Lecture Notes in Computer S

Edited by G. Goos, J. Hartmanis and J. va:

Advisory Board: W. Brauer   D. Gries   J. Stoer

Subhash Bhalla (Ed.)

Information Systems and Data Management

6th International Conference, CISMOD '95
Bombay, India, November 15-17, 1995
Proceedings

Springer

Series Editors

Gerhard Goos
Universität Karlsruhe
Vincenz-Priessnitz-Straße 3, D-76128 Karlsruhe, Germany

Juris Hartmanis
Department of Computer Science, Cornell University
4130 Upson Hall, Ithaca, NY 14853, USA

Jan van Leeuwen
Department of Computer Science,Utrecht University
Padualaan 14, 3584 CH Utrecht, The Netherlands

Volume Editor

Subhash Bhalla
Database Systems Laboratory, University of Aizu
Aizu-Wakamatsu City, Fukushima 965-80, Japan

Cataloging-in-Publication data applied for

Die Deutsche Bibliothek - CIP-Einheitsaufnahme

Information systems and data management : 6th international
conference ; proceedings / CISMOD '95, Bombay, India,
November 15 - 17, 1995. - Berlin ; Heidelberg ; New York ;
Barcelona ; Budapest ; Hong Kong ; London ; Milan ; Paris ;
Tokyo : Springer, 1995
  (Lecture notes in computer science ; Vol. 1006)
  ISBN 3-540-60584-3
NE: Bhalla, Subhash [Hrsg.]; CISMOD <6, 1995, Bombay>; GT

CR Subject Classification (1991): H.2-4, J.1-2, J.4

ISBN 3-540-60584-3 Springer-Verlag Berlin Heidelberg New York

© Springer-Verlag Berlin Heidelberg 1995
Printed in Germany

Typesetting: Camera-ready by author
SPIN 10487157   06/3142 – 5 4 3 2 1 0   Printed on acid-free paper

Preface

The importance of data management and information systems as areas of technology is increasing with more applications. New research and development efforts are being made in mobile databases, real-time databases, visual databases, and knowledge discovery applications. The two areas form the core among the tools for effective exploitation of computer systems.

The 6th international Conference on Information Systems and Management of Data (CISMOD) is being held on November 15-17, 1995 at Bombay in India. The conference aims to bring together researchers and practitioners in the above areas. The conference received close to 60 papers, of which 15 papers were selected through a rigorous refereeing process. I thank all members of the programme committee for their support. Committee members and referees provided detailed comments for nearly all papers, and deserve credit for excellence. I thank all authors who considered CISMOD 95 for making research contributions.

With great pleasure, I thank the invited speakers, Professor J. Biskup, Dr. U. Dayal, Dr. I.S. Mumick, and Dr. N. Terashima for accepting our invitation to give talks and contribute papers to the proceedings. I thank the authors of invited papers, who accepted the invitation to contribute papers and give presentations at the conference. I also thank the tutorial speakers Dr. S.P. Rana, Dr. C. Rolland, and Dr. F. Andres for agreeing to deliver tutorial talks.

The sponsoring organizations and the organizing committee deserve praise for the support they provided. A number of individuals have contributed to the success of the conference. I thank Dr. T. Viswanathan, Dr. P. Dublish, Dr. N. Prakash, Dr. Patkar, Dr. Savanur, and Mr. J. Bhardwaj, for providing continuous support and encouragement.

I have received invaluable support from the University of Aizu. I thank Professor T.L. Kunii, President, and Professor S. Okawa, Head of Department of Computer Software, for making the support available. I also thank all my colleagues at the university for their cooperation and support.

September 1995   Subhash Bhalla

Programme Committee Members

D. Agrawal, Univ. of California
R. Agrawal, IBM Almaden, USA
S. Bhalla, Univ. of Aizu, Japan
J. Biskup, Universitaet Hildesheim
N.bollu, City Univ. of Hong Kong
J.A. Bubenko, SISU, Sweden
A. Cavarero, CNRS, France
X. Castellani, CEDRIC-IIE, France
U. Dayal, HP Labs., Calif., USA
P. De, Univ. of Dayton, USA
P. Dewan, Univ. of N Carolina, USA
P. Dublish, River Run Sof., India
S. Grumbach, INRIA, France
T. Halpin, Uni of Queensland, Austra
T. Ichikawa, Hiroshima Univ., Japan
Rekha Jain, IIM, Ahmadabad, India
H.V. Jagdish, AT&T, USA
S. Jajodia, George Mason Univ.,USA
M. Kitsuregawa,Univ of Tokyo,Japan
M. Kobayashi, Univ. of Aizu, Japan
P. Krishna Reddy, DIT, India
Akhil Kumar, Univ. of Colorado,USA
H.S. Kunii, RICOH, Japan
A.K. Majumdar, IIT Kharagpur,India
Peter McBrien, King's College London
R. Mehrotra, Univ. of Miss-St. Louis
I.S. Mumick, AT&T, USA
N. Parimala, BITS Pilani, India
B. Pernici, Politecnico di Milano,Italy
T.V. Prabhakar, IIT Kanpur, India
B.E. Prasad, Dow Jones Telerate, USA
S. Ram, Univ. of Arizona, USA
S. P. Rana, IBM, USA

M.P. Reddy, Kenan Technologies,USA
C. Rolland, Univ. of Paris, France
M. Saeki, Tokyo Inst. of Tech.,Japan
M. Scholl, Univ of Konstanz,Germany
A. Solvberg, The NIT, Norway
S. Spaccapietra, EPFL-DI-LBD, Switz.
M.V. Sreenivas, River Run Sof., India
Vijay Kumar, Univ. of Missouri,USA
B. Wangler, SISU,Sweden

Additional Reviewers

M. Ahlsen, SISU, Sweden
K. Asada, RICOH, Japan
L.F. Capretz, Univ. of Aizu, Japan
M. Capretz, Univ. of Aizu, Japan
M Cohen, Univ. of Aizu, Japan
Stephen Davis, Univ. of Tokyo
K. Furuse, RICOH, Japan
T. Grust, Univ. of Konstanz
O. Hammami, Univ of Aizu, Japan
Lilian Harada, Fujitsu, Japan
A. Iizawa, RICOH, Japan
P. Johannesson, Stockholm Univ.
A. Kawaguchi, AT&T, USA
M. Nakano, Univ. of Tokyo
C. Nehaniv, Univ. of Aizu, Japan
Y. Ogawa, RICOH, Japan
T. Ohmori,Univ of Elect-C.,Japan
B. Panda, Univ. of Alabama, USA
Indrakshi Ray, George Mason Univ.
Dr. H. Riedel, Univ. of Konstanz
MH Wong, Chinese Univ of H. Kong

Sponsors
Indian National Scientific Documentation Centre (INSDOC), New Delhi,
India

Co-sponsors
IEEE India Council

Table of Contents

Real-Time Database Systems

Performance Evaluation

Knowledge Discovery

Invited Talk

Product and Process Design

Invited Talk

Invited Papers

Conflictfreeness as a Basis for Schema Integration[1]

Love Ekenberg   Paul Johannesson
lovek@dsv.su.se   pajo@dsv.su.se

Logikkonsult NP AB
Jakobsdalsvägen 13
S-126 53 HÄGERSTEN, SWEDEN
and
Department of Computer and Systems Sciences
Royal Institute of Technology and
Stockholm University
Electrum 230
S-164 40 KISTA, SWEDEN
Telefax: +46-8-703 90 25

Abstract

We present a formal framework for the combination of schemas. A main problem addressed is that of determining when two schemas can be meaningfully integrated. Another problem is how to merge two schemas into an integrated schema that has the same information capacity as the original ones, i.e., that the resulting schema can represent as much information as the original schemas. We show that both these problems can be solved by placing a restriction on the schemas to be integrated. The restriction, called conflictfreeness, states that the rules of one schema together with a set of correspondence assertions may not restrict the models of the other schema. We also give decidability and complexity results for the problem of determining conflictfreeness.

Keywords: schema integration, semantic interoperability, conceptual modelling

1. Introduction

Database management systems have been available for more than two decades. They have been used mainly in the form of the hierarchical, network, and relational models. In the mid 1970s the development of semantic database models was initiated. These were introduced primarily as schema design tools, meaning that a schema should first be designed using a high level semantic model and then translated into one of the traditional models for implementation. One advantage of using semantic data models in this context is that it simplifies the integration of different user perspectives. In fact, one of the basic reasons for using a database approach instead of a file approach is that it makes it possible to

---

[1]This work was partly performed within the project SDeLphi at Telia Research AB. It was also partly supported by the NUTEK project SISI.

define a coherent view of the data of an organisation. This view may then be used for serving a number of different user perspectives.

An important part of conceptual design is to integrate various conceptual schemas. The term *schema integration* will, henceforth, be used to refer to this process, which is more precisely defined as "the activity of integrating the schemas of existing or proposed databases into a global, unified schema" [Batini86]. A natural distinction can be drawn between two types of schema integration, namely between view integration and database integration. View integration, which takes place in database design, produces a global conceptual description of a proposed database. Database integration occurs in distributed database management [Özsu90] and produces a global schema of a collection of existing databases.

Research in the area of schema integration has been carried out since the beginning of the 1980s. A comprehensive survey of the area can be found in [Batini86]. Most of the work has been done in the context of the relational model [Biskup86], the functional model [Motro87], or (some extended version of) the ER model, [Larson89], [Spaccapietra92].

It is interesting to note that different approaches to the schema integration problem have been chosen in view integration and database integration. The prevalent approach in view integration has been to derive, more or less automatically, an integrated schema from a set of integration assertions relating equivalent constructs in the views. The integration assertions typically describe set relationships (equality, inclusion, etc.) between the extensions of related entity types [Effel84], [Spaccapietra92], [Johannesson91], [Johannesson93]. In database integration, on the other hand, the common approach has been to provide a restructuring manipulation language that enables a user to build an integrated schema from the original ones, [Dayal84], [Motro87]. The restructuring operators allow a user to modify an object hierarchy (by introducing supertypes or subtypes to existing types) or to modify the attribute structure.

In this paper, an approach using integration assertions was chosen. This approach makes the integration process simpler and easier to understand than the restructuring approach. A main problem addressed in the paper is that of determining when two schemas can be meaningfully integrated. Intuitively, two schemas can be integrated if the rules of one of the schemas together with a set of integration assertions do not restrict the models of the other schema. We formalise this concept using the notion of *conflictfreeness* and show how it can be used to ensure that the merging of two schemas results in an integrated schema with the same information capacity as the original ones. This means that the resulting schema can represent as much information as the original schemas. We show that the problem of conflictfreeness is undecidable, and we outline how it can be addressed for finite domains and determine its complexity properties in this case. An important contribution of the paper is that it shows how previous work on schema integration, which has considered only static and structural aspects, e.g., [Biskup86], can be extended to also handle dynamic aspects of a schema.

The paper is organised as follows. In section 2 we introduce a first-order logic framework for conceptual schemas and define basic notions for schema integration. In section 3 we extend the framework by taking into account dynamic aspects, which are modelled by the event concept. In section 4 we discuss computational considerations for determining when schemas can be integrated, and in the final section we summarise the work and give suggestions for further research.

2. Static Schema Properties

We define the static framework for schema integration. This will be generalised to handle dynamic aspects in section 3. In the definitions below, we assume an underlying *language* L of first order formulas. By a *diagram for* a set S of formulas in a language L, we mean a Herbrand model of S, extended by the negation of the ground atoms in L that are not in the Herbrand model. Thus, a diagram for L is a Herbrand model extended with classical negation.[2]

2.1. Conflictfreeness of Schemas

A *schema* is a finite set S of closed first order formulas in a language L. *L(S)* is the *restriction of L to S*, i.e., L(S) is the set {p | p ∈ L, but p does not contain any predicate symbol, that is not in a rule in S}. The elements in S are called *rules* in L(S). *An integration assertion expressing the schema S_2 in the schema S_1* is a closed first order formula, $\forall x \ (p(x) \leftrightarrow F(x))$, where p is a predicate symbol in $L(S_2)$ and F(x) is a formula in $L(S_1)$.[3] S_i below denotes a schema.

Fig. 2.1 shows a graphical representation of a schema. The labels on the ellipses specify the unary predicate symbols, and those on the arrows specify the binary ones. The graph also depicts domain and range constraints – as an example the arrow between Person and Vehicle represents the formula $\forall x \forall y(\text{owns}(x,y) \rightarrow \text{Person}(x) \wedge \text{Vehicle}(y))$. The arrows labelled ISA represent generalisation relationships, e.g., $\forall x(\text{Car}(x) \rightarrow \text{Vehicle}(x))$.

Fig. 2.2 shows a schema that represents similar information as the previous one. The following is a set of integration assertions that express the schema of fig. 2.2 in the schema of fig. 2.1:

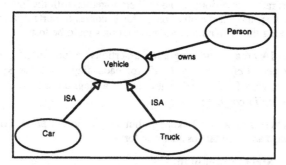

Fig. 2.1

$\forall x(\text{Road\_vehicle}(x) \leftrightarrow \text{Vehicle}(x))$
$\forall x(\text{Owner}(x) \leftrightarrow \text{Person}(x))$
$\forall x(\text{type}(x,'\text{auto}') \leftrightarrow \text{Car}(x))$
$\forall x(\text{type}(x,'\text{lorry}') \leftrightarrow \text{Truck}(x))$
$\forall x \forall y(\text{owned\_by}(x,y) \leftrightarrow \text{owns}(y,x))$

---

[2]For our purposes, this is no loss of generality by the well-known result that a closed formula is satisfiable iff its Herbrand expansion is satisfiable. For a discussion of this expansion theorem and its history, see, e.g. [Dreben79].

[3]We can without loss of generality, in the definitions below, assume that the set of predicate symbols in $L(S_1)$ and $L(S_2)$ are disjoint.

4

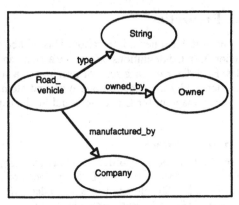

Fig. 2.2

The model theoretic counterpart to the schema concept is the concept of a description. We also define what is meant by two descriptions being isomorphic.

Definition: A *description of a schema S* is the set of all diagrams over L(S) for S. We will also say that the description of a schema S_1 and the description of a schema S_2 are *isomorphic*, if the two descriptions have the same number of elements. A bijection between two isomorphic descriptions is called an *isomorphism*.[4]

Next, we define the concept of conflictfreeness. Intuitively, two schemas are in conflict w.r.t. a set of integration assertions if one of them together with the integration assertions restrict the set of diagrams for the other one. This becomes important later when integrating schemas, obeying the demand that no information should be lost.

Definition: Let IA be a set of integration assertions[5] expressing S_2 in S_1. S_1 *and S_2 are conflictfree w.r.t. IA iff for each diagram* σ *for S_1, there exists a diagram* τ *for S_2, such that* $\sigma \cup \tau$ *is a diagram for IA. Otherwise, S_1 and S_2 are in conflict w.r.t. IA.*

This definition is very similar to [Biskup86], but adapted to our present framework using a first order language and a semantics adequate for that.

2.2. Weak and Strong Dominance

The concept of weak dominance intuitively expresses that the dominating schema, in some sense, contains more information. The dominance concept we introduce is similar to previous concepts of dominance and information capacity equivalence in the literature, [Biskup86], [Hull86], [Miller94]. A major problem with these concepts is that they are too liberal, i.e., there exist schemas that are equivalent according to these concepts although there is no intuitive relationship between them. Our definition of dominance overcomes this problem by being more restrictive than previous definitions. We define

---

[4]This definition is (especially from a practical point of view) very weak and we could instead define isomorphism in terms of the query language, cf. [Miller94], but, as we will see in the sequel, when considering conflictfreeness in the dynamic case and in finite domains, the proposed definition is computationally meaningful.

[5]Note that, by definition, IA also is a schema.

dominance and equivalence not as an absolute relationship between schemas; instead we state that a schema dominates another one only relative to a set of integration assertions that provides an intuitive relationship between the schemas.

Definition: Let IA be a set of integration assertions expressing S_2 in S_1. S_2 *weakly dominates* S_1 *w.r.t. IA* iff there is an isomorphism ξ between the description of S_1 and a subset of the description of S_2, such that for each diagram σ for S_1, $\sigma \cup \xi(\sigma)$ is a diagram for IA.

Thus, the relation of weak dominance between two schemas is transitive. We also will say that two schemas are equivalent if they weakly dominate each other.

Definition: Let IA_1 and IA_2 be two sets of integration assertions expressing S_2 in S_1, and S_1 in S_2 respectively. S_1 *and* S_2 *are equivalent w.r.t.* IA_1 *and* IA_2 iff S_1 weakly dominates S_2 w.r.t. IA_1 and S_2 weakly dominates S_1 w.r.t. IA_2. If $IA_1 = IA_2 = IA$, we will say that S_1 *and* S_2 *are equivalent w.r.t. IA.*

Definition: Let IA be a set of integration assertions expressing S_2 in S_1. S_2 *strongly dominates* S_1 *w.r.t. IA* iff S_2 weakly dominates S_1 w.r.t. IA, and S_1 and S_2 not are equivalent w.r.t. IA.

2.3. Integration of Schemas

We combine two schemas by taking the union of them, and show that the combined schemas weakly dominate each of its components. Thereafter, we refine the definition by eliminating some redundancies.

Definition: The first level combined schema (denoted $\chi_1(S_1, S_2, IA)$) is the schema $S_1 \cup S_2 \cup IA$.

Theorem 1: Let IA be a set of integration assertions expressing S_2 in S_1, such that S_1 and S_2 are conflictfree w.r.t. IA. Then, $\chi_1(S_1, S_2, IA)$ weakly dominates S_1 w.r.t. IA.

The theorem follows immediately from the definitions. Since S_1 and S_2 are conflictfree w.r.t. IA, for every diagram σ of S_1, there exists a diagram τ, such that $\sigma \cup \tau$ is a diagram for $S_1 \cup S_2 \cup IA$. Thus, we can define the function ξ by letting $\xi(\sigma_i)$ be equal to $\sigma_i \cup \tau_i$ for every diagram σ_i of S_1[6]. Clearly, ξ defined in this way is injective since extensions of different descriptions have to be different. The argument that $\chi_1(S_1, S_2, IA)$ weakly dominates S_2 w.r.t. a set of integration assertions is analogous.

---

[6]$\sigma_i \cup \tau_i$ is a diagram for $S_1 \cup S_2 \cup IA$.

Definition: Let IA be a set of integration assertions expressing S_2 in S_1. *The second level combined schema (denoted $\chi_2(S_1, S_2, IA)$) is the schema $S_1 \cup S_2'$, where S_2' is S_2 with all instances of left hand sides in IA exchanged to the corresponding instances of the right hand sides. Thus $L(\chi_2(S_1, S_2, IA))$ is $L(S_1) \cup (L(S_2) - L(IA))$[7].*

Definition: Let IA and S_2' be as in the definition above. *The third level combined schema (denoted $\chi_3(S_1, S_2, IA)$) is the schema $S_1 \cup S_2''$, where S_2'' is $S_2' - \{R \mid R \in S_2' \text{ and } R \text{ is deducible from } S_1\}$.*

Fig. 2.3 shows the third level combined schema for the schemas of figs. 2.1 and 2.2.

As can be seen from the following theorem, $\chi_2(S_1, S_2, IA)$ and $\chi_3(S_1, S_2, IA)$ above, essentially express the same schema as $\chi_1(S_1, S_2, IA)$, but with some redundancies eliminated.

Theorem 2: Let IA be a set of integration assertions expressing S_2 in S_1. Then the schemas $\chi_1(S_1, S_2, IA)$, $\chi_2(S_1, S_2, IA)$ and $\chi_3(S_1, S_2, IA)$ are equivalent w.r.t. IA.

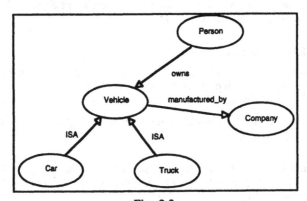

Fig. 2.3

Since $\chi_3(S_1, S_2, IA) \Leftrightarrow \chi_2(S_1, S_2, IA)$[8], σ is a diagram for $\chi_3(S_1, S_2, IA)$ iff σ is a diagram for $\chi_2(S_1, S_2, IA)$. Since the substitution, of right hand side expressions of IA for left hand side expressions of IA in a formula F does not affect the truth value of F, there is a bijection between the description of $\chi_2(S_1, S_2, IA)$ and the description of $\chi_1(S_1, S_2, IA)$. By this, it is also clear that the description of $\chi_1(S_1, S_2, IA)$ and the description of $\chi_3(S_1, S_2, IA)$ are isomorphic. (One isomorphism between them is the combination of the isomorphism above.)

---

[7] I.e., $L(S_1)$ union $L(S_2)$ minus the predicate symbols occurring at the left hand sides of the formulas in IA.

[8] $A \Leftrightarrow B$ denotes that the set A is logically equivalent to the set B.

2.4. Consistency and Entailment

The purpose of the above definitions of conflictfreeness is that no diagram for the schemas, that are combined, should be lost. However, if two schemas are in conflict, it is still interesting to study other aspects of the combination. The weakest requirement is that the combined schema actually expresses anything at all, i.e. that there is a diagram for it.

Definition: Let IA be a set of integration assertions expressing S_2 in S_1. *S_1 and S_2 are consistent w.r.t. IA* iff there exists a diagram for $S_1 \cup S_2 \cup IA$.

Another important issue might be to investigate whether one schema implies another w.r.t. some set of integration assertions. For instance, this is meaningful when comparing a design with its specification, as investigating if a set of requirements expressed in a specification always is fulfilled by a design.

Definition: Let IA be a set of integration assertions expressing S_2 in S_1. *S_1 implies S_2 w.r.t. IA* iff every diagram for S_1 is a diagram for $S_2 \cup IA$.

Thus, it follows immediately that if S_1 implies S_2 w.r.t. IA, then S_1 and S_2 are conflictfree w.r.t. IA.

3. Dynamic Schema Properties

In order to handle dynamics, we extend the definition of a schema to include event rules, that describe possible transitions between different states of a schema, i.e. transitions between schema populations.

3.1. Event Messages, Event Rules, and Events

First, we define the concept of events, event messages, and event rules. Intuitively, an event results in a transition from one diagram to another one, and are a result from an event rule together with an event message. Below, *L* denotes a language and *B* denotes the alphabet underlying the language L.

Definition: An *event message* in L is a vector **u** of constants in the alphabet B.

Definition: An *event rule* in L is a structure <**z**, P(**z**, **w**), C(**z**, **w**)>. P(**z**, **w**) and C(**z**, **w**) are first order formulas in L, and **z** and **w** are vectors of variables in B.[9]

Now, we define a schema to include a set of event rules.

Definition: A *schema* S is a structure <R, ER>, where ER is a set of event rules in L(R).[10] R is a finite set S of closed first order formulas in L. The elements in R are called *rules* in L(R).

Definition: An *event* for a schema <R, ER> is a tuple (σ, ρ), where σ and ρ are diagrams for R, and
  (i) $\rho = \sigma$, or
  (ii) there is a rule <**z**, P(**z**, **w**), C(**z**, **w**)> in ER, and an event message **u**, such that σ is a diagram for P(**u**, **w**) and ρ is a diagram for C(**u**, **w**).

---

[9]The notation A(x,y) means that x and y are free in A(x,y).

3.2. Conflictfreeness

First, we define an analogy to description in the static case. The new concept is a structure consisting of all diagrams to a schema, together with all possible transitions that are possible with respect to the events in the schema.

Definition: The *description of a schema S* is a digraph <M, A>, where M is the set of all diagrams for S, and A, is the set of all events for S.[11]

Below, we will also use the concepts of isomorphic description and subdescription.

Definition: The description <M_1, A_1> of a schema S_1 and the description <M_2, A_2> of a schema S_2 are *isomorphic*, if there is a bijection ξ from <M_1, A_1> to <M_2, A_2>, such that ($\xi(\sigma)$, $\xi(\rho)$) is in A_2 iff (σ, ρ) is in A_1.

As in the static case, this definition is very weak. However, it is still meaningful from a computational point of view.[12]

Definition: A *subdescription of a description* <M, A> is a description <M', A'>, where A' is a subset of A. An *empty description* is a description <M, A>, where M = Ø.

The definition of conflictfreeness in the dynamic case is identical to the corresponding definition in the static case, except for that we have a new definition of the underlying schema concept.

Definition: Let IA be a set of integration assertions expressing S_2 in S_1, and let <M_1, A_1> and <M_2, A_2> be the graph descriptions of S_1 and S_2, respectively. *S_2 and S_1 are conflictfree w.r.t. IA* iff for each diagram σ in M_1, there exists a diagram τ in M_2, such that $\sigma \cup \tau$ is a diagram for IA. Otherwise *S_1 and S_2 are in conflict w.r.t. IA*.

3.3. Dominance and Schema Integration

The definition of weak dominance in the dynamic case is similar to the definition in the static case, but we also require that there should be a (partial) injective function from the dominated schema to the dominating one. This function expresses that the behaviour of the dominated schema is completely reflected in the dominating schema.

Definition: Let IA be a set of integration assertions expressing S_2 in S_1. *S_2 weakly dominates S_1 w.r.t. IA* iff there is an isomorphism ξ between the description <M_1, A_1> of S_1, and a subdescription of the description <M_2, A_2> of S_2, such that for each σ in M_1, $\sigma \cup \xi(\sigma)$ is a diagram for IA.[13]

---

[10]As before, L(R) is the restriction of L to R.

[11]Consequently, events are arcs between elements in M_i.

[12]There is no known algorithm for deciding isomorphism in polynomial time, and (when considering finite domains) it is an open problem whether the problem is NP-Complete [Johnson90].

[13]As in the static case, we can refine the definition of combination to eliminate some redundancies. This can be done very similar to the static case, but a treatment of the details is omitted here.

Now, we can define what it is meant by integrating schemas in the dynamic case, and show that integrated schemas weakly dominates each of its components.

Definition: Let IA be a set of integration assertions expressing S_2 in S_1. The first level combined schema, $\chi_1(S_1, S_2, IA)$, is the schema:
$<R_1 \cup R_2 \cup IA, ER_1 \cup ER_2>$.

Theorem 3: Let IA be a set of integration assertions expressing S_2 in S_1, such that S_1 and S_2 are conflictfree w.r.t. IA. Then, $\chi_1(S_1, S_2, IA)$ dominates S_1 w.r.t. IA.

Assume that the two schemas S_1 and S_2 are conflictfree with respect to a set of integration assertions, IA, and that $<M, A>$ is a description for $\chi_1(S_1, S_2, IA)$. First, we need to show that there is a total injective function ξ, such that $(\xi(\sigma_i), \xi(\sigma_j))$ is in A if (σ_i, σ_j) is in A_1, i.e., the set of all events for S_1. For all events (σ_i, σ_j) in A_1, there are event rules $<z, P(z, w), C(z, w)>$ in ER_1, and event messages u, such that σ_i is a diagram for $P(u, w)$ and σ_j is a diagram for $C(u, w)$, if $\sigma_i \neq \sigma_j$. By the conflictfreeness of S_1 and S_2, there is an extension of σ_i to $\sigma_i \cup \sigma_i'$ and an extension of σ_j to $\sigma_j \cup \sigma_j'$, such that the extensions are diagrams for $R_2 \cup IA$. If σ_i is a diagram for $P(u, w)$, $\sigma_i \cup \sigma_i'$ is a diagram for $P(u, w)$, and if σ_j is a diagram for $C(u, w)$, $\sigma_j \cup \sigma_j'$ is a diagram for $C(u, w)$. Consequently, we can define the function ξ by letting $\xi(\sigma_i)$ be equal to $\sigma_i \cup \sigma_i'$, for every state description σ_i of S_1. It is also obvious that the function ξ has the property that for each σ_i in M_1, $\sigma_i \cup \xi(\sigma_i)$ is a diagram for IA, since $\xi(\sigma_i)$ is a diagram for IA.

As in the static case, also weaker concepts of conflictfreeness could be applicable, and we define what it is meant by two schemas being consistent, and one schema implying another.

Definition: Let IA be a set of integration assertions expressing S_2 in S_1. *S_1 and S_2 are consistent w.r.t. IA* iff $<R_1 \cup R_2 \cup IA, E_1 \cup E_2>$ has a nonempty description.

The definition of that one schema implies another w.r.t. some set of integration assertions, extends the usual concept of consequence in first order logic.

Definition: Let IA be a set of integration assertions expressing S_2 in S_1. *S_1 implies S_2 w.r.t. IA* iff every diagram for R_1 is a diagram for $R_2 \cup IA$, and $A_2 \subseteq A_1$.

As in the static case, we can observe that if S_1 implies S_2 w.r.t. IA, then S_1 and S_2 are conflictfree w.r.t. IA.

3.4. Static and Dynamic Integrity Constraints

The framework we have proposed, is general and can be extended in several directions to include further schema properties. A term often used for a subset of the rules in a schema is static integrity constraints, i.e., constraint of the diagrams for the schema. Static integrity constraints are only first order formulas and are incorporated into schemas as such. Sometimes, we might also be inclined to give restrictions on the set of admissible transitions between different states of the schema as well. We will call these kinds of constraints, dynamic integrity constraints.

Definition: A *dynamic integrity constraint* in a language L is a structure <B(z), A(z)>. B(z) and A(z) are first order formulas in L, such that z is a vector of variables in the alphabet underlying L.

We define a schema to include a dynamic integrity constraint.

Definition: A *schema* S is a structure <R, ER, DI>, where R and ER are as before, and DI is a set of dynamic integrity constraint in L(R). L(R) is the restriction of L to R.

A legal event for a schema is an event that fulfils the dynamic integrity constraints in the schema.

Definition: A *legal event* for a schema <R, ER, DI> is an event (σ, ρ), such that for every dynamic integrity constraint <B(z), A(z)> in DI, if σ is a diagram for B(z), then ρ is a diagram for A(z).

A description of a schema with respect to our new schema concept is slightly different.

Definition: The *description of a schema S* is a digraph <M, A>, where M is the set of all diagrams for S, and A, is the set of all legal events for S.

The concepts of isomorphic description, subdescription, conflictfreeness, and domination are as before, but with the new restriction of the set A_i, i.e., that it consists only of legal events. The definition of schema combination is also very similar to the one above.

Definition: Let IA be a set of integration assertions expressing S_2 in S_1. The first level combined schema, $\chi_1(S_1, S_2, IA)$, is the schema:
$$<R_1 \cup R_2 \cup IA, \ ER_1 \cup ER_2, \ DI_1 \cup DI_2>.$$

By the discussion following theorem 3 above, it should be clear that $\chi_1(S_1, S_2, IA)$, also in our new sense, dominates S_1 as well as S_2 w.r.t. IA, if S_1 and S_2 are conflictfree w.r.t. IA

4. Some Computational Considerations

This section discusses some aspects on the problem of determining conflictfreeness of two schemas.

Since we have no restriction of the formulas in a schema (except that they are first order formulas) the general problems of determining conflictfreeness and weak dominance are undecidable. However, if we restrict our attention to languages with a finite number of constants, we have a problem in propositional logic that is Π_2^P-complete.[14]

This can be realised from the general characterisation of the complexity class Π_k^P and the observation that the criterion of conflictfreeness for two schemas is an expression in second order logic of the form: for all diagrams σ for S_1, there exists an extension of σ to a diagram ρ for $S_1 \cup S_2 \cup IA$.[15]

Despite the problem of determining conflictfreeness is decidable under the assumption of a finite number of constants, it seems that it still is a bit too hard from a computa-

---

[14]The set of NP-complete problems is a subset to the problems in this class.

[15]For a detailed treatment of the different complexity classes, see, e.g., [Johnson90], p. 96.

tional point of view. One way too reduce the computational complexity is to restrict the set IA.

Definition: Let IA be a set of integration assertions expressing S_2 in S_1, and let IA have the additional property that every predicate symbol in $L(S_2)$ occurs on the left hand side in an integration assertion in IA. Then we call IA a *complete set of integration assertions expressing S_2 in S_1*.

Now by theorem 2, the problem of determining conflictfreeness, considering a determined finite domain and a complete set of integration assertions, is clearly a satisfiability problem in first order propositional logic, i.e., an NP-complete problem.[16] Even this seems to be too demanding from a computational point of view, but if we have access to an efficient theorem prover, NP-complete problems could be solved (in most cases) within a reasonable time. For instance, one such theorem prover is built on a method by Gunnar Stålmarck, that has been successfully used for solving satisfiability problems of the kind we get in our context (cf. [Stålmarck95]).

5. Concluding Remarks

When integrating two schemas, a reasonable requirement is that the resulting integrated schema can contain at least as much information as the original schemas. This requirement ensures that no information is lost in the integration process. Further, there must exist a natural mapping between the integrated schema and the original ones. In this paper, we have formalised these requirements by means of the dominance concept. We have also identified a condition, called conflictfreeness, that two schemas and a set of integration assertions must satisfy in order to be mergeable. We have shown that if two schemas are conflictfree w.r.t. a set of integration assertions then the integrated schema does dominate the original ones. Further, we have investigated decidability and complexity properties of the problem of determining conflictfreeness.

The results of this paper are mainly of a theoretical nature and they show that in the general case there is no algorithm for deciding if two schemas can be meaningfully integrated. In our future research, we intend to investigate various restrictions imposed in order to determine conflictfreeness. One line of research is to use efficient theorem provers for propositional logic, [Stålmarck95] to determine conflictfreeness for schemas where the predicates range over finite domains. This assumption is not as restrictive as it might seem at first inspection, since in all schemas occurring in practice the predicates do range over finite domains. Another line of research is to find ways to restrict the forms of permissible integration assertions as opposed to our present approach where arbitrary formulas are allowed in the integration assertions. Further, one might also restrict the forms of rules. In [Johannesson94a], we present some results for determining conflictfreeness in the presence of restricted classes of rules. Still another research direction is to study how schema transformations can be used to remove structural discrepancies between schemas and thereby facilitate the integration process, [Johannesson94b].

---

[16]Strictly speaking, it is not even proved that the (probably) larger class PSPACE $\neq$ NP, and thus we cannot really say that we have decreased the complexity of the problem. However most experts in the field of complexity theory are convinced that this is the case.

Acknowledgements

The authors would like to thank Gunnar Stålmarck and Filip Widebäck, Logikkonsult NP AB for commenting an earlier version of this paper.

References

[Batini86] C. Batini, M. Lenzerini and S. B. Navathe, "A Comparative Analysis of Methodologies for Database Schema Integration", *ACM Computing Surveys*, vol. 18, no. 4, pp. 323-364, 1986.

[Biskup86] J. Biskup and B. Convent, "A Formal View Integration Method", in *International Conference on the Management of Data*, Washington, ACM, 1986.

[Bouzeghoub90] M. Bouzeghoub and I. Comyn-Wattiau, "View Integration by Semantic Unification and Transformation of Data Structures", in *Ninth International Conference on Entity-Relationship Approach*, Ed. H. Kangassalo, pp. 413-430, Lausanne, North-Holland, 1990.

[Dayal84] U. Dayal and H.-Y. Hwang, "View Definition and Generalization for Database Integration in a Multidatabase System", *IEEE Transactions on Software Engineering*, vol. SE-10, no. 6, pp. 628-644, 1984.

[Dreben79] B. Dreben and W. D. Goldfarb, *The Decision Problem: Solvable Classes of Quantification Formulas*, Reading, Mass, Addison-Wesley, 1979.

[Effel84] W. Effelsberg and M. V. Mannino, "Attribute Equivalence in Global Schema Design for Heterogeneous Distributed Databases", *Information Systems*, vol. 9, no. 3/4, pp. 237-240, 1984.

[Hull86] R. Hull, "Relative Information Capacity of Simple Relational Database Schemata", *SIAM Journal of Computing*, vol. 15, no. 3, pp. 856-886, 1986.

[Johnson90] D.S. Johnson "A Catalogue of Complexity Classes", in *Handbook of Theoretical Computer Science: Volume A*, Ed. Jan van Leeuwen, Elsevier, Amsterdam, 1990.

[Johannesson91] P. Johannesson, "A Logic Based Approach to Schema Integration", in *10th International Conference on Entity-Relationship Approach*, Ed. T. Teorey, San Francisco, North-Holland, 1991.

[Johannesson93] P. Johannesson, "A Logical Basis for Schema Integration", in *Third International Workshop on Research Issues in Data Engineering - Interoperability in Multidatabase Systems*, Ed. H. Schek, Vienna, IEEE Press, 1993.

[Johannesson94a] P. Johannesson, "Linguistic Instruments and Qualitative Reasoning for Schema Integration", in *Third International Conference on Information and Knowledge Management*, Ed. N. Adam. Gaithersburg, Maryland, IEEE Press, 1994.

[Johannesson94b] P. Johannesson, "Schema Standardization as an Aid in View Integration", in *Information Systems*, Vol. 19, No 3, 1994.

[Larson89] J. A. Larson, S. Navathe and R. ElMasri, "A Theory of Attribute Equivalence in Databases with Applications to Schema Integration", *IEEE Transactions on Software Engineering*, vol. 15, no. 4, pp. 449-463, 1989.

[Miller94] R. Miller, Y. Ioannidis, and R. Ramakrishnan, "Schema Equivalence in Heterogeneous Systems: Bridging Theory and Practice", in *Information Systems*, Vol. 19, No 1,1994.

[Motro87] A. Motro, "Superviews: Virtual Integration of Multiple Databases", *IEEE Transactions on Software Engineering*, vol. 13, no. 7, pp. 785-798, 1987.

[Spaccapietra92] S. Spaccapietra, C. Parent and Y. Dupont, "Model Independent Assertions for Integration of Heterogeneous Schemas", *The VLDB Journal*, vol. 1, no. 2, pp. 81-126, 1992.

[Stålmarck95] G. Stålmarck, *System for Determining Propositional Logic Theorems by Applying Values and Rules to Triplets that are Generated from a Formula*, US. Patent No. 5 276 897.

[Özsu90] M. T. Özsu and P. Valduriez, *Principles of Distributed Database Systems*, Prentice-Hall, 1990.

Database Schema Design Theory: Achievements and Challenges

Joachim Biskup

Institut für Informatik,
Universität Hildesheim,
D-31141 Hildesheim,
Germany,
biskup@informatik.uni-hildesheim.de

Abstract. Database schema design is seen as to decide on formats for time-varying instances, on rules for supporting inferences and on semantic constraints. Schema design aims at both faithful formalization of the application and optimization at design time. It is guided by four heuristics: Separation of Aspects, Separation of Specializations, Inferential Completeness and Unique Flavor. A theory of schema design is to investigate these heuristics and to provide insight into how syntactic properties of schemas are related to worthwhile semantic properties, how desirable syntactic properties can be decided or achieved algorithmically, and how the syntactic properties determine costs of storage, queries and updates. Some well-known achievements of design theory for relational databases are reviewed: normal forms, view support, deciding implications of semantic constraints, acyclicity, design algorithms removing forbidden substructures. These achievements should be integrated and generalized from (at least) three viewpoints: one unifying framework, embedding in the full design process, interoperability among databases. Specifically, embedding requires to deal with more advanced semantic notions like null values, recursive query languages, complex objects and object identifiers as well as to explore standard internal database structures and operations like access structure and actual constraint enforcement. All conceptual results should be complemented by detailed cost analysis.

1 Introduction

Due to its great importance for database applications, database schema design has attracted a lot of researchers, and accordingly a lot of insight into good schemas has been obtained. On the one side, practical experience suggests to follow some basic design heuristics, which have been ramified into considerable detail. On the other side, theoretical investigations have accumulated many formal notions and theorems on database schema design. Unfortunately, however, theory apparently does not have much impact on practice yet. The purpose of this paper is to improve on this mismatch of theory and practice by

- presenting well-known theoretical results on schema design within a fresh and unifying framework,

- identifying the present shortcomings of database schema design theory as lack of integration and generality, absence of a detailed cost analysis and ignorance of advanced database features, and
- suggesting some directions for future elaboration of database schema design theory.

This paper does not aim at providing a complete survey on well-established and current contributions but at highlighting important examples and worthwhile research directions within an original framework. Accordingly, all references to the literature are to be understood just as hints for further reading.

2   Problem of Schema Design

A database consists of a time independent part, its *schema*, and a time-varying part, its *instance*. At design time, the database administrator *models* the application in hand, let's say using entity-relationship diagrams, and then he *formalizes* the model and declares a database schema, using a data definition language. In the second step the administrator has to decide on the following problems, essentially:

- Which aspects of the application should be *enumerated*, i.e. represented by a time-varying enumeration of ground facts the *formats* of which are statically declared in the schema?
- Which aspects of the application should be *inferrable*, i.e. derivable from the time-varying enumerations, possibly complemented by additional input, by *rules* which are declared in the schema?
- Which aspects should *constrain* the enumerations under updates, i.e. which format-conforming enumerations of ground facts should be considered meaningful in the sense that they satisfy *semantic constraints* which are declared in the schema.

The decisions result in a *schema* that comprises the *formats* for enumerations (the time-varying extensional instances produced over the life time of the database), the *rules* (for intentional views supporting queries), and the *semantic constraints*. Being fixed over the time, the schema statically determines the future dynamic behaviour of the database and, in particular, its usefulness for its end users. Fig.1 illustrates the design and usage of a database and summarizes the terminology introduced so far.

The quality of a schema can be evaluated along two lines of reasoning:

- The schema should *formalize* the application as *faithful* as achievable.
- The schema should allow to execute queries and updates, as far as these operations can be foreseen, as efficiently as possible. From this point of view, schema design can be understood as *optimization at design time*.

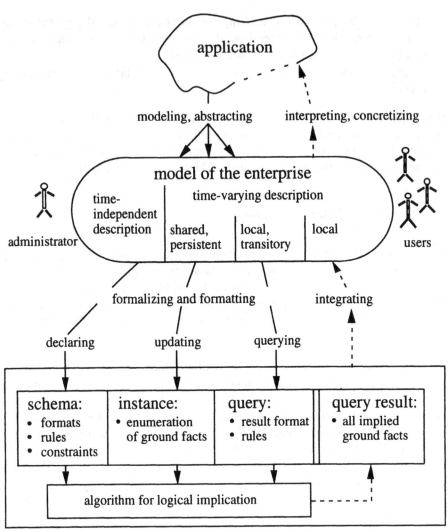

Fig.1. Design and usage of a database

Whether a *faithful formalization* has been achieved or not cannot be evaluated solely based on formal mathematical reasoning. Rather we have to investigate whether the database will successfully provide technical support for communications among those persons who employ the database as end users. Presumably, successful support is based on a common agreement on the following questions: Which *entities* are to be considered basic? Which *relationships* are to be considered basic and how to select from the basic ones those for actual redundancy-free enumerations such that all relationships can be completely inferred? Which actions are to be considered basic?

Optimization at design time, however, can be evaluated in formal mathematical terms by considering

- storage costs (basically determined by the size of the enumerated instances),
- query costs (basically the time complexity of anticipated queries, in particular those that are declared as rules in the schema),
- update costs (basically the time complexity of anticipated insertions and deletions including maintenance of the semantic constraints that are declared in the schema).

3 Design Heuristics

Most guidelines for schema design can be summarized by the following four heuristics:

Separation of Aspects: A declared format should be appropriate to enumerate *exactly one aspect*.

Separation of Specializations: A declared format should be appropriate to conform to *exactly one specialization* of an aspect.

Inferential Completeness: All meaningful aspects that are not enumerated according to a declared format should be inferrable using the query language.

Unique Flavor: Meaningful aspects should be identified and understood by expressing their basic attributes only (and omitting additional context information).

4 Tasks of Design Theory

In order to be helpful in achieving faithful formalizations and in pursuing the design heuristics, the following tasks of design *theory* are due:

- **Task 1:** Formalize the worthwhile semantic requirements and the desirable syntactic properties of schemas!
- **Task 2:** State and prove relationships between the formalized versions of worthwhile semantic requirements and desirable syntactic properties!
- **Task 3:** Find algorithms for deciding on or even achieving syntactic properties of schemas, and prove their correctness and efficiency!

In order to be helpful for optimization at design time, additionally design *theory* should tackle a fourth task:

- **Task 4:** Prove that desirable syntactic properties actually ensure low costs!

Being supplied with appropriate solutions for these tasks, an administrator can effectively benefit from design theory. For, at design time, the administrator essentially has to deal with syntactic material only (supported by Task 3) which must be evaluated with respect to its semantic properties (as stated by Task 1 and Task 2) on the one side and the future operational cost (as stated by Task 4) on the other side.

5 Achievements for Relational Databases

5.1 Notations

For the sake of readability and conciseness we will employ (more or less) standard notations in a somehow sloppy, and sometimes also imprecise, way. In order to study carefully elaborated versions of the notations and of the results the reader should consult the references, in particular the textbooks [Mai83, Ull88, Ull89, PDGvG89, Vos91, MR92, AD93, AHV95, Bis95].

(R_i, X_i, SC_i)	denotes a relation scheme where
R_i	is a relation symbol,
X_i	is a set of attributes (possibly with a range for its values), i.e. a format, and
SC_i	are the local semantic constraints.

A (database) schema comprises relation schemes, rules, and global semantic constraints:

$\langle(R_1, X_1, SC_1), \ldots, (R_n, X_n, SC_n)\|$	relation schemes for extensional enumerations,
$Q_1, \ldots, Q_m\|$	rules (queries) for intensional views,
$SC_{global}\rangle$	global semantic constraints.

Semantic constraints are denoted as follows where X, Y, Y_i, Z are sets of attributes and R_i, R_j are relation symbols:

$X \to Y$	functional dependency,
$X \twoheadrightarrow Y\|Z$ or $\bowtie [X \cup Y, X \cup Z]$	multivalued dependency,
$\bowtie [Y_1, \ldots, Y_k]$	join dependency,
$\Pi_X(R_i) \subset \Pi_Y(R_j)$	inclusion dependency,
SC^+	implicational closure of a set of semantic constraints SC.

5.2 Normal Forms: Separation of Aspects formalized as desirable syntactic property

The first design heuristic, Separation of Aspects, can be rephrased by considering formats and semantic constraints as some kind of structure and requiring that any nontrivial substructure should correspond to, refer to or identify exactly one aspect of the application. Depending on the class of semantic constraints involved we can define different notations of "nontrivial substructure"; but in all cases the notion of "exactly one aspect" is related to the concept of identification of unit pieces of information. In order to formalize the heuristic as desirable syntactic property, normally referred to as "normal form", see Task 1, we favour to express the separation requirement in a negative form: the structure should *not* contain any *forbidden substructures* which might be harmful with respect to the quality measures. Then most algorithms to achieve high quality schemas can conveniently be described as iterated *schema transformations* that stepwise detect and remove forbidden substructures.

The most popular normal forms are listed in Fig.2 giving their names and forbidden substructures, respectively [Cod70, Cod72, Fag77, Del78, Zan76, BBG78, Fag81, Ken83, MR86, BDLM91, DF92]:

name	forbidden substructures
3 NF, third normal form	$Z \rightarrow A \in SC^+$, $A \notin Z$, A nonkey-attribute, (but) $Z \rightarrow X_i \notin SC^+$.
BCNF, Boyce/Codd normal form	$Z \rightarrow A \in SC^+$, $A \notin Z$, (but) $Z \rightarrow X_i \notin SC^+$.
4 NF, fourth normal form	$X \twoheadrightarrow Y \in SC^+$, $Y \not\subset X$, $X \cup Y \subsetneqq X_i$, (but) $X \rightarrow X_i \notin SC^+$.
5 NF, fifth normal form	$\bowtie [Y_1 \ldots Y_k] \in SC^+$, $\bowtie [Y_1 \ldots Y_{i-1}, Y_{i+1}, \ldots, Y_k] \notin SC^+$ for $i = 1, \ldots, k$, (but) there exists j : $Y_j \rightarrow X_i \notin SC^+$.
referential normal form	$\Pi_X(R_i) \subset \Pi_Y(R_j) \in SC^+$, $i \neq j$, (but) $Y \rightarrow X_j \notin SC^+$.
unique key normal form	$X \rightarrow X_i \in SC^+$, X minimal, $Y \rightarrow X_i \in SC^+$, Y minimal, (but) $X \neq Y$.

Fig.2. Normal forms and their forbidden substructures

5.3 View support: Inferential Completeness formalized as worthwhile semantic requirements

The third design heuristic, Inferential Completeness, can be rephrased by considering those aspects of the application which are not explicitly represented by enumerations and requiring that these aspects are completely supported as intensional views by appropriate rules. There are essentially three versions of support: *view instance support, view query support, view update support.*

Restricting our discussion to one-relation views or even so-called universal relation views, we suppose that a database schema of the form
$$DS = \langle \text{schemes for extensional enumerations} | \ | \text{global semantic constraints} \rangle$$
is given, and that some candidate view (or external schema)
$$ES = (\ , U, SC)$$
with set of attributes U and semantic constraints SC should be supported. Then we state the following formal versions of the heuristic as worthwhile semantic property, see Task 1.

- Schema DS provides *view instance support* for ES
  :iff there exists a query Q on DS such that
  $\{instances\ of\ ES\} \subset Q[\{instances\ of\ DS\}]$.

If we have even equality, the view instance support is called *faithful*. In that case, if additionally the supporting query Q is injective on $\{instances\ of\ DS\}$ the view instance support is called *unique*.

- Schema DS provides *view query support* for ES
  :iff for each query P on ES there exists a query P' on DS such that:
  for all instances u of ES there exists an instance $(r_i)_{i=1,\ldots,n}$ of DS such that
  $P(u) = P'((r_i)_{i=1,\ldots,n})$.

Under some rather weak assumptions on the query language we have a fundamental equivalence [AABM82, Hul86, BR88]:

Theorem 1. *DS provides view instance support for ES iff DS provides view query support for ES.*

Based on this equivalence one can easily construct query translations from queries on views to queries on full schemas. For supporting *updates* on views, however, we essentially need that the view instance support is *unique*. For otherwise, well-known as the *view update problem* [BS81, DB82, FC85, Kel86, GHLM93], there is no information available to resolve the ambiguity caused by non-injectivity.

5.4 Syntactic characterization of view support

According to Task 2 the worthwhile semantic requirements of view support should be related to desirable syntactic properties of a schema. The main results available concern universal relation views the supporting query of which is the natural join. For instance we have the following theorems.

Theorem 2. *A schema DS with formats $X_1, \ldots, X_n$ for the extensional enumerations (ignoring local and global semantic constaints of DS) supports a universal relation view $(\ , U, SC)$ by the natural join iff $\bowtie [X_1, \ldots X_n] \in SC^+$.*

Theorem 3. *A schema DS with formats $X_1, \ldots, X_n$ for the extensional enumerations and functional dependencies $F_1, \ldots, F_n$ as local semantic constraints faithfully supports a universal relation view $(\ , U, F)$, where F is a set of functional dependencies, by the natural join if $\bowtie [X_1, \ldots X_n] \in F^+$ and $(\bigcup_{i=1,\ldots,n} F_i)^+ \supset F$.*

More refined results appear for example in [Var82, CM87], and in [Heg94] a rather general theory of schema decomposition is presented.

5.5 Deciding desirable syntactic properties for normal forms and view support

Both heuristics treated so far finally lead to syntactic properties that are basically expressed in terms of implications of semantic constraints. In section 5.2 normal forms, formalizing the Separation of Aspects heuristic, are just defined in these terms, and in section 5.4 view support, formalizing the Inferential Completeness heuristic, has been reduced to these terms. According to Task 3, then, we have to design algorithms to decide implications among semantic constraints and, additionally, to systematically explore all relevant implications.

Here are some prominent examples of results [Arm74, Men79, Bis80, BV84a, BV84b, Var84, Mit83, KCV83, CFP84, CV85, FV84, Var88a, Tha91, BC91]:

Theorem 4. *The implication problem of "$\Phi \in SC$" is decidable for "many important classes" of semantic constraints.*

Theorem 5. *The implication problem "$\Phi \in SC$" is undecidable for the class of "functional and inclusion dependencies".*

In general, the implication problem for semantic constraints is fairly well understood in the relational case: as long as the constraints are "full", i.e. basically no existentially quantified variable occurs positively, we have decidability; otherwise, for "embedded" constraints, we have undecidability due to positively occuring existentially quantified variables that cause the generation of an unlimited number of terms in executing proof procedures.

Theorem 6. *For the class of relation schemes $(\ ,U,F)$ where F is a set of functional dependencies, the problem "$(\ ,U,F)$ is in third normal form" is NP-complete.*

The deep reason for the negative result is that, in this situation, the problem "attribute A appears in a key of relation scheme $(\ ,U,F)$" is already NP-complete; this result in turn is related to the fact that a relation scheme $(\ ,U,F)$ may have exponentially many keys [LO78, JF82, MR83, Kat92, VS93a, DKMST95].

Theorem 7. *For the class of relation schemes $(\ ,U,F)$ where F is a set of functional dependencies, the problem "$(\ ,U,F)$ is in Boyce/Codd normal form" is decidable in polynomial time.*

Indeed, a decision procedure can be based on the following equivalence [Osb78]: Boyce/Codd normal form iff for all $X \rightarrow Y \in F$ with $Y \not\subseteq X: X \rightarrow U \in F^+$.

Theorem 8. *For the class of relation schemes $(\ ,U,F)$ where F is a set of functional dependencies, the problem "$(\ ,U,F)$ is in unique key normal form" is decidable in polynomial time.*

Again, a decision procedure can be based on an equivalence statement [BDLM91]: unique key normal form iff $\{A \mid A \in U \ and \ U \backslash A \rightarrow A \notin F^+\} \rightarrow U \in F^+$.

5.6 Achieving normal forms and view support simultaneously

So far, the Separation of Aspects and the Inferential Completeness heuristics have been treated separately, although, as we have seen in section 5.5., both heuristics lead to related implication problems. According to Task 2, we have to explore the relationship between their formalizations in more detail, in particular whether their formal versions are compatible. As far as we can actually achieve the desirable syntactic properties simultaneously, according to Task 3, we have to design algorithms to obtain them.

The following two theorems are the most well-known examples of results on compatibility.

Theorem 9. *For every (universal relation) scheme $ES = (\ ,U,F)$ where F is a set of functional dependencies, there exists a database schema DS with relation schemes $(\ ,X_1,F_1),\ldots,(\ ,X_n,F_n)$ for extensional enumerations such that:*
i) Schema DS supports ES by the natural join.
ii) Each scheme $(\ ,X_i,F_i)$ of DS is in Boyce/Codd normal form.

Theorem 10. *For every (universal relation) scheme $ES = (\ ,U,F)$ where F is a set of functional dependencies, there exists a database schema DS with relation schemes $(\ ,X_1,F_1),\ldots,(\ ,X_n,F_n)$ for extensional enumerations such that:*
i) Schema DS faithfully supports ES by the natural join.
ii) Each scheme $(\ ,X_i,F_i)$ of DS is in third normal form.

The proofs of these and related theorems are constructive yielding outlines of design algorithms of decomposition and synthesis, respectively [Cod72, Fag77, Fag81, Ber76, BDB79, KM80, LTK81, BK86, SR88, BM87, TLJ90, YÖ92a, YÖ92b]. Such algorithms will be discussed in a more general framework in section 5.11.

5.7 Normal forms ensure low storage and update costs

We have introduced normal forms as desirable syntactic properties formalizing the Separation of Aspects heuristics. According to Task 4, we now justify these normal forms in terms of cost, thus providing formal counterparts to informal motivations of the Separation of Aspects heuristic to avoid so-called "update anomalies".

The benefits of all purely decompositional normal forms in terms of *storage costs* are summarized as follows:

Theorem 11. *A relation scheme $(\ ,U,SC)$ is in decompositional normal form (i.e. BCNF, 4 NF, 5 NF), relative to the class of semantic constraints considered in SC (i.e. functional dependencies, multivalued dependencies, join dependencies)*
iff for each decomposed database schema DS that supports $(\ ,U,SC)$ by the natural join, for each instance of $(\ ,U,SC)$:
size (instance of $(\ ,U,SC)$) $\leq$ size (decomposed instance).

Here size means the number of occurances of constant symbols in the instances. This "folklore theorem" is closely related to a theorem of [VS93b] that characterizes normal forms in terms of data redundancy.

Decompositional normal forms are also helpful to ensure low *update costs* [BG80, Vos88, Bis89, Cha89, HC91, BD93]. As an example, we present a theorem that takes care of functional and inclusion dependencies [BD93]. The theorem characterizes those database schemas that allow maintenance of all semantic constraints by simply checking whether a newly inserted tuple does not violate a key condition.

Theorem 12. *A database scheme DS with relation schemes* $(R_1, X_1, F_1), \ldots,$ (R_n, X_n, F_n) *with functional dependencies as local semantic constraints and inclusion dependencies* I *as global semantic constraints allows* $X \subset X_i$, *for some* i, *as* update object, *i.e.*

i) $X \to X_i \in (I \cup \bigcup_{i=1,\ldots,n} F_i)^+$ *and*

ii)   *for each instance* $(r_1, \ldots, r_i \ldots, r_n)$ *of DS, for each tuple* μ *with*
     $\mu\lceil X \notin \pi_X(r_i)$:
     $(r_1, \ldots, r_i \cup \{\mu\}, \ldots, r_n)$ *is instance of DS*

iff the following properties hold:

iii) R_i *is "not referencing" by inclusion dependencies of* I.

iv) R_i *is in unique key normal form.*

v) R_i *is in Boyce/Codd normal form.*

5.8   Unique Essences: Unique Flavor formalized as worthwhile "semantic" requirement

The fourth design heuristic, Unique Flavor, can be formalized in the framework of designing a so-called universal relation interface for a database schema [MUV84, KKFVU84, Var88b, BB83, BBSK86, BV88, Lev92]. Such an interface should translate queries which are expressed in terms of attributes only (omitting the information about relation schemes) into join paths within the hypergraph structure of the schema. If there are several candidate join paths then, according to the Unique Flavor heuristic, all these candidates should provide essentially the same query answer [BBSK86].

Given a database schema DS a formalized version of this requirement is defined as follows:

$U := \bigcup_{i=1,\ldots,n} X_i$ and $H := \{X_1, \ldots, X_n\}$ describes the *hypergraph* of DS.

$jp : \wp U \to \wp \wp H$, $jp(Y) := \{E \mid E \subset H, \ E \ connected, \ Y \subset \bigcup E, E \ minimal\}$ defines the translation from a set of attributes into *join paths*.

$essence(E, Y) := \{X_i \cap (Y \cup \bigcup_{X_j \in E, X_j \neq X_i} X_j) \mid X_i \in E\}$ is the *essence* of a join path E over a set of attributes Y.

– Finally, Unique Flavor motivates that essences should be unique:
  For all $Y \subset U$, for all $E, F \in jp(Y) : essence(E, Y) = essence(F, Y)$.

Of course, here the property of unique essences is already defined in purely syntactic terms although it is "semantically" motivated.

5.9 Acyclicity: Unique Flavor formalized as desirable syntactic property

The fourth design heuristic, Unique Flavor, can also be rephrased by considering the hypergraph *structure* of a database schema, as defined by the formats, and requiring that the hypergraph is to some degree acyclic [Fag83, BFMY83, BBSK86]. The two most important degrees are listed below by their names and their *forbidden substructures*:

– γ-*acyclic*:

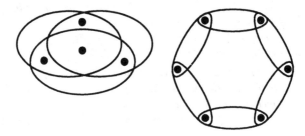

– α-*acyclic*: a nontrivial hypergraph as produced by the GYO-reduction [Gra79, YÖ79] applied to the schema.

As a contribution to Task 2, it turns out that γ-acyclicity syntactically characterizes the "semantic" property of section 5.8 [BBSK86]:

Theorem 13. *DS is γ-acyclic iff DS has unique essences.*

As a contribution to Task 3, the desirable syntactic property of acyclicity can be efficiently decided [YÖ79, Gra79, Fag83, TY84, DM86]:

Theorem 14. *The problems "DS is γ-acyclic" and "DS is α-acyclic" are decidable in polynomial time.*

5.10 Acyclicity ensures low query costs

According to Task 4, the impact of acyclicity as desirable syntactic property on costs should be examined. As suggested by the corresponding worthwhile semantic requirement, the evaluation of view queries that are join paths should be efficient. Indeed, we have the following assertions [Fag83, BFMY83]:

Theorem 15. *DS is γ-acyclic*
iff all *noncartesian join trees are* monotone
    (i.e. for pairwise consistent relations $r_1, \ldots r_n$ all partial results are consistent)
iff projections $\pi_Y(u)$ with $u := \underset{i=1,\ldots,n}{\bowtie} r_i$, $r_i := \pi_{X_i}(u)$,
    can be computed by determining a covering join path and evaluating it
    (i.e. $\pi_Y(u) = \pi_Y(\underset{X_i \in E}{\bowtie} r_i)$ for some $E \in jp(Y)$).

Theorem 16. *DS is α-acyclic*
iff there exists a monotone join tree
    (which, essentially, is determined by the GYO-reduction).

5.11 Design algorithms remove forbidden substructures

Fig.3 summarizes the presented achievements with respect to Tasks 1, 2 and 4. Finally, according to Task 3, a lot of design algorithms for *achieving* desirable syntactic properties have been proposed. The general skeleton of such algorithms can be described as follows [BC86, Bis95]:

- Identify the desirable syntactic properties of particular interest, and define the appropriate worthwhile semantic requirements related to Inferential Completeness.
- Construct an initial database schema that satisfies the semantic requirements.
- While the syntactic properties are still not achieved:
  - detect a forbidden substructure and remove it
  - leaving the semantic requirements invariant.

6 Challenges

6.1 Integration and generalization of results

The achievements of design theory should be integrated and generalized from (at least) three viewpoints:

- one unifying framework,
- embedding in full design process,
- interoperability among databases.

All achievements have to be carried out within *one unifying framework*. Unfortunately, however, previous work often does not exhibit the required homogeneity.

formalize heuristics		
separation	**unique flavour**	**inference**
Task 1. syntactic properties no forbidden substructure: – 3 NF – BCNF – 4 NF – 5 NF – referential NF – unique key NF	*Task 1. "semantic" requirements* – join paths are essentially unique *Task 1. syntactic properties* no forbidden substructure: – γ-acyclic – α-acyclic	*Task 1. semantic requirements* – (faithful) view instance support – view query support *Task 2. syntactic characterization* for FD's, join support: – $\bowtie [X_1,\ldots,X_n] \in SC^+$ – $(\bigcup_i F_i)^+ \supset F$

Task 2. relationships between syntactic properties and semantic requirements

– BCNF and join support are compatible (decomposition)
– 3 NF and faithful join support are compatible (synthesis)
– γ-acyclic iff join paths are essentially unique

optimize at design time		
Task 4. syntactic properties ensure low costs		
storage	**query**	**update**
– normal forms	– γ-acyclic: • join trees are monotone • projection by covering joins – α-acyclic: existence of monotone join trees	– $\begin{cases} \text{not referencing} \\ \text{one key} \\ \text{BCNF} \end{cases}$

Fig.3. A short summary of presented achievements with respect to Tasks 1, 2 and 4.

Two main differences show up. First, there are widely differing assumptions on the inclusion dependencies to be stated as part of the global semantic constraints SC_{global} for a database schema; in fact they range from the universal instance assumption (there exists a relation u: $r_i = \pi_{X_i}(u)$) over assuming pairwise consistency (for all i, j : $\pi_{X_i \cap X_j}(r_i) = \pi_{X_i \cap X_j}(r_j)$) to not assuming any implicit inter-relational constraints through names at all (for all i, j : $X_i \cap X_j = \emptyset$) but requiring to state all such constraints explicitly. Second, there are different assumptions on the formal meaning of semantic constraints and, accordingly, on the implicational closure of constraints; in fact they range from assuming a strictly local scope of constraints to requiring their validity within a so-called representative instance [Hon82, Sag83, GMV86]. Of course both types of differences are related since inclusion dependencies may actually extend the scope of locally declared constraints. Theorem 5 shows that this interaction is nontrivial indeed. Additionally, varying the scope of constraints may force us to consider "embedded" constraints which again may let us run into undecidable implication problems. Finally, the modalized view of integrity constraints, that sees constraints as statements about the contents of the database rather than on some external world [Rei92], should be taken into consideration. My personal impression is that basically we would have sufficient fragmented knowledge to build the wanted unifying framework but this ambitious project still has to be worked out in detail.

All achievements have to be *embedded* into the full design process. Fig.1 suggests a three-step design cycle as sketched in Fig.4. For semantic modelling the entity-relationship approach or some variant or extension [Che76, TYF86, Ste93] is widely accepted. The resulting entity-relationship schema is subject to further modification and final transformation into a concrete database declaration for the target system [RR94, FV95].

The design algorithms mentioned in section 5.11 are to be located within the second step. They aim at conceptual improvements starting with the input from the first step and yielding an appropriate output for further refinement in the third step. Therefore the data model for abstract database schemas has to be powerful enough to represent all relevant notions of semantic modelling, and it should reflect common features of all concrete target systems. Accordingly, on the one side design theory must be consistently generalized to deal with more advanced semantic notions like null values, recursive query languages, complex objects or object identifiers, and on the other side design theory should also explore standard internal database structures and operations like access structures and actual constraint enforcement. These challenges will be further discussed in the next subsections.

Classical schema design is supposed to produce a single centralized conceptual database schema which may support several external views. In particular all efforts of view integration are directed to this goal [BLN86, BC86, Con86, LNE89, Joh94]. Current trends in computing, however, make full integration obsolete but rather require *interoperability among* more or less *autonomous databases*. Therefore all achievements of design theory should be reconsidered and reinter-

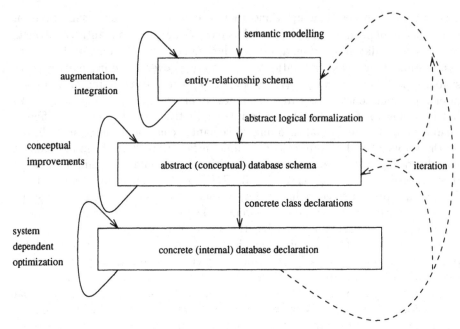

Fig.4. A three-step design cycle

preted from the viewpoint of distributed computing [SPD92, SGN93, MIR93, MIR94, RS94]. Of course, as for the centralized case, the undecidability of basic problems in schema integration have to be circumvented by appropriate restrictions or approximations [Con86, BR88, MIR94]. A particular challenge would be to elaborate a design theory for approaches like the so-called personal model of data [BB88] an implementation of which would consist of a network of many autonomous but closely cooperating databases.

6.2 Null values and the Separation of Specializations heuristic

Formal rephrasements of the Separation of Aspects heuristic are closely connected with Inferential Completeness provided by view support which is based on the natural join or, equivalently, on equality testing and logical conjunction. Sections 5.2 – 5.4 and Fig.3 provide a short introduction for a lot of results. The Separation of Specializations heuristic has not been investigated in similar depth. Clearly, formal rephrasements have to be connected with Inferential Completeness provided by view support which is based on a generalized union and null values or, equivalently, on logical disjunction and an appropriate notion of non-applicability. It should be noted, however, that the general theory of schema decomposition presented in [Heg94] tackles some version of both kinds of view support.

For relational databases specializations mostly formally mean to apply selections whereby the (generalized) union can be used to infer the original relation. For instance a relation r could be specialized into relations $r_1, \ldots, r_k$ such that $r_i := \sigma_{\Phi_i}(r)$ for some appropriate selection formulas Φ_i and $r = r_1 \cup \ldots \cup r_k$. At design time, when there are no enumerated instances yet, the selection formulas should originate from considering semantic constraints. There appears to be two approaches for such considerations.

Firstly, we could relax the meaning of semantic constraints by requiring that they should hold in "most cases" but allowing exceptions as well. Then we can separate the usual case from the exceptional one. The so-called horizontal decomposition according to afunctional dependencies [DP84, PDGvG89, Heg94] is a well-known example. Here, for a given functional dependency $X \rightarrow Y$ and the corresponding afunctional dependency $X \nrightarrow Y$, a relation r is uniquely specialized into the horizontal fragments r_1 and r_2 such that r_1, satisfying $X \rightarrow Y$, contains the "usual" tuples and r_2, satisfying $X \nrightarrow Y$, gathers the "exceptions". This example could be generalized by employing min-max dependencies [GM85] of the syntactic form $X \overset{l}{=} {}^u Y$ which are satisfied by a relation r iff for all tuples $\mu \in \pi_X(r)$ the cardinality of $\pi_Y(\sigma_{X=\mu}(r))$ ranges between l and u. Apparently, a functional dependency $X \rightarrow Y$ is equivalent to $X \overset{1}{=} {}^1 Y$, and an afunctional dependency $X \nrightarrow Y$ is equivalent to $X \overset{2}{=} {}^\infty Y$. Using min-max dependencies we would have many options to specify formally what is identified as "usual" case and what kinds of "exceptions" are allowed. Furthermore, "exceptional" cases of the form $X \overset{u}{=} {}^u Y$ with a small number u can be handled by introducing new attributes $Y_1, \ldots, Y_u$ and converting an implicit set $\{y_1, \ldots y_u\}$ of Y-values into a sequence, i.e. a subtuple $(y_1, \ldots y_u)$ over the new attribute set. Of course, in the context of achieving view support we would need the multivalued dependency $X \twoheadrightarrow Y$ in addition. Finally it is important to observe that actually "usual" cases and "exceptional" ones are treated uniformly by formally specifying pertinent semantic constraints which have the normal (unrelaxed) meaning.

Secondly, we can further extend the expressiveness of semantic constraints beyond restricting the cardinality of nonempty sets of the form $\pi_Y(\sigma_{X=\mu}(r))$, namely by requiring that certain sets are nonempty indeed, or empty indeed, respectively. The syntax of a nonnull dependency [CTFB89] has the form $\sigma_\Phi \rightarrow X \neq \lambda$, where Φ is a selection formula and X a set of attributes; such a constraint is satisfied by a relation r iff for all tuples $\mu \in \sigma_\Phi(r)$ the subtuple $\mu\lceil X$ is defined and unequal to the null value λ. Correspondingly we could consider null dependencies of the form $\sigma_\Phi \rightarrow X = \lambda$ stating that for all $\mu \in \sigma_\Phi(r)$ the X-components of μ are all set to the null value λ. If we interpret λ as non-applicable and if we succeed in determining complementary selection formulas Φ_1 and Φ_2 for a null dependency $\sigma_{\Phi_1} \rightarrow X = \lambda$ and a nonnull dependency $\sigma_{\Phi_2} \rightarrow X \neq \lambda$, then we can employ these constraints to build an specialization hierarchy and the corresponding instance inclusion between the (more general) Φ_1-case and (the more specialized) Φ_2-case.

A lot of work has been done considering the implication problem for semantic constraints if their semantics is defined in terms of relations with null values [Lie79, Vas79, Gra84, AM86, Tha91, LL94], and studying query evaluation for such relations [Bis83, IL84, AKG91, Lev92, Lib95], but the impact for schema design has less been investigated [CM87, Heg94].

6.3 Impact of more powerful query languages for rules

The Separation heuristics are necessarily complemented by the Inferential Completeness heuristic, because any fragmentation of information into its "basic aspects" at design time has to be compensated by providing appropriate means to recover the original information from its fragments at query time. The means at hand to this purpose is just the *query language* used to declare the *rules* of the schema. Clearly, the expressive power of the query language essentially determines which kinds of recovering are achievable and, hence, how fine information can be fragmented. This insight is reflected in the formal definitions of view support (as presented in section 5.3) by their implicit parametrization with respect to the query language under consideration.

For relational databases we have the relational algebra as prototype for a query language. As we have seen, mainly the natural join and, to a lesser extent, the (generalized) union are actually used to define views. For logic oriented databases the relational algebra is strictly extended by recursion thereby offering more powerful options.

A powerful extension of decompositional normal forms for relational database schemas has been outlined in [BC89,Con89], see also [HV88]. The extension is based on the observation that relational normalization removes a join dependency from the declared semantic constraints and inserts the corresponding join query into the declared rules (see Theorem 2 and Theorem 9), and that this transformation is justified by reducing redundancy and hence storage costs (see Theorem 11). Now join dependencies are an example for the rich class of tuple-generating semantic constraints, which in a Horn-clause like language are expressed by possibly recursive sets of clauses the conclusions of which may refer to extensional enumerations. Many of these tuple-generating semantic constraints can be considered as indicating undesirable redundancy, and thus they should be transformed into queries for declared rules. As an example, the requirement that a binary relation should be transitively closed can be expressed as tuple-generating semantic constraint using the well-known Horn-clauses for transitive closure, but we may want to explicitly enumerate merely a skeleton relation from which the transitively closed one can be inferred. For instance, if an application is interested in querying the ancestor relation over some population, we nevertheless would like to enumerate only the parent relation. It appears to me that the basic contributions of [HVC87, HV88, BC89, Con89] need further elaboration, in particular dealing with negation and supporting updates.

6.4 Impact of more powerful structures for formats

Seen from the relational point of view the structures for formats can be made more powerful mainly concerning two issues, namely *complex objects* and *object identifiers.* Complex objects reintroduce sets into the rigidly flat relational world; object identifiers reintroduce explicit references (pointers, abstract addresses) into the purely value oriented relational paradigm. The first augmentation has been pioneered by the nested relational model [PDGvG89], the second one by the RM/T model involving surrogate relations [Cod79]. Their combination is central to object oriented models [AK89, Bee90, ST93, KV93, KLW93] that, in particular, provide the option to refer to a copy of a set by its unique identifier. Of course, the query languages also have to be adjusted to the enhanced options for formats, and therefore the general statements of section 6.3 apply here as well. The design problem for object oriented systems has attracted a lot of researches but most results are still of a somehow heuristic nature. Examples of formal work can be found in [LX93, PTCL93, BMP94].

6.4.1 Pivoting The new features of object oriented models offer a wide range to formalize *relationships between entities* of the real world and, correspondingly, a new type of transformation between those options. In order to *formalize a relationship set*, as modelled by an entity-relationship diagram, we can always canonically simulate the relational approach: for each individual relationship between objects we construct a (relationship) object containing references to the related objects. Accordingly, the class declaration contains a method for each component of the relationship class. For instance, the relationship set of Fig.5a is formalized by the (F-logic) declaration of Fig.5b where U, C, S, T are class designators and *co*, *st*, *te* are method designators; Fig.5c visualizes the structure of a typical instance where term $\square$ stands for an object uniquely identified by the object identifier term.

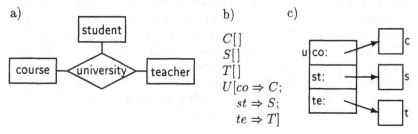

Fig.5. a) a relationship set, **b)** its canonical object-oriented formalization, **c)** the structure of a typical instance

We also need canonical semantic constraints[1] requiring that the component methods are actually defined for every relationship and that a relationship is uniquely formalized:

[complete representation]
$$!- (\exists c)\, x[co \to c] \leftarrow x\colon U$$
$$!- (\exists s)\, x[st \to s] \leftarrow x\colon U$$
$$!- (\exists t)\, x[te \to t] \leftarrow x\colon U$$

[unique representation]
$$!- x_1 = x_2 \leftarrow \quad x_1\colon U[co \to c;\, st \to s;\, te \to t]$$
$$\land \quad x_2\colon U[co \to c;\, st \to s;\, te \to t]$$

Depending on further application-dependent semantic constraints alternative formalizations can be obtained by the schema transformation of pivoting. Roughly speaking, *pivoting* moves information from relationship objects (as introduced by the canonical formalization) to objects representing entities. More precisely, we can select any of the key components of the relationship class as *pivot class* and encode all pertinent information into new methods to be defined for the objects of the pivot class. In some cases the application-dependent constraints indicate that the pivoted version can be further improved. Assuming that T is a key component and that the functional constraint $T \to C$ holds pivoting would transform the situation of Fig.5 into that of Fig.6.

a)
$$C[\,]$$
$$S[\,]$$
$$T[u\_m \Rrightarrow S;\, co\_m \Rrightarrow C]$$

b)

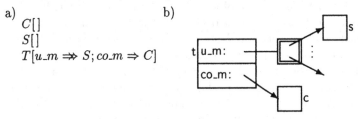

Fig.6. **a)** the pivoted formalization, **b)** the structure of its typical instance

As a formal transformation on (F-logic [KLW93]) database schemas pivoting was introduced and investigated in [BMP94] and [BSMP95] and clearly deserves further studies. Being composed of some kind of nesting, some adaption of decomposition and class removal, pivoting can be generalized in various ways and therefore it is a very powerful transformation for object-oriented schemas which, as a special case, also comprises some sort of class specialization (introducing a new subclass). In order to preserve semantics (in the sense of view support to ensure Inferential Completeness) pivoting is driven by semantic constraints, which also seem to indicate the cases in which pivoting is recommendable as optimization at design time. Roughly summarized:

- Pivoting diminishes the time complexity of enforcing semantic constraints if the left-hand side of a functional constraint corresponds to the pivot class (the objects of which are the goals of the pivoting). In that case, as in classical relational normal forms, the functional constraint becomes a key constraint that can be efficiently maintained under updates.
- Pivoting helps to eliminate two different kinds of redundancy:
  - First by moving properties to more suitable objects.
  - Second by removing objects that turn out to be unnecessary.

---

[1] This F-logic formula [KLW93] should be read as follows: As a semantic constraint it is required (!-): if object x is an instance of class U ($x\colon U$) then ($\leftarrow$) there exists an object c such that for x the method co delivers c ($x[co \to c]$).

6.4.2 Sharing If complex objects and object identifiers are disposable, a design can also employ advanced possibilities of data sharing and, accordingly, a schema transformation to favour explicit object sharing via object identifiers. The cost benefit from object sharing is similar to that resulting from decomposition into normal form in the relational case, as presented by Theorem 11. Basically, the relational case and the object oriented case differ in how the structured data is referred to: by user defined values or by system generated object identifiers. The difference can be illustrated by the following example. For a relation scheme $(R, \{Person, City, Zip\}, \{Person \rightarrow City, City \rightarrow Zip\})$, the two tuples of the instance shown in Fig.7a repeatedly contain the data that the city han has zip code number 511; after decomposing the scheme into BCNF and accordingly projecting the instance, this data is referred to by the user defined value han and thereby shared. In a corresponding object oriented formalization, the instance objects of Fig.8a again repeatedly represent that data; a "normalizing transformation" would create a new object for that data which is referred to by its object identifier as shown in Fig.8b.

a)

R	Person	City	Zip
	miller	han	511
	smith	han	511

b)

R1	Person	City
	miller	han
	smith	han

R2	City	Zip
	han	511

Fig.7. a) repeated data, b) shared via a user defined value

a)   b)

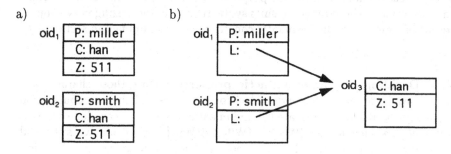

Fig.8. a) repeated data, b) shared via an object identifier

In general, the goal is to avoid repeated occurences of data which indicate a violation of the Separation of Aspects heuristic. Therefore we are looking for the desirable syntactic properties the negations of which characterize "forbidden substructures" (which are the transitive functional dependencies in the relational case) which may result in repeated data, and for corresponding transformations to remove them stepwise. Our own project [BP95] is still somehow preliminary and can only be sketched here:

– Consider *complex values* built up from atomic values by pertinent *constructors*, say for tuple and set forming.
– Define a formal notion that an object *represents* a complex value.

- Identify the (unwanted) possibility of repeated representation of complex values from the *implicational closure* of the declared *semantic constraints*, i.e. identify "forbidden substructures".
- Algorithmically *transform* the schema by introducing a new class appropriately defined for the type of the (possibly) repeated complex values in order to replace the methods used for representing these complex values by a new method that just references a (possibly) shared object of the new class.

Of course, there is room for many variations, according to Tasks 1 and 2 in particular with respect to the selected constructors, the notion of representation, and the considered semantic constraints. According to Task 3 for any variation a lot of specific problems have to be solved: an efficient inference system for the involved semantic constraints, an efficient recognition algorithm for forbidden substructures, and efficient algorithms for transforming schemas including their semantic constraints. And Task 4 concerning costs has to be tackled too.

6.4.3 View support Of course, if we transform an object oriented schema by pivoting or take advantage of data sharing, then we must pay attention to the third design heuristic, Inferential Completeness, in an appropriate form. Fundamental work has been provided, for example, by [AK89, dSAD94] defining a notion of *equivalence* of object oriented schemas and defining a general *object oriented view* mechanism. But for purposes of pivoting and the sharing transformation we need more refined results syntactically characterizing view support in terms of logically implied semantic constraints (as exemplified in section 5.4).

6.4.4 Deciding desirable syntactic properties Once those characterizations are available the corresponding decision problems and their computational complexity should be investigated. Fundamental work has been done on *path functional dependencies* by [Wed92, IW94, vBW94]. But again we still need more refined and more general results about the implicational closure of semantic constraints in object oriented database schemas.

6.4.5 Types and classes In the relational approach the distinction between types and classes is usually blurred although also there at least a rudimentary form would be helpful. In particular, the hypergraph associated with a relational database schema DS should be enriched by a second sort of hyperedges representing the value domains of attributes, i.e. the atomic types of the schema. For instance, in Fig.9 the circle like hyperedges represent relation schemes whereas the rectangle like hyperedge indicates that the attributes Par and Ch have the same domain (say because the attribute Ch is used as a role name for persons acting as child in the birth register given by the three attribute scheme).

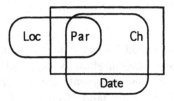

Fig.9. A two-sorted schema hypergraph representing relation schemes and domains

The domain specification can be enriched also by the inclusion dependencies to be enforced. Specification of types and of corresponding inclusion dependencies are also useful for horizontal decompositions the fragments of which have the same type. Of course, in object oriented approaches types are even more important, in particular in order to specify specialization hierarchies [CW85, Oho90, BB92, BS92, FM94, KZ95]. As mentioned before in sections 6.2 and 6.4.1, respectively, *null dependencies* as well as *pivoting* may have an impact on specialization. Although types are implicitly used in practical schema design methodologies (see for instance [RR94]) they are rarely investigated in the design theory (see [Heg94] for an exception).

6.5 Detailed cost analysis

Schema design can be understood as *optimization at design time*, and the Separation heuristics are guided by avoiding redundancy and thereby saving costs. Accordingly, Task 4 of design theory is devoted to prove that low costs are actually achieved. Unfortunately, however, little precise results on costs have been presented. Main obstacles to precise cost statements are to define agreed abstract cost measures and to evaluate not only best case and worst case costs but also average case costs. Concerning storage costs some size measure, as used in Theorem 11, can be useful. But for object oriented database we should distinguish between the occurences of values and the occurences of object identifiers because the usage of object identifiers requires some space overhead for converting them into main storage pointers after object activation. Concerning query costs we have to explicitly specify the anticipated queries and to take access structures into consideration. Concerning update costs we again have to consider access structures and investigate the time complexity of enforcement of semantic constraints [Nic82, MH89, Lip90, JQ92, GW93] and the space overhead due to access structures. It should be clear from the sketchy remarks that in general we can obtain only tradeoff statements within a specific cost model. But also precise partial results focusing on a limited point of view can be helpful. For deductive databases, a first rough outline is presented already in [HV88]. The Performance Framework and Assistant of [Nix94] provides a tool for cost driven design decisions. In [vBvdW92, vB94] the transformation from the conceptual into the internal layer is guided by storage and time costs and their tradeoffs. The case study of [BSMP95] and the cost estimates of [ST95] are also examples of this type of work.

7 Final claim

For the basic relational data model substantial fragments of a database schema design theory are available. Their usefulness for practical schema design tools will finally become clear if these fragments can be appropriately unified and combined. For more advanced data models many details of the design impact of their additional features still deserve further research. Again, the practical usefulness of theoretical contributions will depend on their compatibility and homogeneity.

Acknowledgement: I would like to thank Ralf Menzel and Torsten Polle for valuable discussions and hints.

References

[AABM82] P. Atzeni, G. Ausiello, C. Batini, and M. Moscarini. Inclusion and equivalence between relational database schemata. *Theoretical Computer Science*, 19:267–285, 1982.

[AD93] P. Atzeni and V. De Antonellis. *Relational Database Theory*. Benjamin/Cummings, Redwood City, CA, 1993.

[AHV95] S. Abiteboul, R. Hull, and V. Vianu. *Foundations of Databases*. Addison-Wesley, Reading, MA, 1995.

[AK89] S. Abiteboul and P. C. Kanellakis. Object identity as a query language primitive. In *ACM SIGMOD International Conference on Management of Data*, pages 159–173, 1989.

[AKG91] S. Abiteboul, K. Kanellakis, and G. Grahne. On the representation and querying of sets of possible worlds. *Theoretical Computer Science*, 78:159–187, 1991.

[AM86] P. Atzeni and N. M. Morfuni. Functional dependencies and constraints on null values in database relations. *Information and Control*, 70(1):1–31, 1986.

[Arm74] W. W. Armstrong. Dependency structures of data base relationships. In J. L. Rosenfeld, editor, *Proceedings of IFIP Congress 1974*, pages 580–583. North-Holland, Amsterdam, 1974.

[BB83] J. Biskup and H. H. Brüggemann. Universal relation views: A pragmatic approach. In *Proceedings of the 9th International Conference on Very Large Data Bases*, pages 172–185, 1983.

[BB88] J. Biskup and H. H. Brüggemann. The personal model of data - towards a privacy-oriented information system. *Computers & Security*, 7:575–597, 1988.

[BB92] D. Beneventano and S. Bergamaschi. Subsumption for complex object data models. In *Database Theory—ICDT '92*, number 646 in Lecture Notes in Computer Science, pages 357–375. Springer-Verlag, 1992.

[BBG78] C. Beeri, P. A. Bernstein, and N. Goodman. A sophisticated introduction to database normalization theory. In *Proceedings of the 4th International Conference on Very Large Data Bases, Berlin*, pages 113–124, September 1978.

[BBSK86] J. Biskup, H. H. Brüggemann, L. Schnetgöke, and M. Kramer. One flavor assumption and γ-acyclicity for universal relation views. In *Proceedings of the Fifth ACM SIGACT-SIGMOD-SIGART Symposium on Principles of Database Systems*, pages 148–159, 1986.

[BC86]   J. Biskup and B. Convent. A formal view integration method. In *Proceedings of the ACM SIGMOD International Conference on Management of Data, Washington*, pages 398–407, 1986.

[BC89]   J. Biskup and B. Convent. Towards a schema design methodology for deductive databases. In J. Demetrovics and B. Thalheim, editors, *Proceedings of the Symposium on Mathematical Fundamentals of Database Systems (MFDBS '89)*, number 364 in Lecture Notes in Computer Science, pages 37–52. Springer, 1989.

[BC91]   J. Biskup and B. Convent. Relational chase procedures interpreted as resolution with paramodulation. *Fundamenta Informaticae*, XV(2):123–138, 1991.

[BD93]   J. Biskup and P. Dublish. Objects in relational database schemes with functional, inclusion and exclusion dependencies. *Informatique théorique et Applications / Theoretical Informatics and Applications*, 27(3):183–219, 1993.

[BDB79]   J. Biskup, U. Dayal, and P. A. Bernstein. Synthesizing independent database schemas. In P. A. Bernstein, editor, *Proceedings of the ACM SIGMOD International Conference on Management of Data (SIGMOD '79), Boston*, pages 143–151, New York, NY, 1979. ACM.

[BDLM91] J. Biskup, J. Demetrovics, L. O. Libkin, and I. B. Muchnik. On relational database schemes having unique minimal key. *Journal of Information Processing and Cybernetics EIK*, 27(4):217–225, 1991.

[Bee90]   C. Beeri. A formal approach to object-oriented databases. *Data & Knowledge Engineering*, 5(1990):353–382, 1990.

[Ber76]   P. A. Bernstein. Synthesizing third normal form relations from functional dependencies. *ACM Transactions on Database Systems*, 1(4):272–298, December 1976.

[BFMY83] C. Beeri, R. Fagin, D. Maier, and M. Yannakakis. On the desirability of acyclic database schemes. *Journal of the ACM*, 30:479–513, 1983.

[BG80]   P. A. Bernstein and N. Goodman. What does Boyce-Codd normal form do? In *Proceedings of the 6th International Conference on Very Large Data Bases*, pages 245–259, 1980.

[Bis80]   J. Biskup. Inferences of multivalued dependencies in fixed and undetermined universes. *Theoretical Computer Science*, 10:93–105, 1980.

[Bis83]   J. Biskup. A foundation of Codd's relational maybe-operations. *ACM Transactions on Database Systems*, 8:608–636, 1983.

[Bis89]   J. Biskup. Boyce-Codd normal form and object normal forms. *Information Processing Letters*, 32(1):29–33, 1989.

[Bis95]   J. Biskup. *Grundlagen von Informationssystemen*. Vieweg, Braunschweig-Wiesbaden, 1995. to appear.

[BK86]   C. Beeri and M. Kifer. An integrated approach to logical design of relational database schemes. *ACM Transactions on Database Systems*, 11(2):134–158, 1986.

[BLN86]   C. Batini, M. Lenzerini, and S. B. Navathe. A comparative analysis of methodologies for database schema integration. *ACM Computing Surveys*, 18:323–364, 1986.

[BM87]   J. Biskup and R. Meyer. Design of relational database schemes by deleting attributes in the canonical decomposition. *Journal of Computer and System Sciences*, 35(1):1–22, 1987.

[BMP94]   J. Biskup, R. Menzel, and T. Polle. Transforming an entity-relationship schema into object-oriented database schemas. Technical Report 17/94, Institut für Informatik, Universität Hildesheim, June 1994. Short version in Proceedings 2nd International Workshop on Advances in Databases and Information Systems, Moscow, pages 67–78, 1995.

[BP95]   J. Biskup and T. Polle. Sharing in object-oriented database models. In preparation, 1995.

[BR88]   J. Biskup and U. Räsch. The equivalence problem for relational database schemes. In *Proceedings of the 1st Symposium on Mathematical Fundamentals of Database Systems*, number 305 in Lecture Notes in Computer Science, pages 42–70. Springer-Verlag, Berlin etc., 1988.

[BS81]   F. Bancilhon and N. Spyratos. Update semantics of relational views. *ACM Transactions on Database Systems*, 6(4):557–575, 1981.

[BS92]   S. Bergamaschi and C. Sartori. On taxonomic reasoning in conceptual design. *ACM Transactions on Database Systems*, 17:385–422, 1992.

[BSMP95]   J. Biskup, Y. Sagiv, R. Menzel, and T. Polle. A case study on object-oriented database schema design. Technical report, Universität Hildesheim, 1995.

[BV84a]   C. Beeri and M. Y. Vardi. Formal systems for tuple and equality generating dependencies. *SIAM Journal on Computing*, 13(1):76–98, 1984.

[BV84b]   C. Beeri and M. Y. Vardi. A proof procedure for data dependencies. *Journal of the ACM*, 31(4):718–741, October 1984.

[BV88]   V. Brosda and G. Vossen. Update and retrieval in a relational database through a universal schema interface. *ACM Transactions on Database Systems*, 13(1988):449–485, 1988.

[CFP84]   M. A. Casanova, R. Fagin, and C. H. Papadimitriou. Inclusion dependencies and their interaction with functional dependencies. *Journal of Computer and System Sciences*, 28(1):29–59, 1984.

[Cha89]   E. P. F. Chan. A design theory for solving the anomalies problem. *SIAM Journal on Computing*, 18(3):429–448, June 1989.

[Che76]   P. P.-S. Chen. The entity-relationship-model — towards a unified view of data. *ACM Transactions on Database Systems*, 1(1):9–36, March 1976.

[CM87]   E. P. F. Chan and A. O. Mendelzon. Independent and separable database schemes. *SIAM Journal on Computing*, 16(5):841–851, 1987.

[Cod70]   E. F. Codd. A relational model of data for large shared data banks. *Communications of the ACM*, 13(6):377–387, June 1970.

[Cod72]   E. F. Codd. Further normalization of the database relational model. In R. Rustin, editor, *Database Systems*, number 6 in Courant Institute Computer Science Symposia Series, pages 33–64. Prentice Hall, Englewood Cliffs, NJ, 1972. also in: IBM Research Report RJ909.

[Cod79]   E. F. Codd. Extending the database relational model to capture more meaning. *ACM Transactions on Database Systems*, 4(4):397–434, December 1979.

[Con86]   B. Convent. Unsolvable problems related to the view integration approach. In *Database Theory—ICDT '86*, number 243 in Lecture Notes in Computer Science, pages 141–156. Springer, Berlin etc., 1986.

[Con89]   B. Convent. *Datenbankschemaentwurf für ein logikorientiertes Datenmo-dell.* PhD thesis, Universität Hildesheim, March 1989.

[CTFB89]   M. A. Casanova, L. Tucherman, A. L. Furtado, and A. Braga. Optimization of relational schemes containing inclusion dependencies. In *Proceedings of the 15th International Conference on Very Large Databases, Amsterdam,* pages 317–325, 1989.

[CV85]   A. K. Chandra and M. Y. Vardi. The implication problem for functional and inclusion dependencies is undecidable. *SIAM Journal on Computing,* 14(3):671–677, 1985.

[CW85]   L. Cardelli and P. Wegner. On understanding types, data abstraction and polymorphism. *ACM Computing Surveys,* 17:471–522, 1985.

[DB82]   U. Dayal and P. A. Bernstein. On the correct translation of update operations on relational views. *ACM Transactions on Database Systems,* 8(3):381–416, 1982.

[Del78]   C. Delobel. Normalization and hierarchical dependencies in the relational data model. *ACM Transactions on Database Systems,* 3:201–222, 1978.

[DF92]   C. J. Date and R. Fagin. Simple conditions for guaranteeing higher normal forms in relational databases. *ACM Transactions on Database Systems,* 17:465–476, 1992.

[DKM$^+$95] J. Demetrovics, G. O. H. Katona, D. Miklos, O. Seleznjew, and B. Thalheim. The average length of key and functional dependencies in (random) databases. In G. Gottlob and M. Y. Vardi, editors, *Database Theory—ICDT '95,* pages 266–279. Springer-Verlag, Berlin etc., 1995.

[DM86]   A. D'Atri and M. Moscarini. Recognition algorithms and design methodologies for acyclic database. In P. C. Kanellakis and F. Preparata, editors, *Advances in Computing Research,* volume 3, pages 164–185. JAI Press, Inc., Greenwich, CT, 1986.

[DP84]   P. DeBra and J. Paredaens. Horizontal decompositions for handling exceptions to functional dependencies. In H. Gallaire, J. Minker, and J. M. Nicolas, editors, *Advances in Database Theory,* volume 2. Plenum, New York - London, 1984.

[dSAD94]   C. S. dos Santos, S. Abiteboul, and C. Delobel. Virtual schemas and bases. In M. Jarke, J. Bubenko, and K. Jeffery, editors, *Advances in Database Technology—EDBT '94,* number 779 in Lecture Notes in Computer Science, pages 81–94. Springer-Verlag, Berlin etc., 1994.

[Fag77]   R. Fagin. Multivalued dependencies and a new normal form for relational databases. *ACM Transactions on Database Systems,* 2(3):262–278, September 1977.

[Fag81]   R. Fagin. A normal form for relational databases that is based on domains and keys. *ACM Transactions on Database Systems,* 6(3):387–415, 1981.

[Fag83]   R. Fagin. Degrees of acyclicity for hypergraphs and relational database schemes. *Journal of the ACM,* 30(3):514–550, July 1983.

[FC85]   A. L. Furtado and M. A. Casanova. Updating relational views. In W. Kim, D. S. Reiner, and D. S. Batory, editors, *Query Processing in Database Systems.* Springer-Verlag, Berlin, 1985.

[FM94]   A. Formica and M. Missikoff. Correctness of ISA hierarchies in object-oriented database schemas. In M. Jarke, J. Bubenko, and K. Jeffery, editors, *Advances in Database Technology—EDBT '94,* number 779 in Lecture Notes in Computer Science, pages 231–244. Springer-Verlag, Berlin etc., 1994.

[FV84]   R. Fagin and M. Y. Vardi. The theory of data dependencies - an overview. In *Proceedings of the 11th International Colloquium on Automata, Languages and Programming*, number 172 in Lecture Notes in Computer Science, pages 1–22. Springer-Verlag, Berlin etc., 1984.

[FV95]   C. Fahrner and G. Vossen. A survey of database design transformations based on the Entity-Relationship model. *Data & Knowledge Engineering*, 15(1995):212–250, 1995.

[GHLM93]   J. Grant, J. Horty, J. Lobo, and J. Minker. View Updates in Stratified Disjunctive Databases. *Journal of Automated Reasoning*, 11:249–267, 1993.

[GM85]   J. Grant and J. Minker. Normalization and axiomatization for numerical dependencies. *Information and Control*, 65:1–17, 1985.

[GMV86]   M. H. Graham, A. O. Mendelzon, and M. Y. Vardi. Notions of dependency satisfaction. *Journal of the ACM*, 33(1):105–129, 1986.

[Gra79]   M. H. Graham. On the universal relation. Systems research group report, University of Toronto, 1979.

[Gra84]   G. Grahne. Dependency satisfaction in databases with incomplete information. In U. Dayal, editor, *Proceedings of the 10th International Conference on Very Large Data Bases*, pages 37–45, Singapore, 1984.

[GW93]   A. Gupta and J. Widom. Local verification of global integrity constraints in distributed databases. In *ACM SIGMOD International Conference on Management of Data*, pages 49–58, 1993.

[HC91]   H. J. Hernández and E. P. F. Chan. Constant-time-maintainable BCNF database schemes. *ACM Transactions on Database Systems*, 16(4):571–599, December 1991.

[Heg94]   S. J. Hegner. Unique complements and decompositions of database schemata. *Journal of Computer and System Sciences*, 48:9–57, 1994.

[Hon82]   P. Honeyman. Testing satisfaction of functional dependencies. *Journal of the ACM*, 29(3):668–677, 1982.

[Hul86]   R. Hull. Relative information capacity of simple relational database schemata. *SIAM Journal on Computing*, 15(3):856–886, 1986.

[HV88]   P. Helman and R. Veroff. Designing deductive databases. *Journal of Automated Reasoning*, 4(1):29–68, 1988.

[HVC87]   P. Helman, R. Veroff, and A. Cable. Deductive database design in the presence of updates. Technical Report CS 87-9, University of New Mexico, Albuquerque, 1987.

[IL84]   T. Imieliński and W. Lipski, Jr. Incomplete information in relational databases. *Journal of the ACM*, 31(4):761–791, 1984.

[IW94]   M. Ito and G. E. Weddell. Implication problems for functional constraints on databases supporting complex objects. *Journal of Computer and System Sciences*, 49:726–768, 1994.

[JF82]   J. H. Jou and P. C. Fischer. The complexity of recognizing 3NF relation schemes. *Information Processing Letters*, 14(4):187–190, 1982.

[Joh94]   P. Johannesson. Schema standardization as an aid in view integration. *Information Systems*, 19(3):275–290, 1994.

[JQ92]   H. V. Jagadish and X. Qian. Integrity maintenance in an object-oriented database. In L.-Y. Yuan, editor, *Proceedings of the 18th International Conference on Very Large Data Bases*, pages 469–480, British Columbia, Canada, 1992.

[Kat92]   G. O. H. Katona. Combinatorial and algebraic results for database rela-
            tions. In *Database Theory—ICDT '92*, number 646 in Lectures Notes in
            Computer Science, pages 1–20. Springer-Verlag, Berlin etc., 1992.

[KCV83]   P. C. Kanellakis, S. S. Cosmadakis, and M. Y. Vardi. Unary inclusion
            dependencies have polynomial time inference problems. In *Proceedings
            of the 15th Symposium on Theory of Computing, Boston*, pages 246–277,
            1983.

[Kel86]   A. M. Keller. The role of semantics in translating view updates. *IEEE
            Computer*, 19(1):63–73, January 1986.

[Ken83]   W. Kent. A simple guide to five normal forms in relational databases.
            Communications of the ACM, 26(2):120–125, 1983.

[KKF⁺84]   H. F. Korth, G. M. Kuper, J. Feigenbaum, A. VanGeldern, and J. D. Ull-
            man. A database system based on the universal relation assumption. *ACM
            Transactions on Database Systems*, 9(1984):331–347, 1984.

[KLW93]   M. Kifer, G. Lausen, and J. Wu. Logical foundations of object-oriented
            and frame-based languages. Technical Report TR 93/06, State University
            of New York (SUNY) at Stony Brook, Department of Computer Science,
            NY 11794, April 1993. To appear in Journal of the ACM.

[KM80]   P. Kandzia and M. Mangelmann. On covering Boyce-Codd normal forms.
            Information Processing Letters, 11:218–223, 1980.

[KV93]   G. M. Kuper and M. Y. Vardi. The logical data model. *ACM Transactions
            on Database Systems*, 18:379–413, 1993.

[KZ95]   L. A. Kalinichenko and V. I. Zadorozhny. Type inferencing based on com-
            plete type specifications. In *Proceedings 2nd International Workshop on
            Advances in Databases and Information Systems (ADBIS'95)*, pages 111–
            117, Moscow, 1995.

[Lev92]   M. Levene. *The Nested Universal Relation Database Model*. Lecture Notes
            in Computer Science 595. Springer, Berlin etc., 1992.

[Lib95]   L. Libkin. Approximation in databases. In G. Gottlob and M. Y. Vardi,
            editors, *Database Theory—ICDT '95*, pages 411–424. Springer-Verlag, Ber-
            lin etc., 1995.

[Lie79]   Y. E. Lien. Multivalued dependencies with nulls in relational databases.
            In *Proceedings of the 5th International Conference on Very Large Data
            Bases*, pages 61–66, 1979.

[Lip90]   U. Lipeck. Transformation of dynamic integrity constraints into transac-
            tion specifications. *Theoretical Computer Science*, 76:115–142, 1990.

[LL94]   M. Levene and G. Loizou. The nested universal relation model. *Journal
            of Computer and System Sciences*, 49:683–717, 1994.

[LNE89]   J. A. Larson, S. B. Navathe, and R. Elmasri. A theory of attribute equiv-
            alence in databases with application to schema integration. *IEEE Trans-
            actions on Software Engineering*, 15(4):449–463, 1989.

[LO78]   C. L. Lucchesi and S. L. Osborn. Candidate keys for relations. *Journal of
            Computer and System Sciences*, 17(2):270–279, 1978.

[LTK81]   T.-W. Ling, F. W. Tompa, and T. Kameda. An improved third normal
            form for relational databases. *ACM Transactions on Database Systems*,
            6(2):329–346, 1981.

[LX93]   K. J. Lieberherr and C. Xiao. Formal foundations for object-oriented
            data modeling. *IEEE Transactions on Knowledge and Data Engineering*,
            5(3):462–478, June 1993.

[Mai83]   D. Maier. *The Theory of Relational Databases*. Computer Science Press, Rockville, MD, 1983.

[Men79]   A. O. Mendelzon. On axiomatizing multivalued dependencies in relational databases. *Journal of the ACM*, 26(1):37–44, 1979.

[MH89]   W. W. McCune and L. J. Henschen. Maintaining state constraints in relational databases: A proof theoretic basis. *Journal of the ACM*, 36(1):46–68, 1989.

[MIR93]   R. J. Miller, Y. E. Ioannidis, and R. Ramakrishnan. The use of information capacity in schema integration and translation. In R. Agrawal, editor, *Proceedings of the 19th International Conference on Very Large Data Bases*, pages 120–133, Dublin, Irland, 1993.

[MIR94]   R. J. Miller, Y. E. Ioannidis, and R. Ramakrishnan. Schema equivalence in heterogeneous systems: bridging theory and practice. *Information Systems*, 19(1):3–31, 1994.

[Mit83]   J. C. Mitchell. The implication problem for functional and inclusion dependencies. *Information and Control*, 56(3):154–173, 1983.

[MR83]   H. Mannila and K.-J. Räihä. On the relationship of minimum and optimum covers for a set of functional dependencies. *Acta Informatica*, 20:143–158, 1983.

[MR86]   H. Mannila and K.-J. Räihä. Inclusion dependencies in database design. In *Proceedings of the Second International Conference on Data Engineering*, pages 713–718, Washington, DC, 1986. IEEE Computer Society Press.

[MR92]   H. Mannila and K.-J. Räihä. *The Design of Relational Databases*. Addison-Wesley, Wokingham, England, 1992.

[MUV84]   D. Maier, J. D. Ullman, and M. Y. Vardi. On the foundations of the universal relation model. *ACM Transactions on Database Systems*, 9(2):283–308, June 1984.

[Nic82]   J.-M. Nicolas. Logic for improving integrity checking in relational databases. *Acta Informatica*, 18(3):227–253, 1982.

[Nix94]   B. A. Nixon. Representing and using performance requirements during the development of information systems. In M. Jarke, J. Bubenko, and K. Jeffery, editors, *Advances in Database Technology—EDBT '94*, number 779 in Lecture Notes in Computer Science, pages 187–200. Springer-Verlag, Berlin etc., 1994.

[Oho90]   A. Ohori. Semantics of types for database objects. *Theoretical Computer Science*, 1990.

[Osb78]   S. L. Osborn. *Normal Forms for Relational Data Bases*. PhD thesis, Department of Computer Science, University of Waterloo, 1978.

[PDGvG89]   J. Paredaens, P. DeBra, M. Gyssens, and D. van Gucht. *The Structure of the Relational Database Model*. Number 17 in EATCS Monographs on Theoretical Computer Science. Springer-Verlag, Berlin, 1989.

[PTCL93]   P. Poncelet, M. Teisseire, R. Cicchetti, and L. Lakhal. Towards a formal approach for object database design. In R. Agrawal, editor, *Proceedings of the 19th International Conference on Very Large Data Bases*, pages 278–289, Dublin, Irland, 1993.

[Rei92]   R. Reiter. What should a database know. *Journal of Logic Programming*, 14:127–153, 1992.

[RR94]   A. Rosenthal and D. Reiner. Tools and transformations — rigorous and otherwise — for practical database design. *ACM Transactions on Database Systems*, 19:167–211, 1994.

[RS94]   A. Rosenthal and L. J. Seligman. Data integration in the large: the challenge of reuse. In *Proceedings of the 20th International Conference on Very Large Data Bases*, pages 1–7, 1994.

[Sag83]   Y. Sagiv. A characterization of globally consistent databases and their correct access paths. *ACM Transactions on Database Systems*, 8(2):266–286, 1983.

[SGN93]   A. Sheth, S. Gala, and S. Navathe. On automatic reasoning for schema integration. *International Journal on Intelligent and Cooperative Information Systems*, 2(1), March 1993.

[SPD92]   S. Spaccapietra, C. Parent, and Y. Dupont. Model independent assertions for integration of heterogeneous schemas. *VLDB J.*, 1:81–126, 1992.

[SR88]   D. Seipel and D. Ruland. Designing gamma-acyclic database schemes using decomposition and augmentation techniques. In *Proc. 1st Symposium on Mathematical Fundamentals of Database Systems*, number 305 in Lecture Notes in Computer Science, pages 197–209. Springer-Verlag, Berlin etc., 1988.

[ST93]   K.-D. Schewe and B. Thalheim. Fundamental concepts of object oriented databases. *Acta Cybernetica*, 11(1-2):49–83, 1993.

[ST95]   M. Steeg and B. Thalheim. A computational approach to conceptual database optimization. Technical report, BTU Cottbus, May 1995.

[Ste93]   W. Stein. Objektorientierte Analysemethoden — ein Vergleich. *Informatik-Spektrum*, 16:317–332, 1993.

[Tha91]   B. Thalheim. *Dependencies in relational databases*. Teubner, Stuttgart - Leipzig, 1991.

[TLJ90]   P. Thanisch, G. Loizou, and G. Jones. Succint database schemes. *International Journal of Computer Mathematics*, 33:55–69, 1990.

[TY84]   R. E. Tarjan and M. Yannakakis. Simple linear-time algorithms to test chordality of graphs, test acyclicity of hypergraphs, and selectivity reduce acyclic hypergraphs. *SIAM Journal on Computing*, 13:566–579, 1984.

[TYF86]   T. J. Teorey, D. Yang, and J. P. Fry. A logical design methodology for relational databases using the extended entity-relationship model. *ACM Computing Surveys*, 18(2):197–222, 1986.

[Ull88]   J. D. Ullman. *Principles of Database and Knowledge-Base Systems (Volume I)*. Computer Science Press, Rockville, MD, 1988.

[Ull89]   J. D. Ullman. *Principles of Database and Knowledge-Base Systems (Volume II: The New Technologies)*. Computer Science Press, Rockville, MD, 1989.

[Var82]   M. Y. Vardi. On decomposition of relational databases. In *Proc. 23rd Symposium on Foundations of Computer Science*, pages 176–185, 1982.

[Var84]   M. Y. Vardi. The implication and finite implication problem for typed template dependencies. *Journal of Computer and System Sciences*, 28:3–28, 1984.

[Var88a]   M. Y. Vardi. Fundamentals of dependency theory. In E. Börger, editor, *Trends in theoretical computer science*, pages 171–224. Computer Science Press, Rockville, 1988.

[Var88b]   M. Y. Vardi. The universal-relation data model for logical independence. *IEEE Software*, 5(1988):80–85, 1988.

[Vas79]   Y. Vassiliou. Null values in database management, a denotational semantics approach. In *Proc. ACM SIGMOD Symp. on the Management of Data*, pages 162–169, 1979.

[vB94]   P. van Bommel. Experiences with EDO: An evolutionary database opti-
             mizer. *Data & Knowledge Engineering*, 13(1994):243–263, 1994.

[vBvdW92]   P. van Bommel and T. P. van der Weide. Reducing the search space
             for conceptual schema transformation. *Data & Knowledge Engineering*,
             8(1992):269–292, 1992.

[vBW94]   M. F. van Bommel and G. E. Weddell. Reasoning about equations and
             functional dependencies on complex objects. *IEEE Transactions on Know-
             ledge and Data Engineering*, 6(3):455–469, 1994.

[Vos88]   G. Vossen. A new characterization of FD implication with an application
             to update anomalies. *Information Processing Letters*, 29(3):131–135, 1988.

[Vos91]   G. Vossen. *Data Models, Database Languages and Database Management
             Systems*. Addison-Wesley, Wokingham, England, 1991.

[VS93a]   M. W. Vincent and B. Srinivasan. A note on relation schemes which are in
             3NF but not in BCNF. *Information Processing Letters*, 48:281–283, 1993.

[VS93b]   M. W. Vincent and B. Srinivasan. Redundancy and the justification for
             fourth normal form in relational databases. *International Journal of
             Foundations of Computer Science*, 4:355–365, 1993.

[Wed92]   G. E. Weddell. Reasoning about functional dependencies generalized for
             semantic data models. *ACM Transactions on Database Systems*, 17(1):32–
             64, March 1992.

[YÖ79]   C. T. Yu and Z. M. Özsoyoğlu. An algorithm for tree-query membership of
             a distributed query. In *Proceedings of the 3rd IEEE COMPSAC, Chicago*,
             pages 306–312, 1979.

[YÖ92a]   L.-Y. Yuan and Z. M. Özsoyoğlu. Design of desirable relational database
             schemes. *Journal of Computer and System Sciences*, 45:435–470, 1992.

[YÖ92b]   L.-Y. Yuan and Z. M. Özsoyoğlu. Unifying functional and multivalued
             dependencies for relational database design. *Information Science*, 59:185–
             211, 1992.

[Zan76]   C. Zaniolo. *Analysis and design of relational schemata for database sys-
             tems*. PhD thesis, University of California Los Angeles, Computer Science
             Department, 1976. Technical Report UCLA-ENG-7669, July 1976.

Maintaining Surrogate Data for Query Acceleration in Multilevel Secure Database Systems

Brajendra Panda
Computer and Information Sciences Department
Alabama A&M University
Normal, AL 35762, USA

William Perrizo
Computer Science Department
North Dakota State University
Fargo, ND 58105, USA

Abstract

Various models have been developed for classified data management in a multilevel secure database system. While concurrency control policies have been the focus of these models, the PRISM model was developed to solve the query delay problem in such systems. In order to accelerate multilevel queries, the PRISM model maintains surrogate data in different relations, as opposed to the actual data values, that result in higher record density in a page, thus requiring less number of page I/Os. In order to establish mapping between the surrogate values and the actual data values some auxiliary data structures are maintained. This paper presents the cost of maintaining such surrogate values and corresponding data structures in the model. The result is compared with the maintenance cost of the relations in the SeaView model that has been used as the base model for the PRISM model.

Key Words: Multilevel Secure Database Systems, Mandatory Access Control, Data and User Classifications, Query Processing.

1 Introduction

Many database applications require the use of classified data that needs to be protected against unauthorized accesses. Existing database systems use discretionary access control procedures that give the owner of a data item the right to decide on the access privileges of other users to the same data item. However, this policy does not provide adequate mechanisms for preventing unauthorized disclosure of information. Mandatory security policy, on the other hand, uses formal authorization of users to access data of various sensitivity levels. The Bell-LaPadula model [BEL74] is one of the widely accepted models that offer a set of access control rules to provide mandatory access control policy. This model is defined in terms of subjects and objects. A subject represents an active entity in the system (e.g., a process), whereas an object represents passive data (e.g., a relation, a record, a field, etc.). Every object has a security classification which consists of a hierarchical sensitivity level (e.g., Top Secret (TS), Secret (S), Confidential (C), Unclassified (U)) and a set of categories (e.g., Financial, Administrative, etc.). Every subject has a clearance level. In order for a subject to be granted access to an object, the Bell-LaPadula model imposes the following restrictions:

The Simple Security Policy: A subject can be given a read access to an object only if the subject's clearance level is the same as or higher than the object's classification level.

*The *-Property*: A subject can be given a write access to an object only if the subject's clearance level is the same as or lower than the object's classification level.

However, data at higher levels may be contaminated by allowing lower level subjects to write at higher levels. Therefore, there is a restricted version of the *-Property that permits users to write at their own levels only. These properties protect information from being unveiled directly. Besides, the system must be able to protect sensitive information from disclosure through indirect means, such as covert signalling channels [LAM73]. Covert channels are communication channels that allow malicious subjects to transfer information to low users, thus violating systems security policy.

Various secure database system models ([SMI92], [JAJ91], [HAI91], [DEN87] for example) have been proposed so far which use multilevel security policies. But, these systems have primarily focused on the concurrency control issues. Due to various security constraints in the concurrency control mechanism, the efficiency of the system gets degraded. In most cases, high-level users seeking low-level data suffer the most due to rigid security policies. Since high-level users are most often involved in crucial real-time decision-making policies, it is undesirable that queries initiated by these users be delayed. In [PER93] the authors have proposed a model to enhance the efficiency of a multilevel secure database system. The Protected Information System Manager, or PRISM in short, presented in [PAN95] is an improved model that further optimizes multilevel query performance. The novelty of this approach is to maintain surrogate relations, as opposed to the actual data values, that result in faster accesses by increasing the record density in a page. In order to provide the mapping between the actual data values and the surrogate values, some additional data structures are maintained. Query optimization is achieved by completely reducing the number of base-relation pages read during the recovery process of multilevel relations. A comparison analysis of query execution time in the PRISM model and the SeaView model, since the PRISM model follows the SeaView [LUN90] model, has been presented in [PAN95]. In this paper, the cost of maintaining such surrogate tuples and other required data structures are computed and the result is compared with the maintenance cost of the relations in the SeaView.

The next section outlines the SeaView model, the PRISM model is described in section 3. Section 4 computes the base relation maintenance costs in both the models. The result is shown in section 5, and section 6 concludes the paper.

2 The SeaView Model

In the SeaView model, a relation, R, is represented by the schema $R(A_1, C_1, ..., A_n, C_n, TC)$, where A_1 is the *apparent key* (same as the primary key, as defined by the user) and each C_i is the classification of the attribute A_i. The domain of C_i is the range of classifications for data that can be associated with attribute A_i. The attribute TC in the schema represents the Tuple Class of the record, which is the least upper bound of the attribute classifications in the tuple. In order to prevent from possible covert channels, this model developed the concept of polyinstantiation by allowing users at different levels to insert records with the same primary key value. These records will either have different classifications of the primary key or have different values of any of the non-key attributes. Polyinstantiation arises primarily for two reasons: 1) when a high user believes in a different value of an entity than that provided to a low user and, hence, updates the value; 2) when a low user updates the value of an entity without knowing that a different value exists at a higher level. When polyinstantiation occurs, the apparent key fails to uniquely identify a tuple. SeaView has resolved this problem by using the notion of *full primary key*. A full primary key in a multilevel relation is the combination of the apparent key and classifications of all attribute values in the tuple. Figure 1 shows a multilevel relation MISSILE with three attributes: name, range, and speed of different missiles, with some polyinstantiated records.

Name	Range	Speed	TC
MT1 U	350 U	800 C	C
NT5 U	450 U	800 U	U
NT5 U	480 C	750 C	C
DNT U	400 U	800 U	U
DNT U	450 C	800 U	C
KR1 U	500 U	700 U	U
FD7 C	450 C	850 C	C
KR1 C	400 C	null C	C

Figure 1: A multilevel relation MISSILE.

For security reasons, the multilevel relations are partitioned both vertically and horizontally into single-level base relations and then are stored separately. The decomposition and recovery algorithms of multilevel relations in the SeaView model are presented in [DEN87] and [LUN88]. Figure 2 below shows the base relations, as obtained in SeaView, of the MISSILE relation given in Figure 1.

MISSILE$_{name,u}$ MISSILE$_{range,u,u}$ MISSILE$_{speed,u,u}$

MISSILE$_{name,u}$	MISSILE$_{range,u,u}$		MISSILE$_{speed,u,u}$	
MT1	MT1	350	NT5	800
NT5	KR1	500	DNT	800
DNT	NT5	450	KR1	700
KR1	DNT	400		

	MISSILE$_{range,u,c}$		MISSILE$_{speed,u,c}$	
	DNT	450	NT5	750
	NT5	480	MT1	800

MISSILE$_{name,c}$	MISSILE$_{range,c,c}$		MISSILE$_{speed,c,c}$	
FD7	FD7	450	FD7	850
KR1	KR1	400		

Figure 2: The base relations for the MISSILE relation in SeaView.

3 The PRISM Model

PRISM decomposes all multilevel relations into standard single level base relations, in a similar manner as SeaView does, which are then stored separately at different levels. These base relations are obtained by vertically (attribute wise) and horizontally (according to different classification levels) fragmenting the original multilevel relation. The multilevel relation can be obtained by taking the outer join of the corresponding base relations that are required by the user. However, the decomposition process does not occur every time a multilevel relation is built from the base relations unless the multilevel relation have been updated. Even then, only the level that gets affected by the update will invoke the decomposition process which is invisible to lower levels. Every access to any of the base relations at any level is protected by a reference monitor at that level, and each of these reference monitors enforce mandatory security. Therefore, a subject will be denied access to any data in an underlying relation unless the subject satisfies the security constraints.

For the purpose of query acceleration, we maintain some auxiliary data structures that were initially introduced in [PER91] and later modified in [PER93]. We utilize each primary key relation (pk-relations) as a table, which we call the Domain Value Table (DVT), where the relative record number (RRN) of a primary key value is used as the value identifier (vid we denote in short) of that value. Instead of storing the primary key and the non-key attribute values in the base relations, unlike the SeaView, we store the corresponding vid values along with the attribute values. This structure serves two primary purposes. First, this reduces the size of each record, thus increasing the number of records per page and decreasing the number of pages per relation. Secondly, the vid list of each base relation helps in constructing a bit vector (referred to as Domain Vector or DV) for that base relation. A DV helps in determining the presence or absence of a value in a relation's joining attribute (i.e., the primary key attribute, in this situation, for each base relation). The presence of the vid value is indicated by a 1-bit and the absence is indicated by a 0-bit in the corresponding position in the DV. The DVs, along with either clustered or non-clustered indices, eliminate the need to access tuples which do not participate in the join. The correspondence between a value and its position in the DV, at each level, is provided by the primary key relation and the relative record numbers of its tuples. Thus, the number of bits in the DVs, at a given level, is the same as the number of records in the primary key relation that exists at that level. The reader should note that the DVs for the primary key relations will consist of all ones and hence need not be maintained. Again, for each base relation, we maintain an index (B^+ tree preferably) that uses the vid values as the keys to obtain their record numbers in the relation. We denote this type of indices as vid indices. The size of this structure is also much smaller than the regular index structure for each relation, since, the vid size (which is of type small integer) is significantly less than the size of the primary key that is usually used as the key of the index. Figure 3 illustrates the base relations of the multilevel relation MISSILE as stored in the PRISM model and Figures 4.1 and 4.2 present the auxiliary data structures required for the base relations for query acceleration purpose.

MISSILE$_{name,u}$	MISSILE$_{range,u,u}$		MISSILE$_{speed,u,u}$	
MT1	1	350	2	800
NT5	4	500	3	800
DNT	2	450	4	700
KR1	3	400		

MISSILE$_{range,u,c}$		MISSILE$_{speed,u,c}$	
3	450	2	750
2	480	1	800

MISSILE$_{name,c}$	MISSILE$_{range,c,c}$		MISSILE$_{speed,c,c}$	
FD7	1	450	1	850
KR1	2	400		

Figure 3: The base relations for the MISSILE relation in PRISM.

$$\text{DV.MISSILE}_{range,u,u} = 1111 \qquad \text{DV.MISSILE}_{speed,u,u} = 0111$$
$$\text{DV.MISSILE}_{range,u,c} = 0110 \qquad \text{DV.MISSILE}_{speed,u,c} = 1100$$
$$\text{DV.MISSILE}_{range,c,c} = 11 \qquad \text{DV.MISSILE}_{speed,c,c} = 10$$

Figure 4.1: The DVs for the base relations.

MISSILE$_{name,u}$		MISSILE$_{range,u,u}$		MISSILE$_{speed,u,u}$	
vid	rec#	vid	rec#	vid	rec#
1	1	1	1	2	1
2	2	2	3	3	2
3	3	3	4	4	3
4	4	4	2		

MISSILE$_{range,u,c}$		MISSILE$_{speed,u,c}$	
vid	rec#	vid	rec#
2	2	1	2
3	1	2	1

MISSILE$_{name,c}$		MISSILE$_{range,c,c}$		MISSILE$_{speed,c,c}$	
vid	rec#	vid	rec#	vid	rec#
1	1	1	1	1	1
2	2	2	2		

Figure 4.2: The vid indices for the base relations.

3.1 Decomposition of a Multilevel Relation in PRISM

In this section, we present the decomposition formula in the PRISM in detail. As described earlier, the primary key relations are created by storing only the apparent primary key attribute(s) at each level. Each of these pk-relations is indexed as a regular data-space index and, hence, access to any of the records is done as in any regular relation. Next, each non-key attribute present in the multilevel relation is fragmented according to different levels present in it and then stored along with the *vid* of the corresponding key value. The attribute relations are indexed as vid indices instead of data-space indices. The use of a vid index and the vid values, as explained earlier, results in faster access because of the smaller size of the structure. The decomposition algorithm is described formally below.

Let A_1 be the apparent primary key of the multilevel relation, R. For each class, x, one Primary Key Relation, $R_{1,x}(A_1)$, is created by taking the values of A_1 for which $C_1 = x$. Such a relation is stored at level x. For every non-key attribute, A_i, an Attribute Relation, $R_{i,x,y}(A_1,A_i)$, is created by storing the values of $(vid(A_1),$ $A_i)$, for each class x,y such that x and y are the classifications of the apparent key and the non-key, respectively, i.e., $C_1 = x$, and $C_i = y$ (the value $vid(A_1)$ is taken from relation $R_{1,x}(A_1)$). Such a relation is classified as y and stored at level y. During the insertion into the attribute relation, the presence of duplicate is checked and eliminated if such a tuple exists.

3.2 Recovery of a Multilevel Relation in PRISM

Let us assume $[a,b]$ as a sublattice of the security lattice that needs to be accessed by a user to answer the query, where the system low level $\leq a \leq b \leq$ the level of the user. When the user does not explicitly specify the values of a and b, then $a =$ the system low level and $b =$ the level of the user are assumed. As an alternative, these values could also be $a = b =$ the level of the user, as considered in the Smith-Winslett Model [SMI92]. For the rest of this work, we deal with levels that are in the interval $[a,b]$ only.

In this algorithm, before performing the outer join operation to answer a multilevel query, a Query Vector is constructed for every level, x, and is denoted by QV_x. The number of bits in QV_x is the same as the number of entries in the primary key relation at level x. A bit is set to 1 in QV_x at position i, if the corresponding key value at the relative record number i in the key relation participates in the query. If the query does not involve any selection at level x, the QV_x would be entirely 1-bits. Otherwise, the base relations having the participating attributes are read at each level, x, and the selected vid positions in QV_x are set to ones. If there is more than one attribute involved in the selection criteria, then the smallest participating base relation is read, the selected vids are searched in the index of the next smallest participating base relation to avoid reading non-participating pages, and then the selection criteria are applied to further reduce the number of vids. By continuing this process for all participating base relations, a fully reduced list is obtained and the query vector is built.

Next, for each primary key attribute set at level x, a Polyinstantiated Domain Vector, $PDV_{x,y}$, is created in the following way for each level y such that $x \leq y$.

1. Let $PDV'_{i,x,y}$ be the vector obtained by logically ORing the domain vectors: $DV.R_{i,x,z}$ for all z, where $x \leq z < y$ and i is the attribute number required in the output. If $i = 1$, i.e., the primary key attribute is being considered, then $DV.R_{i,x,z} = DV.R_{i,x}$ for $z = x$, and $DV.R_{i,x,z}$ is a zero vector (i.e., a vector with all zero bits) for $z > x$.

2. Next $PDV_{i,x,y}$ is constructed by ANDing each $PDV'_{i,x,y}$ with the corresponding $DV.R_{i,x,y}$. The positions of 1-bits in this vector denote the positions of those vids at level x, that have at least one polyinstantiated element in the ith attribute up to level y.

3. $PDV_{x,y}$ is constructed by ORing all $PDV_{i,x,y}$s, which represents the vids having polyinstantiated elements in their records that are visible to users up to level y.

Next, for each level, x, the vids that do not participate in the query and/or do not have any polyinstantiated elements visible up to level y are filtered out. The vector that represents such information is obtained by logically ANDing QV_x with $PDV_{x,y}$ and is denoted by $PQV_{x,y}$. The readers should note that QV_x represents the key values at level x that participate in the query, irrespective of whether or not their corresponding records have any polyinstantiated elements.

Before carrying on any build or probe phase of joins, a table, called Select Omit Table, denoted by SOT_x at each primary key level x, is built. The number of columns in each SOT is the same as the number of attributes needed in the output, and the number of rows is the exact number of records that would appear in the output. Each element in such a table at level x consists of two components: the first one gives the address of a tuple, and the second one represents the attribute classification y of the base relation, $R_{i,x,y}$, where the tuple appears.

To construct SOT_x, QV_x is taken first and scanned to find the position of 1-bits in it. The position of such a bit indicates that the corresponding key position would appear in the result. To find out the record number of such a key value in attribute relation, R_i, that is required in the output, the corresponding position in the domain vector, $DV.R_{i,x,x}$, is searched. If the bit is one, then the vid is used as the search key in the vid index of $R_{i,x,x}$ to find the record position, and the (record address, x) pair is entered in SOT_x under column i. Otherwise, $DV.R_{i,x,z}$, where z is the next higher level of x in the security lattice, is checked. The search continues up to level b. If a 1-bit is detected, the corresponding index is checked, and the (record address, z) pair is entered in SOT_x under ith column. Next, $PQV_{x,y}$ is scanned for the presence of 1-bits. But this time, the search starts from the DV of base relation $R_{i,x,y}$ and, if not found, continues downward in the security lattice until the DV of $R_{i,x,x}$ is searched.

After constructing SOT_x, for a given x, the records are retrieved from the base relations in the following way. If element (n,z) appears under the ith column, the record with address n is retrieved from the relation $R_{i,x,z}$, and then the build and probe phases of the join are performed. After completion of a record, the tuple class attribute is added by taking the least upper bound of all attribute classifications.

We claim that this algorithm does not allow a flow of information violating the system's security policies. The data structures we maintain in this method are classified according to the associated primary key relations; thus, they do not signal information downward. For further explanation, readers are referred to [PAN95].

4 Cost of Maintaining Base Relations

In this section, the costs of maintaining different base relations are analyzed. The three different primary costs involved in maintaining base relations are the cost of inserting a new record, the cost of deleting an existing record, and the cost of updating attribute value(s) of an existing record. However, as an update to a record is assumed to be the deletion of the record followed by the insertion of the modified record, the update cost will not be considered as a separate case. An update that results in polyinstantiated elements or changing a null value of a non-key attribute to a non-null value in a multilevel relation will affect the corresponding attribute relation as an insertion of a new record into it. But, this does not affect the primary key relation, since the primary key value of the record still remains the same. Similarly, the deletion of a polyinstantiated element or modification of a non-null value to a null value in the multilevel relation results in the deletion of the corresponding record in the corresponding base relation without affecting the primary key relation. Therefore, in the case of the PRISM, this type of insertion or deletion does not affect the length of the DVs, it only modifies the bit values in the DVs. A rather different case would be when a tuple polyinstantiation occurs or a new record is inserted into the multilevel relation. In such a situation, the primary key relation as well as some attribute relations need to be updated (new records are inserted). Similarly, deletion of a polyinstantiated tuple or any record from the multilevel relation results in deletion of records from the primary key relation and attribute relation(s). In this type of insertions and deletions, the PRISM updates the length and bit-values of DVs. Therefore, the two cases for each model that need to be discussed are: insertions or deletions that affect the attribute relation(s) only, and insertions or deletions that affect both the primary key and attribute relation(s).

4.1 Update Costs in the SeaView Model

The insertion costs in the SeaView model are estimated for both the cases as follows:
Case 1 (Insertion into an attribute relation only):
In this case, a record with the primary key and the attribute value need to be inserted into the corresponding attribute relation, and then the index needs to be

updated. The cost of inserting the record into a base relation involves the cost of reading the last page from the disk, moving the new record into it, and then writing the page into the disk. This costs $C_1 = 2 * T\_IO + T\_move$, where T_IO is time to perform a random page I/O operation, and T_move is time to move a tuple in main memory. Updating the index involves reading two pages (one from the second level and one from the third level) from the index, adding a node at the third level, and writing the modified page back to the secondary storage. This costs $C_2(best) = 3 *$ $T\_IO + T\_move$. In the worst case, writing an entry in one of the pages in the third level of the index may result in updating all the pages in the second and third levels of it. Thus, the worst case cost $C_2(worst) = 2 * m * T\_IO + n * T\_move$, where m is the total number of pages in the second and third levels of the index, and n is the number of entries in these pages (also the same as the number of records in the attribute relation) after the update. The value of m is calculated as $n/FO + n/(FO * FO)$, where FO is the average fanout structure in the index node. The total insertion cost in case1, therefore, is $C_{insert\_1} = C_1 + C_2(best)$ in the best case, or $C_{insert\_1} = C_2 + C_2(worst)$ in the worst case.

Case 2 (Insertion into both primary key and attribute relations):

The cost in this case is the cost estimated in case 1 plus the additional cost of inserting the record into the primary key relation and updating the index of the primary key relation. Thus, $C_1 = 4 * T\_IO + 2 * T\_move$, $C_2(best) = 6 * T\_IO + 2 * T\_move$, and $C_2(worst) = 2 * (m_1 + m_2) * T\_IO + (n_1 + n_2) * T\_move$, where m_1 and m_2 are the total number of pages in the indices (except the page containing the root node) of the primary key and attribute relations, respectively, and $n1$ and n_2 are the number of entries in these indices, respectively, after the update is made. The total insertion cost in case 2 is $C_{insert\_2} = C_1 + C_2(best)$ in best case, or $C_{insert\_2} = C_1 + C_2(worst)$ in worst case.

The deletion costs in the SeaView model are estimated as follows:

Case 1 (Deletion from attribute relation only):

In order to delete a record from an attribute relation, the corresponding index must be searched first for the record address, then the entry must be deleted from the index, and finally, the record would be deleted (or marked as deleted) from the base relation. The cost in this case, after analyzing the same way as in the case of insertion, is estimated to be $C_{delete\_1} = C_1 + C_2$, where $C_1 = 2 * T\_IO + T\_move$, the best case value of C_2, i.e., $C_2(best) = 3 * T\_IO + T\_move$, and the worst case value of C_2, i.e., $C_2(worst) = 2 * m * T\_IO + n * T\_move$. It must be noted that in case 1, the insertion cost is the same as the deletion.

Case 2 (Deletion from both primary key and attribute relations):

The record in this case needs to be deleted from both primary key and attribute relation indices, and from both the relations. But prior to deletion, the index must be searched in order to obtain the record address in each relation. The total cost in this case is the same as the insertion cost as estimated in case 2, i.e., $C_{delete\_1} = C_{insert\_2}$.

4.2 Update Costs in PRISM

In the PRISM model, the update costs are estimated in a similar way as the SeaView model. However, one point to note here is that the indices maintained for different attribute relations are the vid-indices, not the regular data-space indices, as required in the SeaView model. This results in more record entries in each node of the index, thus, requiring fewer page accesses than that of the SeaView model. The extra cost needed in the PRISM, besides the costs required in the SeaView model, is the cost of updating the DVs for the base relations. In the best case, a bit is simply flipped from zero to one (in case of insertion) or from one to zero (in case of deletion) in the existing DV. This would incur a cost of T_updt. In the worst case, the vector needs to be updated by appending a vector of length one word to the DV, or by deleting the last word from the DV. Again, it might be necessary to rearrange the bit values in the entire vector. Thus, the worst case cost is $(|DV| + 1) * T\_updt$, where $|DV|$ is the size of the domain vector in terms of words, and T_updt is the time to update a vector of length one word. Adding these costs to the formulas derived in the previous section, the final costs of inserting or deleting a record in the PRISM are as follows:

Case 1: best case total cost: $5 * T\_IO + 2 * T\_move + T\_updt.$

worst case total cost: $2 * (m + 1) * T\_IO + (n + 1) * T\_move +$
$$(|DV| + 1) * T\_updt.$$

Case 2: best case total cost: $10 * T\_IO + 4 * T\_move + 2 * T\_updt.$

worst case total cost: $2 * (m_1 + m_2 + 2) * T\_IO + (2 + n_1 + n_2) * T\_move +$
$$(|DV_1| + |DV_2| + 1) * T\_updt,$$

where DV_1, and DV_2 represent DVs for the primary key and attribute relations, respectively.

Table 1: Relation Update Parameter Values

T_IO	25 msec	m	381 for SeaView
T_updt	1.2 μsec	n	60,000
T_move	20 μsec	m	210 for PRISM
m_1	633 for SeaView	m_1	351 for PRISM
m_2	381 for SeaView	m_2	210 for PRISM

5 Comparison Results

Applying the formulas derived in sections 4.1 and 4.2, the insertion/deletion costs in the SeaView model and the PRISM model are calculated in this section. The values of m, n, m_1, and m_2 for the SeaView model and for the PRISM are given in Table 1 below. In calculating n, the percent of null values was taken as 40 percent. The total number of records in a primary key relation was fixed at 100,000. It is to be noted that when percent of null values is 0.4, then there will be 60,000 records on the average in each attribute relation. Therefore, the SeaView model required three levels of indices for each attribute relation, whereas the PRISM required only two levels. The number of records in a base relation is assumed to be the number after the insertion or deletion is performed.

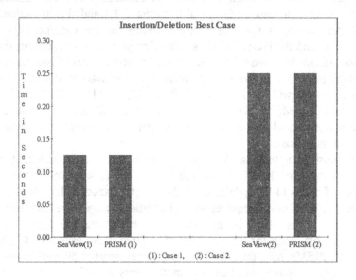

Figure 5: The Best Case Scenario for Insertion/Deletion

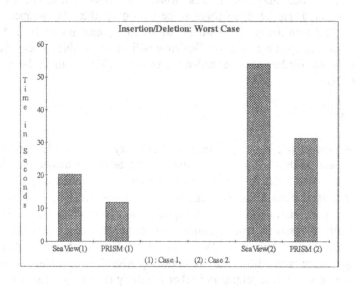

Figure 6: The Worst Case Scenario for Insertion/Deletion

Figure 5 shows the best case scenario for insertion/deletion in both the models. The first pair of bars represents the cost in seconds for inserting/deleting a record into an attribute relation only (Case 1 in the previous section). The second pair of bars represents the same for both primary key and attribute relations (Case 2 in the previous section). The exact values for the PRISM is only a few micro seconds more than that of the SeaView model. These values were as follows: SeaView(1) = 0.12504 seconds, PRISM(1) = 0.1250412 seconds, SeaView(2) = 0.25008 seconds, and PRISM(2) = 0.2500824 seconds, where the number in the parentheses denotes the case as described in section 3.2.1 and also in section 3.2.2.

Figure 6 shows the worst case scenario for insertion/deletion into the SeaView model and the PRISM models. The first pair of bars represents the update costs in seconds for attribute relation only, and the second pair of bars represents the same for both primary key and attribute relations. The exact values in seconds were as follows: SeaView(1) = 20.30002, PRISM(1) = 11.7522712, SeaView(2) = 54.00004, and PRISM(2) = 31.3560824. This large difference was due to the fact that the SeaView model had more number of pages that need to be updated in the worst case in its indices.

It must be noted that for updating each node in the index, it costs 1.2 microseconds more for the PRISM than that of the SeaView model. However, when the number of nodes to be updated exceeds 159 (the fanout value for the SeaView model), the SeaView model requires another additional page to be read, updated, and written back. Thus, there is an increase of twice the I/O time, totaling 50 milliseconds. But, when the total number of updated nodes exceeds 287 (the fanout value of the PRISM), then the PRISM also requires another 50 milliseconds due to the above reason. Hence, the SeaView model beats the PRISM again. However, when the number of nodes to be updated exceeds 378, the SeaView model requires three pages to be read and written back, while the PRISM still requires only two. The PRISM would require three pages to be read only after the number of nodes exceeds 574, but then the SeaView would require four pages to be read and written. And after that, the update cost in the SeaView will never be able to beat that of the PRISM since the number of page read-write in the SeaView will be far more than that of the PRISM.

6 Conclusion

The PRISM model was developed to solve the query delay problem in multilevel secure database models. This model follows the SeaView model that has been developed as a joint effort by the SRI International and Gemini Computers with an objective to achieve class A1 of the secure systems classes [DoD85]. We assume that the SeaView model maintains indices for quicker access into its base relations. The PRISM maintains a special type of indices, called the vid indices, that is based on a different data structure known as DVs. Since the DVs are bit vectors they require less storage space and allow fast execution [PAN95]. In order to show the maintenance cost of the relations and other auxiliary data structures, a cost model has been developed by taking various parameters in a multilevel database system environment. The results have been compared with that of the SeaView model. We have shown that although PRISM uses some additional data structures, their maintenance does not add further to the cost. Rather, when an update affects only one page of the index, the update time is almost the same as that of the SeaView model. Moreover, in the worst case scenario, when the update affects all the pages of the index, updating a record in the SeaView model takes much longer than updating a record in the PRISM.

References

[BEL74]   D. E. Bell and L. J. LaPadula, "Secure Computer Systems: Mathematical Foundations and Model," Technical Report, The Mitre Corp., 1974.

[DEN87]   D. Denning and T. Lunt, R. Schell, M. Heckman, and W. Shockley, "A Multilevel Relational Data Model," Proceedings of the IEEE Symposium on Security and Privacy, p 220-234, Oakland, CA, April 1987.

[DoD85]   Department of Defense, "Department of Defense Trusted Computer System Evaluation Criteria," National Computer Security Center, December 1985.

[HAI91]   J. Haigh, R. O'Brien, and D. Thomsen, "The LDV Secure Relational DBMS Model," Database Security IV, editors S. Jajodia and C. Landwehr, North-Holland, Amsterdam, p. 265-279, 1991.

[JAJ91]   S. Jajodia and R. Sandhu, "Toward A Multilevel Secure Relational Data Model," Proceedings of the ACM SIGMOD Conference, p. 50-59, Denver, CO, May 1991.

[LAM73]   B. W. Lampson, "A Note on the Confinement Problem," Communications of the ACM, Vol. 16, No. 10, p. 613-615, October 1973.

[LUN88]   T. F. Lunt, R. R. Schell, W. R. Shockley, and D. Warren, "Toward a Multilevel Relational Data Language," Proceedings of the IEEE Symposium on Research in Security and Privacy, p. 72-79. 1988.

[LUN90]   T. F. Lunt, D. E. Denning, R. R. Schell, M. Heckman, and W. R. Shockley, "The SeaView Security Model," IEEE Transactions on Software Engineering, Vol. 16, No. 6, June 1990.

[PAN95]   B. Panda, and W. Perrizo, "Query Execution in PRISM and SeaView: A Cost Analysis," Proceedings of the 1995 ACM Symposium on Applied Computing, Nashville, TN, February 1995.

[PER91]   W. Perrizo, J. Gustafson, D. Thureen, D. Wenberg, and W. Davidson, "Domain Vector Accelerator (DVA): A Query Accelerator for Relational Operations," Proceedings of the 7th International conference on Data Engineering, Kobe, Japan, 1991.

[PER93]   W. Perrizo and B. Panda, "Query Acceleration in Multilevel Secure Database Systems," Proceedings of the 16th National Computer Security Conference, Baltimore, MD, September 1993.

[SMI92]   Smith, K. and Winslett, M., "Entity Modeling in the MLS Relational Model," Proceedings of the 18th VLDB Conference, Vancouver, British Columbia, Canada, 1992.

Calibration
of a DBMS Cost Model
with the Software Testpilot

F. Andrès[1], F. Kwakkel[2], M.L. Kersten[2]

1. Ifatec/Euriware
   12/14 Rue du Fort de St Cyr
   Montigny Le Bretonneux
   78182 St Quentin en Yvelines, France
   Email: Frederic.Andres@euriware.fr

2. CWI, Kruislaan 413
   1098 SJ Amsterdam
   Netherlands
   Email: {kwakkel,mk}@cwi.nl

Abstract

Relational database systems come with a query optimizer to plan the execution order of individual operators using a cost model of the underlying system. Despite the large body of research on optimization rules and cost approximation, little effort has been invested in the design of effective techniques for their validation. In this paper we discuss a method to calibrate cost models using the Software Testpilot, a performance assessment tool. The approach is based on taking an initial cost model and to refine it using the capabilities of the underlying DBMS kernel using automatically generated experiments. Apart from simplifying a time consuming task, it improves the effectiveness of the query optimizer by being tuned towards its hardware platform. The technique is illustrated for a novel parallel DBMS, called DBS3.

1   Introduction

Performant relational systems strongly depend on their query optimizers to find a good execution plan. Therefore, they come with an efficient algorithm to explore the space of candidate plans and a cost model to predict resource consumption and response time. The cost model is built around statistics on the actual database population and formulae that capture the cost of the DBMS algorithms and data structures. The former requires (semi- automatic) intervention by the DBA to update the statistics at regular time intervals. The latter is frozen during system design or system installation.

The cost model represents a, necessarily, inaccurate picture of the DBMS processing capabilities for several reasons. First, traditional query optimizers [KBZ86,Selinger79] try to avoid the worst query response time. They do not locate the best execution plan, because the search space of alternatives is too large to be explored for each query separately. Second, potential parallelism and load balancing are difficult to take into account at query compile time. They increase the cost model complexity and necessarily delay part of the optimization problem to query execution time. Finally, a DBMS can run on many platforms for a variety of application domains. This makes cost model calibration a recurring activity. This would not be much of a problem if the desired cost model can be tuned by running a simple benchmark program to adjust the formulae's coefficients. This is not the case, however, since extended relational database systems aimed at complex query processing require both cost model enhancements and benchmarks that reflect the properties of the new application domains.

This paper describes an effective technique to calibrate a DBMS cost model. Its novelty is an automatic procedure for cost model calibration using a probabilistic benchmark and a sophisticated software performance assessment tool, called the Software Testpilot. The probabilistic benchmarking techniques proposed in [Turbyfill89] lend themselves for an automatic tuning procedure, because -unlike adhoc benchmarks- they cover a wide spectrum of possible database characteristics. Another advantage is their use of standard data generators and work loads prescriptions.

Yet, a benchmark alone is not sufficient to calibrate a cost model. For, in general, many points in the envisioned database work space should be inspected for tuning the cost formulae coefficients. A software tool that automates this (manual) task and that supports tuning at minimal cost is in high demand. Our Software Testpilot[1] is a significant step in this direction.

The automatic calibration procedure, result of a collaboration with Bull, has been validated with a re-calibration of the cost model for the advanced DBMS DBS3 [Bergsten91]. It is focused on formulae for the relational operations and its primary access methods. This experiment indicates that automatic calibration is feasible and that our laborious earlier attempts have been accurate to extent that estimated values of majority of the validation queries are within a 10% error of the observed values. This result was obtained by the procedure in 18 hours of processing with the Software Testpilot.

This paper is organized as follows. In Section 2 we place our work in the context of performance evaluation for query optimization. Section 3 introduces the problem of model calibration using part of the DBS3 cost model and the calibration database. Section 4 explains the Software Testpilot and its role in the calibration process. Section 5 describes the calibration experiments and the results obtained. We conclude with open issues and tracks for future research.

2 Performance evaluation for query optimization

The basis for query optimization is obtained through analytical modelling and performance evaluation of prototype systems. A traditional approach [Kooi80,Swami89] is to view cost estimation through performance evaluation as the following abstract problem: Given a query execution plan QEP, a target confidence rate CR for the estimates, an execution model EM and a cost function CF, produce an error $E \leq 100 - CR$ where EM_Q is the subset of the execution model used to evaluate the plan and to compute its cost. Any solution to this problem can be characterized by selecting 1) a cost model and, therefore, 2) an evaluation strategy, and 3) a statistic space.

1. The work reported here has been funded by ESPRIT-III Pythagoras project (P7091)

The cost model encodes the execution model of the target DBMS regarding the relational operations, access methods, data transfer strategies, parallelization strategy, etc. The evaluation strategy is used to propagate the estimates through a graph representation of the query execution plan and to browse the statistic space until the target confidence rate is reached. The choice of either one affects the other. For example, if a cost model with accuracy of 50% is used then the evaluation strategy might be limited to a single query graph scan. More scans would not improve the quality of the response time being estimated. On the other hand, a large statistic space can be browsed with a finer cost model. This is of particular importance for database administration tools, which amortize the cost over many sessions.

Most cost models for (parallel) DBMSs are built for optimal response time of isolated queries. They are derived from the seminal work on System-R [Selinger79] and, generally, extend its query cost model:

$$\text{Cost} = \text{resource\_consumption} + W \times \text{response\_time}$$

where W is a system dependent parameter [Hong92]. Recent research works integrate data flow dependencies and resource contentions to better deal with the characteristic encountered in shared nothing database systems [Andres93,Ganguly92,Wilschut95].

Several degrees of extensibility and flexibility can be observed in state-of-the-art of cost models. A first class contains systems with a fixed model, such as exemplified by Postgres[Stonebraker91], in Starbust[Pirahesh92], and in Gral [Becker92]. By their nature, these query optimizers require a lot of expertise from the designer and modification of the algorithms is rather complicated. The second class of cost models support varying models, such as in Exodus [Graefe87] or in Genesis [Batory87]. A cost model calibration tool simplifies their construction by provision of the required information at compile time.

Cost model calibration or tuning have been investigated by several researchers [Mackert86, Christodulakis84]. A prime goal is to provide DB and DBMS designers a (semi-)automatic process to predict the impact of applications on the DBMS architectures foreseen. The automatic process itself is either a comparison between measurements and cost estimates for a great number of queries[Salza92], or a prelude for the query optimizer generator to determine coefficient [Du92]. Our proposal extends this track by careful selection of interesting points for further investigation automatically.

Automatic tuning is also needed in performance evaluation for database administration tool, such as DBA*Expert[Goldgewicht92] and it can be used to calibrate an index management tool. Combination of a performance model with database and application requirements for the Software Testpilot provides a mechanism to speedup physical database design.

Research on query optimizer calibration relies on techniques for DBMS performance assessment. In particular, tools for multi-environment performance assessment, such as [Gaede81, Kersten93, McDonnell92, Salza92]. These systems provide easy comparison of different systems and they represent a significant improvement over prior work in this field. For example, the use of specification languages (e.g. MBUS, TSL, WSL) simplify the description of the target environments for evaluation and they trade "the cost to build a performance analysis task off" against the "cost of ownership over the life of a system".

3 Calibration of a DBMS cost model

In this section, we introduce part of the cost model for query optimization used in the DBS3 prototype to illustrate the calibration technique. Its performance evaluation is based on the DBMS architectural model, data profiles, and the query execution plan algorithm. The architectural model is captured by a few analytical formulae frozen during system design. The data profiles reflect information contained in the system's data dictionary. It is obtained using the DBA*expert design tool. The query plan generator uses a heuristic search procedure to find a good execution plan [Lanzelotte93]. As such, this model is representative for the approach taken in extended relational systems.

3.1 Model assumptions

The cost model of DBS3 is based on general characteristics of relational systems, i.e. the cost formulae are generic and they can be re-used in other implementations. The system specific properties are captured as much as possible in the formulae coefficients. The key parameters are the tuple size, complexity of the predicate evaluation, and primary data access times. The tuple size determines the storage requirements and transport cost involved. The number of predicates to be evaluated and the number of projection attributes determine the CPU processing cost, which strongly depend on the calibration database properties. The cost model is a further simplified by considering linear formulae over the relation cardinality only.

A critical term in processing cost estimation is to determine the portion of a conjunctive predicate that should be evaluated against a tuple to determine its role in the answer set. We use the function $eval()$ defined in [Whang87], which provides a probabilistic estimate on the number of terms evaluated in a conjunctive expression. The main idea is that if a term in the sequence is false then the remaining predicates need not to be evaluated. Assuming a probability of 0.5 that a term fails, we have:

$$eval(expr) = \frac{2^{card(expr)} - 1}{2^{card(expr) - 1}}$$

where card(expr) denotes the number of terms in the predicate.

3.2 A cost model

For the sake of clarity in the presentation of the calibration method, we focus, only in this paper, on the select and join operations in a cost model for an extended relational system. They can be described by simple, analytical cost formulae with their selectivity obtained from data profiles kept in the data dictionary. Moreover, the target metric is the completion time often called elapsed time, i.e. the time to complete the request as perceived by the user. Response time calculation (the first answer received by the user) requires a more complex model, which is beyond the scope of this paper. The performance functions and parameters are summarized in Table 1.

The first cost formula describes a sequential scan for a memory and disk-based DBMS. The scan is divided into three cost components: scan initialization, tuple loading, and identifying tuples satisfying the predicate. The coefficient of the cost function depend on the DBMS architecture, such as its interpretative or compile time optimization method and the mechanism to control data flow. Fine tuning them is the aim of our calibration procedure.

The join formulae used are a straight forward description of the nested-loop and the index-join algorithms. They are based on the selection cost formulae with specific modifications to reflect the way they access input relations. For a nested loop join, all tuples of both relations R and S are loaded and the join is performed in memory. Therefore, the cost formula is the sum of four parts: nested- loop initialization, operand loading, and comparison of all tuple combinations. The index join algorithm incurs a possible different start-up and inner relation access cost. For, the matching tuples are located through an index lookup rather than a sequential scan. In both algorithms matching the candidates is equally expensive.

Variable	Description	Type
R,S	Database Relations	factor
P	Predicate	factor
$size(A)$	Size of relation A	function
$eval(P)$	Estimated number of terms actually evaluated	function
$Cost_{\{r/i\}scan}(A)$	Cost to scan relation or index A	cost-function
$Cost_{select}(A,B,P)$	Cost to select tuples in AxB for predicate P	cost-function
$Init_{\{ss/is/nl/ij\}}$	Cost to setup and scan/join for the cost models	parameter
$Cost_{\{i/r\}next}$	Cost to process next tuple in relation or index	parameter
$Cost_{pred}$	Cost component to balance function eval(P)	parameter
$Cost_{ss}(A,P)$	Cost of sequential scan of relation A with selection predicate P	cost-function(response)
$Cost_{is}(A)$	Cost of a primary index scan of relation A	cost-function(response)
$Cost_{nl}(A,B,P)$	Cost of a nested loop join for relation A and B with selection predicate P	cost-function(response)
$Cost_{ij}(A,B,P)$	Cost of a index join (index scan on A) for relations A and B with selection predicate P	cost-function(response)

Table. 1. Variables and functions for the cost models

Cost model for a *sequential scan*:

$$\text{Cost}_{rscan}(R) = \text{Cost}_{next} \times \text{size}(R)$$
$$\text{Cost}_{ss}(R, \text{Pred}) = \text{Init}_{ss} + \text{Cost}_{rscan}(R) + \text{Cost}_{Pred}(R, \text{Pred})$$
$$\text{Cost}_{Pred}(R, \text{Pred}) = \text{Cost}_{pred} \times \textit{eval}(\text{Pred}) \times \textit{size}(R)$$

Cost model for an *index scan*:

$$\text{Cost}_{is}(R) = \text{Init}_{is} + \text{Cost}_{iscan} \times \textit{size}(R)$$
$$\text{Cost}_{iscan}(R) = \text{Cost}_{inext} \times \textit{size}(R)$$

Cost model for a *nested loop join*:

$$\text{Cost}_{nl}(R, S, \text{Pred}) = \text{Init}_{nl} + \text{Cost}_{rscan}(R) + \textit{size}(R) \times \text{Cost}_{rscan}(S)$$
$$+ \text{Cost}_{select}(R, S, \text{pred})$$

$$\text{Cost}_{select}(R, S, \text{Pred}) = \text{Cost}_{pred} \times \textit{eval}(\text{Pred}) \times \textit{size}(R) \times \textit{size}(S)$$

Cost model for an *index join* (index scan on S):

$$\text{Cost}_{ij}(R, S, \text{Pred}) = \text{Init}_{ij} + \text{Cost}_{rscan}(R) + \textit{size}(R) \times \text{Cost}_{iscan}(S)$$
$$+ \text{Cost}_{select}(R, S, \text{Pred})$$

3.3   Calibration database

To calibrate a cost model, the user should select a representative database and query set. Taking a real-world database with a subset of its predominant queries leads to an optimizer highly tuned towards a particular application environment. Due to the sake of generality of the method presented in this paper, the AS3AP Benchmark[Turbyfill89], i.e. a probabilistic database benchmark, has been choosen as a basis for our study due to its functional query set. Choosing a probabilistic database benchmark helps in fine-tuning a query optimizer for a broad spectrum of queries, such as encountered in ad-hoc interactive environments.

The benchmark contains a single relation schema over 10 attributes that reflect the attribute domains supported by the DBMS. The tuple size is equal to 100 bytes. The database used in our experiments contained one million tuples. The range of values for an attribute is determined by its type and by the size of the relation. Relations are indexed on the attribute key using a hash function. Secondary keys are represented by btree. Due to tight space constraints, the experiments described in this paper are necessary sketchy, so we chose a subset of the benchmark queries.

4 Calibration using the Software Testpilot

In this Chapter we introduce the Software Testpilot and its role in the calibration process, i.e. finding accurate parameters for the cost formulae. Section 4.1 gives a brief overview of the Software Testpilot as a performance assessment tool. Section 4.2 illustrates programming a search space for performance assessment of the sequential scan. Section 4.3 describes how the system is used as a calibrator.

4.1 The Software Testpilot: a performance assessment tool

The Software Testpilot [Kersten93] is a tool designed to aid the DBMS engineer and user to explore a large workload search space to find the slope, top, and knees of performance figures quickly. The approach taken is based on specifying the abstract workload search space, a small interface library with the target system, and a functional description of the expected performance behavior. Thereafter, the Software Testpilot selects workload parameter values and execute the corresponding DBMS experiments, such that the performance characteristics and quality weaknesses are determined with minimal cost (i.e. time).

The input file for the Software Testpilot, is called a test suite (illustrated in Fig. 1.). The work space is described by *factor* and *response* variables. They are identified by a name, an underlying type, and value range constraints. The interface with the target system is captured by the *action* statements, which relate a user-defined action with the factor and response variables. The expected performance behavior is described with a *hypothesis*, a (linear) model over factor and response variables.

Identifiers starting with a capital denote variables, others denote a property type. For instance, "tablesize:Size" means the variable "Size" is of type "tablesize". Type definitions should be described elsewhere in the test suite. The type definitions of factors describe the range and type of the factors. The factor tablesize has for instance an integer type and a range between 1 and 1000K tuples.

The Software Testpilot, after initialization of the target system, selects a point in the space spanned by the factors for experimentation. The point chosen depends on the hypothesis given and results of previous experiments. Given a point of interest, it constructs an experiment to bring the target system in the required state and to execute the action to obtain the response value. This process is repeated until the required confidence level has been attained or the system is stopped by the user.

The output of the Software Testpilot are the measurement results of all experiments against the target DBMS, the new value of the model parameters, and the confidence level attained for the hypotheses given. The confidence levels are fed back into the Software Testpilot, where they are used to decide if more experiments are required to proof the hypotheses or if they should be remodelled. The experiment results are visualized to monitor progress.

factor => [	name:	tablesize,
	state:	yes,
	type:	integer,
	interest:	midpoint,
	apply:	optimized,
	unit:	tuples,
	scale:	ratio
	range:	[0 .. 1000K]].
factor => [	name:	predicate,
	statefull:	no,
	type:	symbolic,
	value:	generate(predicate),
	apply:	random,
	scale:	nominal].

parameter => [	name:	init_sscan,
	type:	integer,
	unit:	us].
parameter => [	name:	cost_rnext,
	type:	integer,
	unit:	us].
parameter => [	name:	pred_balance,
	type:	integer,
	unit:	us].
response => [	name:	scan_time,
	unit:	us,
	type:	integer].

Fig. 1. Partial Software Testpilot test suite

4.2 The test suite for a sequential scan

The cost formula for the sequential scan is expressed in the Software Testpilot test suite with a *hypothesis* description (Fig. 2.). The hypothesis defines a cost model for a sequential scan with selection predicate. It directly captures the analytical model described in the previous section. The factors are the table size and the selection predicate. The measured response metric is the time to scan the table. The parameters to be solved are *init_sscan*, *cost_rnext* and *pred_balance*.

hypothesis => [	name:	user_hypothesis,
	factors:	[tablesize:Size, predicate:Pred],
	response:	[scan_time: Time],
	parameters:	[init_sscan: Init, cost_rnext:Next, pred_balance:Balance],
	model:	Time = Init + Next * Size + Balance * eval(Pred) * Size].

Fig. 2. The hypothesis for the sequential scan

The interface with the target system (i.e. DBS3) is described using *action* objects. An action object describes a vector through the search space, such as a state change in the target system (e.g. filling a relation) or a response measurement experiment (e.g. tuple selection). Other actions may define the connection, disconnection and, in case of target system failure, restore operations of the DBMS. An action contains a slot to identify a Prolog interface primitive, which deals with the peculiarities of the interface between Software Testpilot and target system.

A hypothesis has a related action that describes the measurement of its response variable(s). The following action describes the measurements of the *scan_time* response variable. The AS3AP relation we work with is called 'tenpct'.

```
action => [   name:     scan_select_relation,
              factors:  [tablesize:X, predicate:Predicate],
              response: [scan_time:Time],
              action: sql_query(select * from tenpct where Predicate, Time)].
```

Fig. 3. An Action object for the sequential scan

4.3   Calibration of the sequential scan

During exploration of the abstract workload search space, the Software Testpilot measures points that have a high probability of (dis)qualifying the hypotheses. When sufficient data is gathered, the best fit parameter values for the model are calculated using a singular value decomposition algorithm (SVD). This algorithm locates the best parameter values using χ^2 minimization. It accepts a function that is a linear combination of the parameters and partial functions and a data set. It calculates analytically the best-fit parameter (P_1 ... P_n) values and their standard deviations as follows. To find the best-fit parameters for the cost model equation, the equation must be expressed as linear system as follows:

$$F(x; P_1, P_2 ... P_n) = P_1 \times G_1(x) + P_2 \times G_2(x) + ... + P_n \times G_n(x)$$

where

$F(x; P_1, P_2 ... P_n)$	the response equation for the cost model
x	the cost model factor
P_i	a cost model parameter
G_i	a cost model sub function for parameter P_i

The measurements are denoted in a data set: $\{ \left(x_i, y_i \right) \ (1 \leq i \leq m) \}$, where x_i is the i-th chosen factor value, y_i the i-th measured response value and m the number of experiments.

Since the cost model describe here does not match the linear system, it can handle only one factor (i.e. x), we have split the hypotheses two parts. The first hypothesis describes the expected behaviour of the DBMS with a fixed selection predicate and calibrates the paramaters *init_sscan* and *cost_rnext* . The second hypothesis re-uses the parameter values and calibrates the *pred_balance* parameter using queries with varying selection predicates. For the first hypothesis we have the following SVD sub-functions: $G_{init\_sscan}(Size) = 1$ and $G_{cost\_rnext}(Size) = Size$. For the second hypothesis, the SVD sub-function is $G_{pred\_balance}(Size, Pred) = eval(Pred) * Size$. The other terms *init_sscan* and *cost_rnext * tablesize* are constants.

5 Practical Calibrations

In this section, we describe the procedure to calculate the coefficients of the cost model introduced in Section 3.

5.1 The target parallel DBMS

DBS3[1] is a parallel relational database system implemented on a shared everything architecture [Bergsten91]. Its main characteristics are as follows. First, DBS3 uses a compilation approach (compiled C code) to obtain a fine grained control of the execution. Second, DBS3 optimizes and parallelizes queries into self-scheduling parallel programs that support inter- and intra-operator parallelism, as well as pipeline parallelism. Third, the underlying execution model is based on a distributed store model, which allows an efficient mapping of the code produced onto several classes of parallel architectures. The calibrated system is running on an Encore MULTIMAX 520 multiprocessor (Mach 2.5). This machine is configured with 10 NS32532 processors (8.5 Mips, each having 256 Kb cache memory), 96 Mb main memory and 1 Gb disk storage. Processors, memory and I/O boards are interconnected by a 100 Mb/s bus.

5.2 Calibration procedure

To calibrate the cost model, the Software Testpilot takes a test suite with the parameterized cost model, type definition of all variables and the interface with the target DBS3. Then queries are posed onto the DBMS and response values are measured.

The processing cycle for the Testpilot is a repetition of flights during which information on the performance relationship(s) is gathered. This process stops when all interesting points have been inspected, when a fatal-error occurs, or when the user terminates further exploration of the workload search space.

A flight starts with selection of a target point in the test space where measurements should be taken. From the large collection of possible points the Testpilot identifies a few candidate that provides the best information to (dis) qualify the hypotheses applicable within the focus of interest.

Before the measurements can be taken, the Testpilot should move the DBMS into the required state, that is, a table should be created and filled. The action(s) to reach this state is (are) inferred from the *factor* and *target* attributes of action descriptions. Often, several feasible plans exist and a choice should be made to avoid time consuming plans.

---

1. This design and the implementation have been done in the ESPRIT Project EP2025 EDS.

For instance, consider a test suite to investigate tuple selection from table sizes between 0 and 100 tuples. To (dis)qualify a linear hypothesis quickly it could investigate table sizes in the order: 0, 100, 50, 25, 75, 13, 38 etc. That is, table sizes lie far away from those already inspected. However, the corresponding flight plans are also expensive. For, after the first tuple selection 100 tuples are inserted. Then again one tuple is selected and 50 tuples are deleted to bring the database to a state of 50 tuples, etc..

A way out of this dilemma is to consider several candidate points and to apply the cheapest flight plan. For instance, if three candidate points are handled together then the execution order becomes: 0, 50, 100 .. 75, 25, 13, etc. The first batch (0,100,50) is sorted on the cost (=unit steps) to reach the point. Then the next three points are generated (25,75,13) and sorted. Now 75 is the 'cheapest' point because its distance to 100 is the minimum of the three points. The execution order becomes: 75,25,13 etc.

If we carry this idea further and allow the Testpilot to generate all candidate points at startup then the sequence becomes: 0, 1, 2, 3 etc.. This is something we do not want either, because it provides only slow insight into the shape and validity of the performance functions. Therefore, the Testpilot uses a buffer of candidates and gives the user a choice between an emphasis on the execution time, the 'cost' of the flight plan to reach that point, and emphasis on selecting interesting points, i.e. the 'weight' of a point.

5.3   Results

In this section, we present the results of the experiments. The results are presented in two steps. First, we use single relation queries to determinate coefficient values. Second, we validate the estimates provided by the cost model using the actual measurements for join queries.

5.3.1   Determination of parameter values

The actual queries used for the automatic tuning procedure are given in Figure 4. The metric used is the elapsed time. For DBS3, each measurement was done after flushing all the buffers to eliminate the buffering effects. Each query is issued 10 times and the testpilot provides updated coefficient values.

```
SELECT * FROM tiny
SELECT * FROM updates WHERE key <= 10
SELECT * FROM hundred WHERE key <= 100
SELECT * FROM hundred WHERE key <= 1000
SELECT int FROM updates WHERE key <= 1000
SELECT key FROM updates WHERE key <= 100
```

Fig. 4. Calibration Queries

From these experiments, the coefficient values can be deduced using the least-square fitting algorithm, which minimizes the error. The results are summarized in

Table 2. We notice that the initialization procedures are almost the same. A comparison

Parameter	Description	Cost [ms]
$\text{Init}_{\{ss/is/nl/ij\}}$	Cost to setup and scan/join for the cost models	22
Cost_{rnext}	Cost to process next tuple in relation	0.28805
Cost_{inext}	Cost to process next tuple in index	0.032
Cost_{pred}	Cost component to balance function eval(P)	0.005

Table. 2. Calculated parameter values in ms

between the measurements results and the cost model is reported in Figures 5, 6,and 7 which illustrate the high precision of the method. The horizontal axes show the relation sizes, the vertical axes give the elapsed time. The maximum error is about 10 % which proofs a good corroboration between the model and the system behaviour.

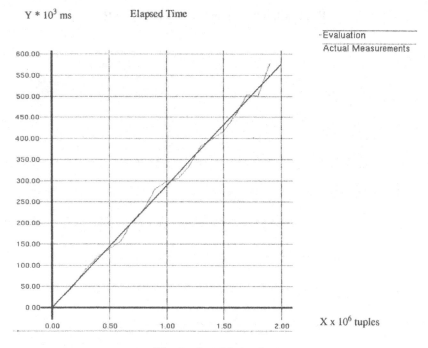

Fig. 5. select * from tiny

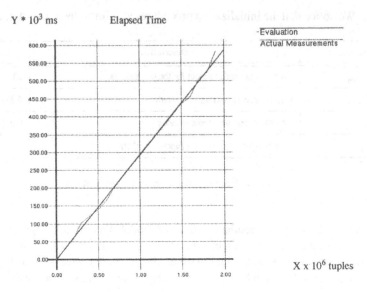

Fig. 6. select * from updates where key <= 10

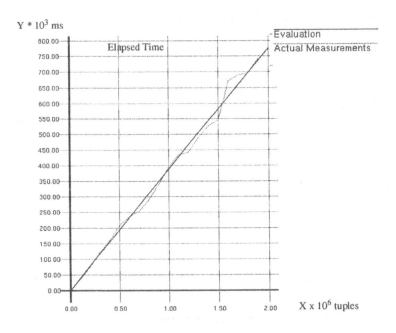

Fig. 7. select * from hundred where key <= 100

5.3.2 Validation of parameter values

We compared the measurements of a set of complex queries taken from AS3AP to the cost formulae based on the previous parameter values. Figures 8 and 9 show the comparison between the measurements and the evaluation with the query : *select * from uniques, hundred where uniques.key = hundred.key.* In figure 8, an index is used on the attribute key *hundred*. In figure 9, no index is used. The result of the validation shows that in more than 85 % of the cases the value observed was within the band of 10 % error from the cost model prediction. Further, in all the other cases, memory swapping phenomenon contribute to the majority of the error measured.

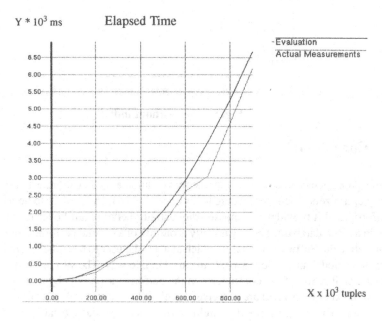

Fig. 8. Join with index

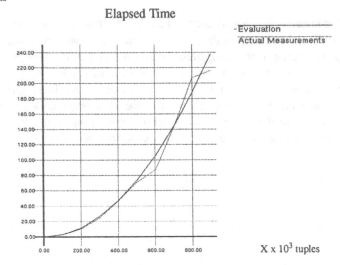

$Y * 10^3$ ms

Elapsed Time

Evaluation
Actual Measurements

$X \times 10^3$ tuples

Fig. 9. Join without index

6   Conclusion

In this paper, we have described a procedure to automatically calibrate a cost model for query optimizers. Such a procedure helps in porting DBMS software to novel hardware platforms and it provides a database administrator better control over resource allocation in a given database. The key novelty is our use of a general performance assessment tool, called the Software Testpilot, which enables the user to describe an analytical performance model and system aspects in a declarative way. The system then helps in automatic calibration of the coefficients in the performance model by selection and execution of experiments that are perceived highly informative to (dis) proof the model assumptions. The system supports a multi-environment approach to calibrate cost models so it can be used for different kinds of environments (various applications,different information systems). Therefore it has been proven effective in the calibration of the query cost model of a modern parallel DBMS.

In particular, the tuning procedure has been used to partially calibrate the query optimizer of the parallel DBMS called DBS3 at Bull. The estimates produced by the cost model were - for a majority of queries- within a 10 % error bound. This information was obtained by running the Software Testpilot for 18 hours in unattended mode. The results and experience obtained form is a good starting point for more elaborate calibration studies on other aspects of the DBS3 prototype.

Based on our experience with DBS3, we have applied the method to other commercial DBMSs. For example, we have extended the automatic procedure to support Oracle where effective tools for database administration are in high demand. We are planning to explore calibration issues of Multimedia DBMSs where large objects are being ma-

73

nipulated. Accurate cost models for such a complex problem domain are still lacking. Given the promising experience gained so far, there is a world of related research issues that can be taken up.

Acknowledgments

We would like to thank Jihad Boulos and Luc Augarde for helpful contributions and Patrick Casadessus for providing us a system support for the DBS3 server at Bull. We also are grateful to Jean Le Bihan (Ifatec) for his encouragement to extend this research.

References

[Andres93]F. Andres "A Multi-Environment Cost Evaluator for Parallel DBMSs" PhD Thesis, University of PARIS VI, 1993.

[Batory87] D. Batory, "Extensible Cost Models and Query Optimization in Genesis", IEEE Database Engineering, 1986,10:4.

[Becker92] L. Becker, and R. H. Guting " Rule-based Optimization and Query Processing in an Extensible Geometric Database System", ACM Transaction on Database Systems, 17 (1992), pp 247-303.

[Bergsten91] B.Bergsten, M. Couprie,and P. Valduriez "Prototyping DBS3, a Shared-Memory Parallel Database System", in Proc.1st Int. Conf. on Parallel and Distributed Information Systems, Miami, Florida,1991.

[Christodulakis84] S. Christodulakis "Implications of Certain Assumptions in Database Performance Evaluation", ACM Trans. on Database syst., 9 (1984), pp. 163-186.

[Du92] W. Du, R. Krishnamurthy, and M. Shan "Query Optimization in Heterogeneous DBMS", in Proceedings of the 18th VLDB conference, Vancouver, British Columbia, Canada, 1992.

[Gaede81] S. L. Gaede "Tools for research in Computer Workload Characterization and modeling", in experimental Comp. Performance and Evaluation, North holland, Amsterdam, 1981.

[Ganguly92] S. Ganguly, W. hasan, and R. Krishnamurthy "Query Optimisation for parallel Execution" in Procs. ACM SIGMOD 1992, pp 9-18.

[Goldgewicht92] W. Goldgewicht "DBA*Expert", report, Bull, France, 1992.

[Graefe87] G. Graefe and D. Dewitt "The EXODUS Optimizer Generator" In Procs of ACM-SIGMOD International Conference on Management of Data, 187, pp 160-172.

[Hong92] W. Hong and M. Stonebraker "Optimization of Parallel Query Execution plans in XPRS" in Proceeding of Int. Symposium on Parallel and Distributed Information Systems,1992, pp 218-225.

[KBZ86] R. Krishnamurthy, H. Boral, and C. Zaniolo "Optimization of non-recursive Queries", in Proc. 12th Int. Conf. of Very Large DataBases, Kyoto, Japan,1986, pp 128-137.

[Kersten93] M.L. Kersten, and F. Kwakkel, "Design and Implementation of a DBMS Performance Assessment Tool",In Proceedings of the 4th Int. DEXA Conference, Prague, Czech republic, 1993, pp 265-276.

[Kooi80] R. P. Kooi "The Optimization of Queries in Relational Databases" PhD Thesis, Case Western Reservie University, 1980.

[Lanzelotte93] R.S.G. Lanzelotte,P. Valduriez, and M.Zait "Optimization of multiway join queries for parallel execution", in Proc. 19 th Int. Conf. on Very Large Data Bases, Dublin, Ireland, 1993, pp 493-504.

[Mackert86] L. Mackert and G. Lohman "R* optimizer validation and performance evaluation for local queries", in ACM SIGMOD International Conf. on Management of Data, 1986, pp. 84 -95.

[Mcdonell92] K. Mcdonell "Benchmark Frameworks and Tools for modelling the Workload profile", in Proceedings of the 6th Int. Conf. on Modelling Techniques and Tools for Computer Performance Evaluation, 1992.

[Pirahesh92] H. Pirahesh, J. M. Hellerstein, and W. Hasan ,"Extensible/Rule Based Query Rewrite Optimization",in Procs. ACM SIGMOD, 1992, pp 39-48.

[Salza92] S. Salza and R. Tomasso " A modelling Tool for the performance analysis of relational database applications"in Proc. 6th Int. Conf. on Modelling Techniques and Tools for Computer Performance Evaluation, 1992.

[Selinger79] P. Selinger, M. Astrahan, D. Chamberlin, R. Lorie and T. Price,"Access path Selection in a Relational Database Management Systems", in Procs. ACM SIGMOD, 1979, pp 23-34.

[Stonebraker91] M. Stonebraker, and G. Kemnitz, "The POSTGRES next-generation database management system", Comm. of ACM, Special Section on Next-Generation Database Systems, 1991, 34(10):78.

[Swami89] A. Swami "Optimization of Large Join Queries", PhD Thesis, Stanford University, June 1989, Stanford CS Report STAN-CS-89-1262.

[Turbyfill89] C. Turbyfill, C. Ori, and D. Bitton "AS3AP - An ANSI Sequel Standard Scalable and Portable Benchmark for Relational Database Systems", DB Sotfware Corporation, 1989.

[Whang87] K-Y Whang "Query Optimization in OBE/QBE A Memory-Resident Domain relational Calculus Database Systems", Tech. Rep. RC11571 IBM Research Division, March 1987.

[Wilschut95] A. N. Wilschut, J. Flokstra, and P. M.G. Apers "Parallel Evaluation of multi-join queries" , in Procs. ACM SIGMOD, 1995, pp 115-126.

'Favourite' SQL-Statements - An Empirical Analysis of SQL-Usage in Commercial Applications

Richard Pönighaus

Adress: Cottagegasse 10, A-1180 Vienna, Austria; e-mail: poenigh@wu-wien.ac.at; phone&fax: +(43) (1) 4702751
Lecturer at Vienna University of Economics and Business Administration, Institute of Information Processing and Information Economics, Department of Applied Computer Science.

Keywords: Database Languages; Empirical Study; Commercial Applications; Database Benchmarks; Statement Complexity; Software Quality Assurance

Abstract. An empirical study investigates usage of SQL in commercial applications of three large Austrian companies. Based on 38,000 statements we analyse the practical meaning of the DML-Part of SQL language constructs. A cost-efficient method of data collection for IBM-DB2 environments is described. We also propose a simple complexity scheme for classifying SQL statements and apply it to our data. Some of our findings are compared with SQL-features used in standard database benchmarks. Since empirical but non-laboratory results in this area are rare this study may be of general interest to the database community.

1 Motivation

Knowing which SQL-constructs are frequently used in real-world applications is desirable for a number of reasons. For instance, this type of information is necessary to derive a workload mix for a generally applicable database benchmark. Currently, these information are either taken from one particular type of real-world application (e.g. for the TPC-suite [10], [11], [12] and Set Query Benchmark [8]) or, as in the case of AS³AP benchmark [13], [14] are based on individual experience rather than thorough empirical analysis of several applications (for an overview of database benchmarks see [6]). Other areas where it is advantageous to know the 'favourite' SQL are the teaching of database programming and optimizer design. Clearly - given the ever growing size of standard SQL - one should concentrate on the language constructs most frequently used. Moreover, one can infer programming styles and language knowledge from the presence or absence of certain constructs and use this information to guide the teaching of SQL.

Currently the database literature only provides little information on SQL-usage in the 'real' world. Some empirical studies on SQL have been done. But most of them are laboratory studies done with student users [7]. [7] did a survey study on usage of SQL in Singapore's IT industry based on a mail questionnaire.

Our paper deals with the issue of determing 'favourite' SQL-constructs from real-world, commercial applications. All four data manipulation statements provided by SQL are considered: SELECT, INSERT, UPDATE, and DELETE. The paper is based on some of the main results of [9], Chap.3.

Section 2 describes the methodolgy used for data collecting and analysing. This methods allows to completely evaluate all embedded SQL-Statements which are in use on a given IBM-DB2 system. The author had the opportunity to apply this method to four IBM-DB2 environments of three Austrian companies. Section 2 also gives a first look to the data. Section 3 analyses SQL-constructs for queries and update operations. Using a statement parser developed by the author we will count syntax elements and give one-dimensional frequency distributions for SQL-constructs of interest. In Section 4 we propose a simple complexity scheme based on the search condition of a particular SQL-statement and apply it to our data. To give examples of the usefulness of empirical data some results of Sects. 3 and 4 are related to standard database benchmarks (Wisconsin Benchmark [5], AS^3AP Benchmark [13], [14], Set Query Benchmark [8] and the sample implementations of TPC Benchmark A, B, C [10], [11], [12]) and statements found in the database literature.

2   Data Collection Method and Analysis Tools

When talking about usage of SQL we have to distinguish between

- embedding of SQL in host languages like C, COBOL, PL/1 etc. and
- SQL as interactive query language via special interfaces like (SQLPLUS-/ORACLE or SPUFI/DB2).

The goal of any *global statement analysis* should be to gain statement texts of each *executed* SQL statement embedded as well as interactive. This type of information can be obtained by using specialized monitors or starting traces. This method has the advantage that the *execution frequency* is measured; interactively and dynamically executed SQL-statements are recorded, too. However, besides the necessity of a monitor, this measurement degrades performance. Moreover, rather long traces are necessary to obtain statistically significant results. Having in mind to get representative results it is also important to choose a *portable solution* between different companies. But the availability of monitors or the possibility to start traces depends on the respective environment under study.

Therefore the author decided to restrict himself to *static embedded SQL* and the measurement of *coding frequency* but not execution frequency. Under this restriction a very cost-efficient method for data collection is possible, i.e. using meta information which is automatically stored and managed in catalog tables by a RDBMS. The RDBMS IBM-DB2 V2R2 manages about 30 system catalog tables (SYSIBM.SYS*) [4]. One of these catalog tables - SYSIBM.SYSSTMT - stores each static SQL-statement embedded in an application program bound at the respective DB2-System[1]. Querying meta information from catalog tables can be done using standard SQL, too; this is a big advantage of RDBMSs. In particular, the following SQL-statement can be used to *extract all SQL queries and update statements from a particular IBM DB2-enviroment*:

```
SELECT B.PLNAME, B.STMTNO, B.NAME, B.SEQNO, B.TEXT
  FROM SYSIBM.SYSSTMT A, SYSIBM.SYSSTMT B
    WHERE
        A.SEQNO=0
    AND (   A.TEXT LIKE '%SELECT %'
        OR A.TEXT LIKE '%INSERT %'
        OR A.TEXT LIKE '%UPDATE %'
        OR A.TEXT LIKE '%DELETE FROM %' )
    AND A.NAME=B.NAME
    AND A.PLNAME=B.PLNAME
    AND A.STMTNO=B.STMTNO
    AND A.SECTNO=B.SECTNO
ORDER BY B.PLNAME,B.NAME,B.STMTNO,B.SEQNO;
```

where NAME refers to the DBRM (Database Request Module), PLNAME is the name of the application plan, PLCREATOR is the authorization id of the plan creator, and STMTNO is the statemment number in the application source program. These columns are used to uniquely identify a statement. Since SQL-statement texts longer than 254 characters are distributed over several rows of SYSIBM.SYSSTMT, this query was embedded in a C-program that reassembled split statements.

The statement texts thus obtained have to be syntactically analysed to extract language constructs used; i.e, one has to develop a SQL-parser that, instead of generating object code, records a particular structure of the syntax tree. Based on the syntax diagrams in [4], which were translated into BNF, the author developed a lexical analyser and parser using the MS-DOS public domain products FLEX (for lexical analysis) and BYACC (for parsing). For the DML-statements under study, 109 terminal and 76 nonterminal symbols as well as 285 rules were used. The generated C-program had approximately 2,400 lines of code and implemented a finite automaton with 494 states.

2.1   Data and Companies

Since the cost for a participating company could be reduced to apply the above mentioned single SQL-statement to their catalog table SYSIBM.SYSSTMT, each of three asked Austrian companies could be 'won' to participate in this study. Table 1 gives an overview on our data basis and shows the empirical distribution of the four DML types under study.

Company A is an Austrian bank, which provided its development (A-dev) and production DB2 environment (A-prod). Companies B and C are Austrian service industry businesses. The largest source of DML-statements was the development

---

[1] Starting with IBM-DB2 V2 Release 3 another catalog table called SYSIBM . SYS-PACKSTMT is available which stores statements belonging to so called packages. Please refer to the respective documentations.

Table 1. Data basis / empirical distribution of SQL statements grouped by companies

Col%	A-dev	A-prod	B	C	Total %	Total nr.
SELECT	72.52	73.90	48.14	73.69	68.09	26131
- Cursor SELECT/SELECT in (%)	(38.19)	(47.73)	(46.58)	(80.86)	(40.81)	(10664)
- Singleton SELECT/SELECT in (%)	(61.81)	(52.27)	(53.42)	(19.14)	(59.19)	(15467)
DELETE	4.4	8.82	9.71	5.39	5.73	2200
INSERT	11.94	7.03	31.87	8.72	15.25	5853
UPDATE	11.13	10.25	10.28	12.2	10.93	4194
Total %	72.38	7.3	18.67	1.64	100.00	
Total	27779	2801	7167	631		38378

system of company A, followed by company B and A's production environment. Company C was at the very beginning of DB2 use. Data from company A were collected in March 1992; companies B and C data were collected in 1990.

A-prod and A-dev. Since 1989 the author has been cooperating with company A, as database consultant, closely. Interpretations of data related to company A given in the following sections are based on the author's above mentioned experience. For safety and performance reasons company A uses two logically and physically separated IBM mainframes. The development machine, on which A-dev is running, is used as development and test environment. The production machine, on which A-prod is running, is exclusively used for the 'real business' of the bank. Most of the programs developed and tested on A-dev will go into production A-prod later on. Statements extracted from A-prod reflect the early SQL usage of application programmers in company A; statements extracted from A-dev show the SQL usage of the now more experienced and advanced application programmers in company A.

3 Study Results

In the following we will analyse the data on a *syntactical level*. We do not make any analysis of semantics of a particular statement. For evaluation purposes we show frequency distributions of some interesting SQL-constructs.

3.1 Coding Frequency of SQL Statement Types

As could be expected in all four environments, the SELECT statement is the most frequently coded statement type (see Tab. 1). In company B the fraction of INSERT statements is considerably larger at the expense of SELECT statements. A closer analysis of the INSERT statements in B revealed that 83.1% of all

INSERT statements are multiple-record INSERT statements with a subselect-clause of the special form

```
INSERT INTO report (atti,..)
        SELECT colfunct(attj),.. FROM relati .. GROUP BY ..
```

with many column functions and expressions. This statement type was used to fill a reporting database. It is interesting to note that if multiple-record INSERT statements are counted as SELECT statements all companies have a nearly identical fraction of SELECT statements between 72.7% and 74.9%.

3.2 SELECT Statements

Embedded SQL differentiates between cursor SELECT statements and non cursor SELECT statements. Cursors are needed for multiple-row selects. Table 1 shows that the majority of all SELECT statements are singleton SELECT statements. That means, that in about 60% of all cases SELECT statements return at most one row.

Projections. In principle the length of projection lists is unlimited. SQL offers the possibility to explicitly define attributes needed or to use the *-notation giving full details of all tables named in the FROM-clause.

*The *-Notation.* The *-notation is a comfortable alternative specifying projection, since it saves keystrokes. Using embedded SQL there is a potential danger; e.g. the meaning of '*' may change if the program is rebound and columns are added to a table (see [2],p.78f). Table 2 shows that *-notations are coded by companies with different intensity. The usage of *-notations (more than 8% of all SELECT statements) in company B seems to be a serious problem. In contrast to company C and A-prod we find a very low fraction of *-notations (0.8%) in company A-dev. Comparing A-prod with A-dev we can confirm an improved program quality for new applications, which is probably based on a higher degree of experience in using SQL.

Explicitly Defined Projection Lists. As can be seen in Tab. 2, projection lists are normally short. Remarkable is the structural change from A-prod to A-dev. The average length changed from 6.4 elements to 2.6 elements. Reasons for these significant smaller projection lists may be:

- better table design (smaller tables, higher degree of normalization),
- qualified specification of those attributes really needed in an application program.

In certain situations shorter projection lists may have a performance advantage (better access paths (e.g., index-only scans instead of non-index-only scans) may be chosen by an RDBMS-optimizer). Therefore the author supposes that programmers in company A are more concerned with performance related questions, now.

Table 2. Length of projection lists measured as number of elements

Col Pct Cum. Col%	A-dev	A-prod	B	C	Total %	Total
*	0.79	2.70	8.35	3.20	1.99	534
	0.79	2.70	8.35	3.20	1.99	
1	55.29	36.96	32.74	24.40	50.24	13514
	56.08	39.66	41.08	27.60	52.23	
2	24.00	13.67	12.83	14.20	21.50	5783
	80.08	53.33	53.92	41.80	73.73	
3	5.27	5.89	10.96	8.60	6.13	1648
	85.35	59.22	64.88	50.40	79.86	
4	3.21	6.74	6.98	5.40	4.04	1086
	88.56	65.96	71.86	55.80	83.89	
5	3.68	5.04	6.05	6.60	4.16	1118
	92.24	71.00	77.91	62.40	88.05	
6-10	4.18	11.60	13.80	11.00	6.18	1663
	96.42	82.60	91.71	73.40	94.23	
11-15	1.39	6.43	5.17	12.40	2.51	675
	97.82	89.03	96.88	85.80	96.74	
16-20	0.91	4.09	0.97	6.00	1.27	342
	98.72	93.12	97.84	91.80	98.01	
> 20	1.28	6.88	2.16	8.20	1.99	534
	100.00	100.00	100.00	100.00	100.00	
Total%	76.78	8.27	13.09	1.86	100.00	
Total	20651	2224	3522	500		26897
Mean	2.58	6.25	3.97	8.17	3.17	
Mean ($X >= 1$)	2.60	6.43	4.33	8.44	3.23	

Aggregate Functions / Grouping. Table 3 depicts the frequencies each of the five built-in column functions of SQL are appearing at least once per statement. Our analysis confirms a high relevance of aggregate functions for commercial applications. However, we can find some significant differences in the usage of a particular function. While COUNT seems to be the most important function (unweighted average over all companies is about 5%) - somewhat surprising - MIN and especially AVG do not seem to be worth mentioning. Note that Companies A and C do not use AVG at all.

Based on our study we can state that the aggregate functions (COUNT, SUM) used in the Set Query Benchmark (see for example [8]) have a good coverage of commercial reality. While companies A and C only make moderate use of GROUP BY company C uses this construct in 7.18% of all queries. This seems to be related to the fraction of queries using the column function SUM (7.35%). HAVING seems to be too complicated to use.

Table 3. Aggregate functions and grouping (frequencies with which each construct appears at least once per statement)

function	A-dev	A-prod	B	C	unwgt. avgerage
column functions					
MIN	0.13	0.54	0.20	0.40	0.32
MAX	1.77	5.85	1.76	1.60	2.74
SUM	0.50	1.35	7.35	1.60	2.70
AVG	0.00	0.00	0.23	0.00	0.06
COUNT	3.42	10.21	3.75	2.80	5.04
all column functions	5.82	17.95	13.29	6.40	10.86
grouping					
GROUP BY	0.08	0.04	7.18	1.00	2.08
HAVING	0.01	0.04	0.09	0	0.04

Joins. Usage of joins can be measured by counting the number of tables in the FROM-clause of a query. Table 4 and the corresponding Fig. 1 show a homogeneous distribution for Company B and C and company A-prod: the main parts of queries are selections on a single table. Only between 2.88% and 5.6% of the SELECT statements are joins. If joins are used they are normally 2-way joins. The most complex joins are a 4-way join in A-prod, a 5-way join in B and only a 3-way join in C.

From the author's point of view the low usage of joins may have different reasons:

- Intended conservative usage of SQL in the implementation phase of RDBMSs
- Unfamiliarity with usage of joins and therefore uncertainty related to their performance
- Adaption of programming techniques from conventional data management systems or flat files
- Low level of normalisation

A completely different structure of usage can be found for A-dev. In this more advanced environment nearly 30% of all queries are joins. 8% of all queries are at least 3-way joins. Comparing A-dev with the other three environments we observe a clear trend to higher complexities of coded statements. Beside a more and more decreasing relevance of the above mentioned reasons against joins the user requirements related to the information quality of new commercial applications may increase.

Ordering. In the context of embedded SQL the analysis of ORDER BY - usage has to be restricted to cursor SELECT statements which do not contain a FOR UPDATE OF cursor clause. As mentioned above, only 40% of all queries use cursors. Table 5 shows the important role of ordering in commercial applications.

Table 4. Usage of simple SELECT statements versus n-way joins; number of tables or views in the FROM-clause

Col Pct Cum. Col%	A-dev	A-prod	B	C	Total %	Total
1	70.05	97.12	96.00	94.40	76.13	20478
	70.05	97.12	96.00	94.40	76.13	
2	21.64	2.61	3.27	5.20	17.35	4667
	91.68	99.73	99.26	99.60	93.49	
3	5.47	0.18	0.45	0.40	4.28	1152
	97.15	99.91	99.72	100.00	97.77	
4	1.91	0.09	0.26	0.00	1.51	405
	99.06	100.00	99.97	100.00	99.28	
5	0.39	0.00	0.03	0.00	0.30	81
	99.45	100.00	100.00	100.00	99.58	
6	0.35	0.00	0.00	0.00	0.27	73
	99.80	100.00	100.00	100.00	99.85	
7	0.02	0.00	0.00	0.00	0.01	4
	99.82	100.00	100.00	100.00	99.86	
8	0.02	0.00	0.00	0.00	0.02	5
	99.85	100.00	100.00	100.00	99.88	
9	0.13	0.00	0.00	0.00	0.10	26
	99.97	100.00	100.00	100.00	99.98	
10	0.03	0.00	0.00	0.00	0.02	6
	100.00	100.00	100.00	100.00	100.00	
Total %	76.78	8.27	13.09	1.86	100.00	
Total	20651	2224	3522	500		26897
Mean	1.43	1.03	1.05	1.06	1.34	

The fraction of queries containing an ORDER BY lies between 26.43% (A-Dev) and 53.51% (C).

It seems to be a *serious weakness* of most of the standard database benchmarks that they do not cover this important SQL feature explicitly. Wisconsin Benchmark [5], AS^{3}AP Benchmark [13], [14], Set Query Benchmark [8] and the sample implementations of TPC Benchmark A and TPC Benchmark B given in [10], [11] are lacking in providing any ORDER BY clause! Only the sample implementation of TPC Benchmark C [12] provides SELECT statements with ORDER BY-clauses.

It is interesting to note that the fraction of ORDER BY-clauses decreases from 48.18% in A-prod to 'only' 26.43% in the more advanced environment A-dev. From a performance-oriented point of view this may be favoured, since sorting is normaly costly in terms of CPU-usage and response time. Assuming a similar application structure the author supposes that programmers and systems analysts are now more critically examining whether ordering of data is really needed.

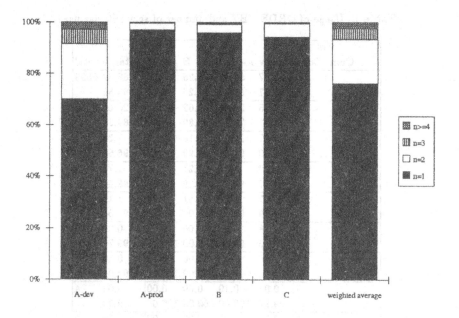

Fig. 1. Usage of simple SELECT statements versus n-way joins

3.3 Search Condition Elements by SQL-Type

This section deals with some important syntax elements which are common to all four SQL-DML statement types. The 'highest' common syntax construct is the *search condition*. As comparison measure we will use the frequencies with which each of the syntactical elements appears at least once per statement. Using embedded SQL some subclasses of statements are of no interest for syntactical analyses (see Tab. 6). In the following those statements will not be considered.

No Search Condition. The usage of a search condition is optional for all considered statement types. As Tab. 7 shows there are remarkable differences for the fractions of statements which do not use any search condition. While global UPDATE statements ($< 1\%$) are rare, the fraction of global DELETE statements (12%) is astonishingly high. Remarkable is the 25% fraction of multiple record INSERT statements which are specified without a search condition. As mentioned above in our data multiple record INSERT statements are mainly used for filling of reporting databases. Obviously there is a high need for overall evaluations.

The important global DELETE statement is not covered by any of the standard database benchmarks. From the author's point of view it would be most interesting to compare performance figures for this special statement type. Since there are different techniques used to implement global DELETE statements the author expects extremely different performance results.

Table 5. Usage of ORDER BY and number of sort criterias used

Col Pct Cum. Col%	A-dev	A-prod	B	C	Total %	total
0	73.57	51.82	63.26	46.49	68.82	6599
	73.57	51.82	63.26	46.49	68.82	
1	17.75	24.40	30.62	19.59	20.48	1964
	91.32	76.22	93.89	66.08	89.30	
2	5.77	15.68	4.10	29.24	7.34	704
	97.09	91.90	97.99	95.32	96.64	
3	1.49	4.15	1.21	3.22	1.77	170
	98.57	96.05	99.19	98.54	98.41	
4	1.02	2.80	0.81	1.17	1.17	112
	99.59	98.86	100.00	99.71	99.58	
5	0.18	0.62	0.00	0.29	0.20	19
	99.76	99.48	100.00	100.00	99.78	
6	0.16	0.42	0.00	0.00	0.16	15
	99.93	99.90	100.00	100.00	99.94	
7	0.04	0.10	0.00	0.00	0.04	4
	99.97	100.00	100.00	100.00	99.98	
8	0.03	0.00	0.00	0.00	0.02	2
	100.00	100.00	100.00	100.00	100.00	
Total %	70.86	10.04	15.53	3.57	100.00	
Total	6795	963	1489	342		9589
Mean	0.40	0.86	0.46	0.94	0.48	
Mean ($X >= 1$)	1.52	1.78	1.24	1.75	1.52	

Table 6. Used/interesting SQL-statement subtypes for comparative analysises

SQL-statement type	syntacticaly 'interesting'	syntactically 'uninteresting'
SELECT	all	
INSERT	multiple record inserts	single record inserts
UPDATE	searched updates	positioned updates
DELETE	searched deletes	positioned deletes

Boolean Operators (AND/OR). Table 7 shows the overwhelming importance of the boolean operator AND compared with OR. It is interesting that for SELECT statements, INSERT statements and DELETE statements the relative frequencies of OR is at the same level (2.5%). Conjunctions are much more frequent than disjunctions, especially for UPDATE statements. Search conditions for UPDATE statements seem to be much more restrictive than those for other statement types.

Table 7. Syntax elements in search conditions by SQL-types (frequencies with which each syntactical element used appears at least once per statement)

Syntax element	SELECT	INSERT	UPDATE	DELETE
no predicate	5.12	25.77	0.98	12.03
conjunction/disjunction				
AND	72.61	41.82	55.35	51.22
OR	2.42	2.54	0.14	2.45
$\frac{AND}{OR}$	30.00	16.46	395.36	20.91
comparison operators				
=	91.90	55.65	97.19	85.82
¬ =	4.15	10.27	0.64	0.90
<=	2.10	0.16	-	3.49
>=	1.31	0.04	0.20	0.60
<>	0.07	-	-	-
>	1.42	6.50	0.20	0.70
¬ >	0.03	-	-	-
<	0.65	-	-	2.15
¬ <	0.37	-	-	-
BETWEEN	0.86	17.56	0.03	0.05
NOT BETWEEN	-	-	-	-
LIKE	4.22	0.32	-	0.35
NOT LIKE	0.03	0.16	-	-
IS (NOT) NULL	0.39	-	0.27	0.20
IN	6.47	5.23	1.93	0.95
subqueries involving predicates				
IN	0.39	1.23	-	0.85
EXISTS	0.25	0.12	-	-
simp.comp.op	1.20	-	-	-
SOME,ANY,ALL	-	-	-	-
Total	1.81	1.26	-	0.85

Comparison Operators. We find a dominating preference for using the equal-operator. Also, we see a remarkable fraction of statements using the 'not equal'-operator. The powerfull range predicate BETWEEN is often used in multiple record INSERT statements. However there was not a single statement using NOT BETWEEN. The pattern matching predicate LIKE is used preferably for SELECT statements. The IN-predicate has some importance to SELECT statements and INSERT statements. Interestingly, the fraction of statements using the NULL-predicates is very low.

None of the standard benchmarks covers IN and LIKE, which are the second and third most frequently used comparison operators for SELECT statements.

Predicates Involving Subqueries. The usage of predicates involving subqueries is relatively low. As to be expected most frequently subqueries appear in SELECT statements (1.81%). Within these SELECT statements subqueries involved by simple comparison operators are most frequently observed. In INSERT statements and DELETE statements the IN-operator is the dominating or the only predicate involving subqueries. The problematic SOME, ANY, ALL-options for subqueries involved by simple comparison operators (see for example the discussion about the high error potential of these constructs in [3], p191f.) are never used.

4   Statement Complexity

In our preceding analysis we always analysed exactly one SQL-construct at a time. Now we propose a complexity scheme which allows a classification of SQL-statements according to their search conditions. As in Sect. 3.3 only the interesting statement subtypes are considered.

[7] used a simple complexity scheme for their study of SQL usage in Singapore's IT industry. This study was based on a mail questionnaire survey. The complexity levels are simply determined by counting the numbers of relations, attributes or predicates in a SELECT statement (see Tab. 8). Thanks to our statement parser's ability to automatically evaluate statement characteristics, we can apply a more elaborated complexity scheme to our data.

Table 8. Complexity scheme used by Lu et al. [7] to classify SELECT statements

syntaxelement / complexity level	low	moderate	high
No. of relations	1-2	3-4	≥ 5
No. of attributes	1-4	5-6	≥ 7
No. of predicates	1-2	3-4	≥ 5

[15] proposed a frame for a systematically increasing statement complexity based on [1]. This frame was used by [16] for synthetical benchmarks of RDBMSes. Four levels of conditions were defined:

- *atomic conditions* as simple comparison on a single column,
- *item conditions* as disjunctions (OR) of atomic conditions of the same column,
- *record conditions* as conjunctions (AND) of two item conditions,
- *query conditions* as disjunctions (OR) of record conditions.

However, not all types of search conditions are captured by this scheme. For instance, range queries, i.e. conjunctions of two atomic conditions, are not contained in the classification.

Thus, the author used a complexity scheme that is summarized in Tab. 9. For a more detailed motivation of this scheme refer to [9], pp103-107. Basically this scheme is built on the operators used in a predicate, the boolean operators used to combine predicates and the number of logical table accesses necessary to evaluate a search condition.

Table 9. Complexity scheme for SQL-statements

complexity class	description
simple table accesses (logically only one access to exactly one table is neccessary)	
1	no search condition
	with search condition
	one predicate (no conjunction/disjunction)
2	one equal-predicate
3	one simple predicate with any of the comparison operators of table 7
	more than one predicate (with conjunctions/disjunctions)
	only equal-predicates
4	only AND
5	as class 4, but at least one OR
	at least one simple comparison operator of class 3
6	only AND
7	as class 6, but at least one OR
complex selections (logically more than one access to tables is neccessary)	
8	equi-joins and additional predicates of class 4
9	subqueries, and all other types of joins not included in class 8

4.1 Complexity and SQL-Statement Types

Figure 2 shows the emprical distribution of statements classified by their complexity. Even if it is scientifically not correct to apply averages on *ordinal values* we will additionally give the mean values and standard deviations of complexities.

The mean complexity (standard deviation) is 4.4 (2.2) for SELECT statements, 3.1 (1.7) for INSERT statements, 3.1 (1.0) for UPDATE statements and 3.1 (1.4) for DELETE statements. On this basis the following conclusions can be drawn:

- The mean complexity of SELECT statements is the highest. Note that statement-specific *additional complexities* like ORDER BY and UNION are not considered.
- The mean complexity of UPDATE statements is the lowest.
- Most of the statements just use equal-predicates eventually conjuncted by AND.

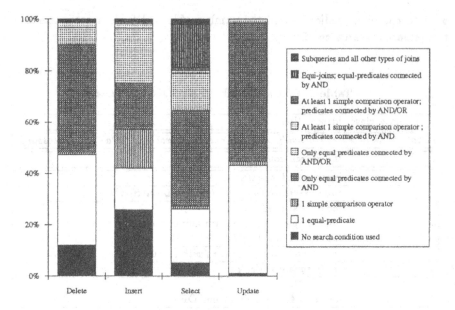

Fig. 2. Complexity and SQL-statement types

- If joins are formulated, normally additional predicates are simple (equal predicates conjuncted by AND)
- The fraction of complex joins and statements including subqueries is low (maximal 2% for selections).
- Notable exceptions of the overall pattern can be seen for multiple record INSERT statements, where a higher fraction of full table SELECT statements, significantly more range queries and disjunctions can be observed. As mentioned above these queries are used to fill a reporting database.

4.2 Complexity and Companies

Figure 3 shows the distribution of complexity for our four analysed companies. The mean statement complexity (standard deviation) is 4.2 (2.2) for company A-dev, 4.0 (1.6) for A-prod, 3.4 (1.3) for B and 4.0 (1.7) for C.

- The range of mean complexities between companies ($4.51 - 3.73 = 0.78$) is significantly lower than the respective range between statement types ($4.56 - 3.15 = 1.41$).
- Note that company A-Prod and C have a nearly identical complexity pattern.
- The comparison of A-dev with A-prod may be used to get an idea of the effects of experience with relational database usage and SQL:

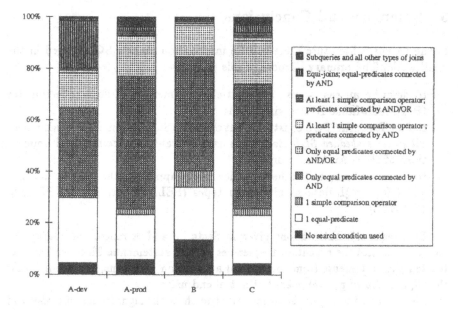

Fig. 3. Complexity and companies

1. The fraction of simple statements increases from A-prod to A-dev (complexity class 2)
2. The fraction of *simple* joins (complexity class 8) increases from 1.7% (A-prod) to 18% (A-dev).
3. The fraction of subqueries and complex joins (complexity class 9) decreases from 3.8% (A-prod) to 1.6% (A-dev).

These opposite trends may be explained by a improved programming quality in company A:

1. More complexity class 2 statements indicate that primary keys are used more often for data positioning, i.e. the relations are better designed and the most efficient data access is possible.
2. The increase in simple joins indicates more normalization and a better table design as well as the fact that the simple 'replacement' of traditional flat files by relations with the same structure is vanishing.
3. As subqueries often can be replaced by simple joins with increased performance, the decrease of these constructs could be attributed to increased knowledge about performance of SQL.

5 Summary and Conclusions

In the database literature there is little information on how SQL is used in the 'real' world. The primary purpose of this paper was

- to describe an easy-to-use and cost-efficient method of data collection for IBM's mainframe DB2 environments,
- to find 'favourite' SQL constructs out of 38,000 SQL statements, which reflect *all static embedded SQL statements* used in real-world, commercial applications of three Austrian companies,
- to propose a complexity scheme, which can be applied to the search-conditions of all four SQL data manipulation types (SELECT, INSERT, UPDATE, DELETE).

The frequency distributions given in Sects. 3 and 4 reflect the *coding frequencies* but not the execution frequencies of SQL statements. Since we have extracted the statements from commercial application programs, the results show the *SQL-usage of programmers* but not of end users.

Throughout the paper the author tried to show the significance of real-world statistics exemplarily. Some of the findings were related to SQL features used in standard database benchmarks (Wisconsin Benchmark, AS^{3}AP Benchmark, Set Query Benchmark and the sample implementations of TPC Benchmark A, B, C given in [10], [11], [12]):

- It was found that explicit ordering of data using the ORDER BY-clause is one of the most important requirements of commercial applications. ORDER BY-clauses are only covered by the TPC-C Benchmark.
- Aggregate functions play an important role in commercial applications. The Set Query Benchmark, which uses the two most frequently coded aggregate functions COUNT and SUM, has a good coverage of commercial reality.
- By far the most frequently used comparison operator is the equal-operator.
- The second and third most frequently coded comparison operators are IN and LIKE. None of the above mentioned benchmarks cover those two SQL-constructs.
- We found a astonishingly high fraction of global DELETE statements (about 12% of all DELETE statements). This statement type is not covered by any of the above mentioned benchmarks, too. It would be most interesting to compare performance figures for this important statement type.

Comparing company A's production and development system we could get an impression how SQL usage may change over time. Many indications of an improved programming quality in A-dev were shown. The interpretations of differences between A-prod and A-dev given in Sects. 3 and 4 are based on a long-term, close cooperation of company A with the author.

C.J.Date and Colin J. White discuss the DML part of SQL in Chapts. 6-8 of their text book [2]. Based on the findings, especially those results derived in Sect. 4, we may agree with their conclusion [2], p.137:

91

'This brings us to the end of our detailed discussion of the four data manipulation statements of SQL, namely SELECT, INSERT, UPDATE, and DELETE. Most of the complexity of those statements (what complexity there is) resides in the SELECT statement; once you have a reasonable understanding of SELECT, the other statements are fairly straightforward, as you can see. In practice, of course, the SELECT statement is usually pretty straightforward as well.'

References

1. A . F . Cardenas, "Evaluation and Selection of File Organization - A Model and System", *Com. ACM*, Vol. 16 No. 9, 1973.
2. C . J . Date., Colin J . White, *A Guide to DB2. Second Edition*, Addison-Wesley Publishing Company, Reading, Mass., 1987.
3. C . J . Date, *A Guide to the SQL Standard*, Addison-Wesley Publishing Company, Reading, Mass., 1987.
4. *IBM Database 2 Version 2. Reference Summary. Release 2*, IBM form no. SX26-3771-1.
5. D . J . DeWitt, "The Wisconsin Benchmark: Past, Present, and Future", in [6].
6. Gray (ed.), *The Benchmark Handbook for Database and Transaction Processing Systems*, 2nd edition, San Mateo California, 1993.
7. H . Lu, H . C . Chan, K . K . Wei, "A Survey on Usage of SQL", *ACM-SIGMOD*, Dec.1993.
8. P . E . O'Neil, "The Set Query Benchmark", in [6].
9. R . Pönighaus, *Performance-Prognosen für relationale Datenbankmanagementsysteme*, Diss., Vienna University of Economics and Business Administration. Dec. 1993.
10. Transaction Processing Performance Council (TPC), "TPC Benchmark^{TM} A. Standard Specification - Revision 1.1" in [6].
11. Transaction Processing Performance Council (TPC), "TPC Benchmark^{TM} B. Standard Specification - Revision 1.1" in [6].
12. Transaction Processing Performance Council (TPC), "TPC Benchmark^{TM} C. Standard Specification - Revision 1.0" in [6].
13. C . Turbyfill, *Comparative Benchmarking of Relational Database Systems*, Diss., Cornell Univ. 1987.
14. C . Turbyfill, C. Orji, D. Bitton, " AS^{3}AP: An Ansi SQL Standard Scaleable and Portable Benchmark for Relational Systems" in [6].
15. S . B . Yao, A . R . Hevner, *A Guide to Performance Evaluation of Database Systems*, Editor: D . R . Benigni, National Bureau of Standards Special Publication 500-118, 1984.
16. S . B . Yao, A . R . Hevner, H . Y . Myers, "Analysis of Database System Architectures Using Benchmarks", *IEEE Transactions on Software Engineering*, Vol. 13 No.6, 1987.

Static Allocation in Distributed Objectbase Systems: A Graphical Approach *

Subhrajyoti Bhar and Ken Barker

Advanced Database Systems Laboratory
Department of Computer Science
University of Manitoba
Winnipeg R3T 2N2, Canada
{bhar,barker}@cs.umanitoba.ca

Abstract. In recent years, object oriented (OO) models have emerged as preferred data models for a wide range of application domains. This is primarily due to the capability of the OO-models to manage and represent information of any arbitrary complexity. Distributed Objectbase Systems (DOBS) combines the benefit of distributed processing and the power of manipulation and abstraction of complex information to provide a powerful environment for distributed computing that is not available in conventional distributed database systems. Application performance and processing cost in a DOBS are greatly influenced by communication overhead involved in accessing nonlocal data and making remote method invocations. Efficient distribution design is mandatory to ensure optimal performance of the distributed system at minimum cost. Distribution design in DOBS involves fragmentation of the objectbase and allocation of the resulting class fragments between the nodes of the network.

We adopt a top-down approach for the distribution design for a DOBS and assume that the global conceptual schema is partitioned into a set of class fragments. We focus our attention to optimal allocation of the fragments between the sites of the network subject to a set of constraints. Our entity of allocation is a class fragment. The allocation scheme defines the local conceptual schema at every site.

The problem of allocation has been addressed for distributed file systems (DFS) and distributed (relational) database systems (DDBS). The additional complexity introduced by the object model has only recently been investigated with no result appearing that address the specific problem of *allocation*. This paper addresses the problem of object allocation in DOBS. A general allocation taxonomy is defined based on data model, degree of redundancy and design objective, and different allocation models are classified based on this taxonomy. An allocation model for DOBS is formulated and an algorithm is presented for static nonredundant allocation in DOBS.

* This research was partially supported by the Natural Science and Engineering Research Council (NSERC) of Canada under an operating grant (OGP-0105566).

1 Introduction

Optimal distribution of data and programs across a computer network is a major design problem that directly impact performance. An optimal distribution design enhances application performance by reducing communication overhead due to references to nonlocal programs or data in a cost efficient way. Distribution design involves two logical steps: (1) fragmentation and (2) allocation. Fragmentation refers to clustering programs and data into groups or "fragments" based on *locality of access* exhibited by applications. The resulting fragments are distributed across the network using an allocation scheme. The problem of allocation has been examined for distributed file systems (DFS) [LL83, GS90, Wah84, Chu69, DF82] and for distributed (relational) database systems (DDBS) [ÖV91, CP84, MR76, CY89, Ape88, SCW83]. The additional complexity introduced by the object environment has only recently been investigated with no results appearing that address the specific problem of *allocation in Distributed Objectbase System*. Benefits obtained from an optimal allocation scheme in a distributed system include: (1) Placing fragments "close to" the site where it is used results in significant reduction in communication cost due to intersite communication across the network. Minimizing network communication also reduces network delay which enhances system performance (ie. throughput, response time and turnaround time); (2) Level of concurrency can be increased by allowing fragment replication which in turn improves throughput and response time. Further, copies of fragments placed at different sites improves fragment availability and system reliability; and (3) Using some load balancing technique, utilization of shared resource (eg. processors, communication channels, I/O devices *etc.*) can be improved and congestion avoided or minimized by distributing the "load" across the network.

An *objectbase* is a collection of objects grouped into a finite number of classes. Objects with common attributes and methods belong to the the same class. Encapsulation bundles together attributes (data values) and the methods (procedures) that manipulate them. Inheritance allows reuse and incremental redefinition of new classes in terms of existing ones. A distributed objectbase system (DOBS) is a collection of local objectbases located at different local sites, interconnected by a communication network [EB94].

A distributed objectbase system (DOBS) is a collection of local objectbases, one at every site. The sites are interconnected by a communication network. At every site the *local conceptual schema (LCS)* defines the logical organization of class objects used by applications executing at local sites. Each *LCS* is mapped to a *logical internal schema (LIS)* which defines the physical organization of class objects stored at local site. The global objectbase, which is aggregation of all local objectbases, is defined by a *global conceptual schema*. Applications executing at every local site view the organization of class objects as defined by the *external schema* at that site. A *global directory (GD)* provides global mapping between different organizational views of class objects distributed across the network and provides location transparency to applications.

The following assumptions are made : (1) The distributed environment is homogeneous with respect to data models for the local objectbases; (2) The network provides reliable, fault-free communication for interaction between nonlocal objects; (3) The set of applications executing on the distributed objectbase does not change with respect to time, location and access behavior; and (4) Class fragments or class objects do not migrate from one site to another.

1.1 The Object Model

The objectbase is a collection of encapsulated objects. An object contains a set (possibly empty) of attributes (data values) that defines its state and a set (possibly empty) of methods [2] (procedures) to access and manipulate them. The attributes and methods are bundled together to form an encapsulated object. Any access to the attributes of an object is permitted only by executing any of the methods belonging to that object. Objects are grouped into classes. Objects belonging to a class have in common all attributes and methods defined for that class. Inheritance allows object in "different" classes to share methods. Thus 'subclasses' are generated by inheriting attributes and/or methods from one or more (in the case of multiple inheritance) of other classes (called superclasses). However, a subclass may contain additional attributes and methods that are not inherited from other class. The overall inheritance hierarchy of the objectbase is captured in a *class lattice* with a 'root class' as the ancestor of every other class. Each class has a class identifier Cid and every object has a globally unique object identifier Oid. An object is represented as a 4-tuple $< Cid, Oid, A, M >$, where Cid is the class identifier of the class containing the object, Oid is the object identifier and A and M are the set of attributes and the set of methods contained by the object respectively.

We assume the objectbase is partitioned into a set of class fragments by using a fragmentation scheme. The fragmentation algorithm is applied to every class of the objectbase to generate "class fragments" for that class [EB93, EB94]. The fragments generated are correct with respect to disjointness, completeness and reconstructibility. However, the type of fragmentation applied is orthogonal to the allocation problem.

Therefore, the problem addressed in this paper can be stated as: Given a set of class fragments of finite size, a communication network with interconnected nodes, each having its own storage capacity and processing power and a set of applications executing at different sites, design an allocation scheme for optimally distributing the fragments between the nodes of the network so that the operating cost is minimum and performance is optimal.

1.2 Motivation

The problem of allocation has been addressed for distributed file systems [LL83, GS90, Wah84, Chu69, DF82] and distributed relational database systems [OV91, CP84, MR76, CY89, Ape88, SCW83]. Previous research work demonstrate the

---

[2] Methods are also known as behavior.

difficulties in obtaining an optimal allocation leads to computational intractable solutions [Esw74]. 'Near optimal' solutions are obtained in most cases by applying heuristic techniques based on the nature of the specific problem.

Distributed file system solutions are not applicable in distributed database environment because the data is more tightly integrated and is managed by a single system that is responsible for the entire database. File allocation means individual files are independent units of distribution. Application do not see strong relationships between files so modeling the application's behavior in a distributed file system is comparatively simple.

A file access in a distributed environment refers to either a remote access or a local read/write depending on the location of the file. Therefore, the allocation model only considers accesses to a single file, so multi-file accesses required in a distributed database are ignored (consider, for example, a relational join). Therefore, files are allocated independently so only resource and performance constraints need to be considered in the model.

Database and object-oriented systems introduce additional complexity. Since relational database system transactions often involves operations on several relations, possibly located at different sites, the modeling of application behavior is more complex. Application behavior is typically modeled by determining the portions of the data accessed together. This behavior is reflected in the formation of *fragments* where data is closely affined[3]. Fragmentation can occur horizontally (collecting together groups of tuples accessed together), vertically (collecting together attributes that are accessed together), or hybrid (applying horizontal and vertical fragmentation) [ÖV91, EB93, EB94].

Object-oriented system are more complex because of the inherent relationships between objects. The object data model is capable of capturing relationships of arbitrary complexity so modeling their relationships and interactions is very difficult. Karlapalem and Navathe [KKM92] identify the difference between relational and object-oriented environment in terms of interactions between data and methods. In distributed database, the data and programs that access them are disjoint so the allocation scheme applies to data only. An object-oriented environment must account for the encapsulation of methods and the data referenced, so method invocation plays a major role in allocating objects. Hence, the allocation problem needs to address placement of data and method *together*. Moreover, as method in one object can invoke another method in the same or different object, the dependencies between methods must also be considered.

The distributed object model must consider all relevant constraints to achieve a complete solution to the allocation problem.

1.3 A Taxonomy for Allocation

The allocation models for distributed systems can be broadly classified based on (1) type of data model, (2) degree of redundancy and (3) design objective. The

---

[3] Affinity is essentially the likelihood and/or frequency that an application will reference two attributes or tuples.

taxonomy illustrated in Figure 1 identifies different allocation models in each class. The influence of each of the characterizing parameters is briefly explained below.

- Type of Data Model: A data model defines the data structures and the operations to be performed on them. The goal of an allocation scheme is to distribute data and programs to facilitate application execution. Thus a strong relationship exists between an allocation model and the data model for which it is designed. Distributed systems are classified based on the data model as (1) distributed file system (DFS), (2) distributed database (relational) system (DDBS) and (3) distributed objectbase system (DOBS). Accordingly, the allocation problems are classified as file allocation problem (FAP), database allocation problem (DAP) and object allocation problem (OAP) respectively.

- Degree of Redundancy: Introducing additional copies or replicas of individual data units increases data availability and fault tolerance. On the other hand, multiple copies occupy more storage space which is a limited resource in distributed system. Moreover, the cost of performing updates increases linearly [CA80] with the number of copies maintained. A judicial trade-off is required to decide the optimal number of copies of each data unit to maintain. The complexity of the distribution design will depend on the degree of replication. While read-only queries can be served by the 'nearest' available copy of the requested data thereby reducing response time but maintaining consistency between all copies becomes substantial overhead.

- Design Objective: Allocation, as part of distribution design, is a combinatorial optimization problem with a specific objective function that is subjected to maximization or minimization with respect to a set of constraints. Formulation of the objective function determines the nature of the optimization model. Accordingly, the allocation models are classified as 'cost models' or 'performance models'. In cost models, the objective function is formulated with various cost factors (eg. communication cost, storage cost, processing cost etc.) involved in allocation and the overall cost is subjected to minimization with respect to a set of constraints, to obtain an optimal allocation scheme. Performance issues are either ignored or they are included in the cost model as constraints. The performance model performs optimization on a set of performance metrics (eg. throughput, response time, etc.) to arrive at an allocation scheme that meets the upper bound for cost. Ideally however, an optimal allocation should meet both cost and performance requirements simultaneously. The corresponding allocation model is referred to as 'cost & performance model' in the taxonomy illustrated in Figure 1.

The rest of the document is organized as follows: Section 2 a review of background and related work on allocation in distributed systems is provided. We adopt a heuristic approach for solving the allocation problem in DOBS with a nonredundant cost model. Section 3 presents a heuristic algorithm based on graphical approach for DOBS nonredundant cost model. We conclude in Section 4 with concluding remarks and future research directions.

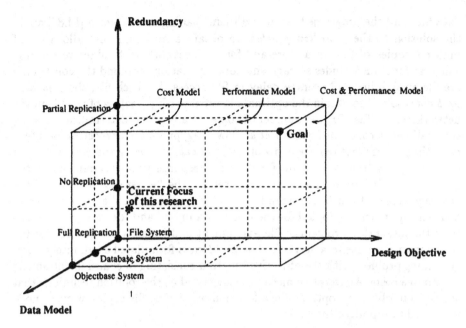

Fig. 1. A Taxonomy for Allocation Models

2 Related Work

FAP and DAP are combinatorial optimization problems that have been studied by many researchers over a long period of time. Both cost and performance models are proposed for either of the two problems.

2.1 FAP/Cost Models

Chu [Chu69], Casey [Cas72], Wah [Wah84], Levin and Morgan [LM77], Laning and Leonard [LL83] and Ghosh *et al.* [GMM92] present cost models for FAP. The proposed models differ in the formulation of the objective function and the solution approach. Casey [Cas72] presents a "cost graph" model to determine the optimal number of copies of a single file that should be placed in a network. Wah [Wah84] shows isomorphism between FAP and the 'warehouse location problem' in operations research (OR) to justify the applicability of OR techniques to solve FAP. Application of branch and bound method is demonstrated as a solution technique for FAP. Chu [Chu69] models the communication traffic between the nodes of a network as M/D/1 queueing process and formulates the FAP as a 0/1 integer programming problem with the storage and communication costs as the components of the objective function to be minimized. File availability and response time are included as constraints and the nonlinear problem is transformed to linear by adding additional constraints. Levin and Morgan [LM77] present a FAP model that captures the interaction between the data files and the programs that access them. Laning and Leonard [LL83] apply the solution to the *p-median problem* to obtain a minimum cost allocation of multiple copies of files in a store-and-forward network with adaptive routing. The cost function includes storage and communication costs and the constraints are file availability and average delay in file access. For each file, the *p-median problem* is solved to obtain the optimal placements of "p" copies of the file in the network. The value of p, the number of copies of any file, is determined based on the availability constraint. Ghosh *et al.* [GMM92] present FAP/cost models that meet the performance requirements of the applications. Delay analysis is done on query by query basis to account for different response time requirements of the queries. Two FAP/cost models are proposed for the "worst case delay" and the "average delay". Branch and bound method is employed for the former whereas heuristic approach is adopted for the latter. Mahmoud and Riordon [MR76] consider the integrated problem of file assignment and capacity assignment for the communication links in a network. The problem is formulated as an integer programming problem with the objective function consisting of communication and file storage costs. An iterative method is suggested as the solution technique that attempts to obtain an optimal file allocation by varying file copies at every node using "add-drop algorithm".

2.2 FAP/Performance Models

FAP/performance models are studied by Ramamoorthy *et al.* [RC70], Arora [AG73], Chen [Che73] and Kleinrock [Kle76]. Ramamoorthy *et al.* [RC70] apply known results from memory hierarchy research to solve the FAP for minimum file access time under certain assumptions. Files are assumed to be divisible across the nodes of the network and queueing delays and communication overheads are ignored. These assumptions, being quite restrictive, do not allow general applicability of the solutions. Arora [AG73] extends the model to include execution time, access time and data transfer time in the objective function. Techniques for solving transportation problems in operations research is employed to find an optimal solution. Queueing network models must be considered to account for queueing delays in FAP. Chen [Che73], Buzen [Buz71], Hughes [HM73] and Piepmeir [Pie75] present file allocation schemes using queueing models for a client/server network. A closed queueing model of a central server (star) network is considered by Buzen [Buz71] and Hughes [HM73] where a constant degree of multiprogramming ensures a constant number of file requests in the network. The objective function is system throughput which is subjected to maximization. Only general guidelines are provided to accomplish the file allocation heuristically.

Chen [Che73] formulates a queueing model for an open central server network. File request frequencies and execution times are assumed to be exponentially distributed and standard results for M/M/1 queue are applied to obtain file request distributions that minimizes average response time. While allocating the files to match the distribution exactly, files are assumed to be divisible and nodal capacity constraints are ignored. Buzen [BG74] and Piepmeir [Pie75] extend these results to an M/G/1 queuing model.

Kleinrock [Kle76] suggests an iterative technique that finds near optimal file allocation for an open general topological network starting from a "guess allocation" used as an initial file assignment. For given link capacities and file request frequencies the objective is to minimize file response time. Jones [JFK78] presents a goal-programming formulation of the allocation problem that minimizes operating and storage costs of all the file copies subject to a capacity constraint. Overhead involved in performing update operation with multiple copies is considered. An iterative method is suggested that improves an initial allocation in each step.

2.3 DAP/Cost Models

Cornell and Yu [CY89], Gavish and Pirkul [Gav87, GP86], Ceri, Navathe *et al.* [SCW83] and Apers [Ape88] present DAP cost models for allocating single copies of every relation in a distributed relational database system. Ceri, Navathe *et al.* [SCW83] model the logical database schema as a directed graph with "relations" as nodes and their semantic relationships as edges joining the nodes. The global schema is partitioned into independent subsets which are placed independently. Sacca and Wiederhold [SW83] present an allocation scheme for nonredundant relational database to evenly distribute the transaction processing load between the sites. Apers [Ape88] argue to show that optimal allocation of database fragments would depend on the execution order of the database operations which is determined by the query processor. A method is proposed that performs partial allocation of data in iterative steps based on scheduling information provided by the query processor. Gavish *et al.* assume that transaction exhibit "strong locality of reference" in making placement decisions for processors and databases in a distributed banking network. The allocation problem is formulated as an integer programming problem and heuristic procedure and branch and bound method are suggested as the solution techniques. Cornell and Yu [CY89] identify interdependency between access plan selection[4] and the assignment of database relations to different sites of a distributed database system. Relations are not fragmented and are assigned to a site as a whole. A fully connected network with equal communication capacity on each link is assumed. An integer programming problem is formulated with the communication cost as the objective function which is subjected to minimization. Yu, Siu *et al.* [YSLC83] present an adaptive algorithm for allocation of database files in a star network. Update costs are ignored as only one copy of each file is considered. Starting from an arbitrary initial file distribution, the adaptive process computes online the "gain" of placing a file at any particular site. For each query being processed, the "gain" is evaluated for all possible file allocation of the files being accessed by the query. The final allocation is obtained by placing every file at a site where

---

[4] Typically access plan selection is performed by query processor [Ape88].

the corresponding gain is maximum. The nonredundant DAP model is extended for replicated allocation by including update costs in the cost function.

In a distributed system, network topology and distribution of data are closely related. Design of a distributed system should exploit the coupling between data distribution and program and network topology to arrive at a design that provides optimal system performance at minimum cost. Irani and Khabbaz [IK82] and Chen and Akoka [CA80] address the integrated problem of design of communication network and distribution of database files. Fisher and Hochbaum [FH80] and Chieu and Raghavendra [CR90] address the problem of allocation of multiple copies of a database in a distributed network.

Fisher and Hochbaum [FH80] present a DAP model that makes placement decisions for copies of databases by trading off the access cost, which is reduced by additional copies, against the cost of storing and updating the additional copies. Chieu and Raghavendra [CR90] address the problem of finding the optimal number of database copies and their locations to optimize the communication cost due to remote accesses between pairs of network sites. It is assumed that the local storage capacity at each site is large enough to hold a copy of the entire database and there are at least three copies of the database present in the network.

2.4 DAP/Performance Models

Cornell and Yu [CY89] propose heuristic technique to minimize response time under the constraints of available storage and processing power at every node. An open queueing model is formulated where each node is modeled as an M/M/1 queue. Ciciani *et al.* [CDY90] examine the performance trade offs of data replication in a distributed system with respect to additional overheads due to multiple data copies.

2.5 Distribution Design in DOBS

Karlapalem *et al.* [KKM92] discuss various issues involved in distribution design of object-oriented database systems. They identify the characteristic features of the object-oriented model that make the problem of distribution design more complex in comparison to its peer problem in relational database system.

They argue that the critical issues for the distribution design for an object-oriented system are : (1) Level of transparency, (2) Method invocation strategy, (3) Application processing semantics and (4) structural and behavioral properties of the objectbase. They identify the goal of a distribution design of an object-oriented system as: (1) to reduce irrelevant data access by application and (2) to reduce the amount of inter-site data transfer.

Ezeife and Barker [EB93, EB94] present algorithms for obtaining horizontal and vertical fragments by optimally partitioning the global objectbase. The fragmentation schemes are designed to achieve improved performance by reducing irrelevant data access, minimizing data transfer, increasing concurrency, and reducing irrelevant data replication.

They assume that the distributed database consists of local databases at individual sites each having its own external and internal schema. Based on application access frequency and application execution pattern, the fragmentation schemes are suggested that reduce processing costs by minimizing communication overhead and irrelevant data access. A taxonomy is presented that classifies different models based on method type and the type of the attributes accessed. The different models suggested are: (1) simple attribute/simple method, (2) simple attribute/complex method, (3) complex attribute/simple method and (4) complex attribute/complex method. Horizontal[5], and vertical[6] fragmentation schemes for different class models are presented.

The problem of allocation has not been investigated extensively in the object oriented environment. Although a few articles have appeared in literature that discuss various issues and constraints of the problem no algorithm or solution strategy has been suggested. Karlapalem, Navathe and Morsi [KKM92] identify the importance of fragmentation and allocation of object base for minimizing irrelevant data access and nonlocal data references as a part of distribution design. Characteristics of object oriented system that distinguishes it from a relational system are listed and their impact on distribution design are discussed.

3 Static Allocation

This section presents algorithms for static allocation in DOBS. The algorithms are based on the nonredundant cost model for DOBS defined in the allocation taxonomy in Figure 1 presented in Section 1. The allocation problem is NP complete for distributed file system (DFS) and for distributed relational database system (DDBS) so with the additional complexity introduced by the object model, it is highly unlikely that an optimal, polynomial time solution to the allocation problem for a DOBS could be obtained. This indicates that heuristic approaches need to be adopted to obtain a "near optimal" solution in a reasonable time. Therefore, the algorithm proposed here is based on heuristic. An initial allocation is obtained as a starting solution which is then subjected to optimization based on heuristic technique. Section 3.1 formally describes the static nonredundant allocation model for DOBS with assumptions explicitly stated and definitions used in the model clearly presented. The allocation problem in DOBS is formally described with respect to this model in Section 3.2. Section 3.3 presents an algorithm for generating an initial allocation which can be used as a starting solution for the heuristic optimization based on graphical approach described in Section 3.4.

---

[5] Horizontal fragments are generated by partitioning the set of instance objects of a class.

[6] Vertical fragments are generated by partitioning the set of attributes and methods of a class.

3.1 The DOBS Allocation Model

This section formally describes the static, nonredundant allocation model for distributing class fragments in a DOBS. The allocation problem is then formally defined based on this model. We proceed by stating explicitly the assumptions made and the definitions required to formally present the model:

1. The objectbase is partitioned into a finite set of class fragments. Each fragment has a finite size which is sum of the sizes of its constituent objects or attributes or methods.

2. We assume that the fragmentation algorithm that generates class fragments is correct so the class fragments are disjoint, complete and reconstructible [EB93].

3. The network provides reliable and fault-free communication between sites by message passing. The network topology and the routing scheme do not change with time.

4. Applications do not migrate between sites and the frequency of executing any application at any particular site is constant.

5. Applications are always invoked from their local site. In other words, users invoking any application are assumed to be located at the same site where the application resides.

6. Objects do not migrate between classes and no new class is added or any existing class deleted. Therefore, the objectbase schema remains unchanged.

The following additional assumptions are made to keep the complexity of the model manageable: (1) Communication cost is the dominating factor so it is subjected to minimization. Therefore, only the communication costs due to non-local references are included in the 'cost function'; (2) Identical communication cost are assumed for all references in either direction between any pair of fragments. Therefore, the "interfragment reference count" is taken as the metric to measure the "interfragment reference cost" between any two fragments; (3) 'Remote invocation' of an application is treated identically to its local invocation, that is the cost of any remote invocation is equivalent to local execution of the same application; (4) The objectbase is nonredundant. In other words, no class objects or class fragments are replicated and there is one and only one copy of each of them in the distributed objectbase; (5) Number of class fragments to be distributed (m) is large compared to number of network sites (n); (6) Local processing power at every site is sufficient to handle transaction processing load; and (7) The communication links connecting the sites have sufficient capacity to support the transaction traffic.

3.2 The Problem

The formal description of the problem requires a precise definition of several terms.

Definition 3.1 An *objectbase* $\mathcal{OB} = \{f_1, f_2, \ldots, f_m\}$ is a set of m class fragments generated by some fragmentation algorithm. Each fragment f_i has a finite size v_i. ■

Definition 3.2 *Method Attribute Usage* $MAU(m_j^i, a_k^i)$ *of a method* m_j^i *of a class* C_i *for an attribute* a_k^i *in the same class* C_i is the average number of accesses made to attribute a_k^i by every invocation of method m_j^i . ∎

Definition 3.3 *Method Method Usage* $MMU(m_j^i, m_k^l)$ *of a method* m_j^i *of a class* C_i *for another method* m_k^l *in any class* C_l is the average number of accesses (invocations) made to method m_k^l by every invocation of method m_j^i . ∎

An application q_k accessing the distributed objectbase is a sequence of method invocations on an object or on a set of objects. The invocation of method j of class C_i is denoted by m_j^i.

Definition 3.4 The *Method Invocation Sequence* $MIS(q_k)$ *of application* q_k denoted $\{m_{j1}^{i1}, m_{j2}^{i2}, \ldots, m_{jn}^{in}\}$ where each m_j^i refers to a method invocation by q_k. ∎

Definition 3.5 *Application Site Affinity* $ASA(q_i, s_j)$ *between an application* q_i *and a site* s_j is the number of times the application q_i is executed at site s_j. ∎

Definition 3.6 *Application Fragment Affinity* $AFA(q_i, f_j)$ *between an application* q_i *and a fragment* f_j is the number of references made by application q_i to fragment f_j. ∎

The following definitions are adapted from [EB93, EB94] :

Definition 3.7 *Method Attribute Reference* $MAR(m_j^i)$ *of a method* m_j^i *of a class* C_i is the set of all attributes of the class C_i referenced by the method m_j^i. ∎

Definition 3.8 *Method Method Reference* $MMR(m_j^i)$ *of a method* m_j^i *of a class* C_i is the set of all methods of any class referenced by the method m_j^i. ∎

Definition 3.9 *Access frequency of an application* is the number of accesses an application makes to "data". If $Q = \{q_1, q_2, \ldots, q_q\}$ is the set of applications accessing the objectbase, $acc(q_i, d_j)$ indicates the access frequency of application q_i on data item d_j where "data" item d_j can be a fragment, an object, or an attribute or method of an object. ∎

The problem of allocating object fragments in a distributed system can be formally stated as follows:

Given a set of n network sites $S = \{s_1, s_2, \ldots, s_n\}$ with respective storage capacities $D = \{d_1, d_2, \ldots, d_n\}$ and a set of m fragments $F = \{f_1, f_2, \ldots, f_m\}$ where each fragment f_i is of size v_i and $m \gg n$, the **goal** is to find a mapping from F to S such that

(i) The sum of the storage requirements of all the fragments assigned to a particular site must not exceed the total available storage at that site.

(ii) The number of non-local references is minimized.

To keep the complexity of the problem manageable, the issues of load balancing, network delay, and system reliability are not considered.

3.3 Generating a Starting Solution

This section presents an algorithm for generating an initial feasible allocation for the model described in the previous section with respect to the storage constraints at every site. The initial allocation is used as the starting solution by the optimization procedure described in Section 3.4.

For any feasible solution to be useful, the total available storage at all sites taken together must be large enough to hold all the fragments. In other words, $\sum_{i|d_i \in D} d_i \geq \sum_{i|f_i \in F} v_i$, where v_i is the size of fragment f_i.

This condition is necessary but not sufficient. At every site, the assigned fragments may not produce an "exact fit" in the local storage so some amount of storage is left unused. The total waste of storage at all individual sites may be large enough to hold another fragment, but since the available storage is not contiguous, the fragment cannot be allocated without splitting. Thereby, requiring additional storage due to internal fragmentation of local storage spaces. Since we are adopting a "top-down" design approach, we need to make a 'reasonable' estimate of the total storage requirement to obtain a feasible allocation for a given set of fragments.

We assume that an estimate [7] of total storage, large enough for a feasible allocation, is available as input to Algorithm 2 that generates an initial feasible allocation. An initial allocation is obtained by applying an algorithm that makes placement decisions for the fragments based on the available storage at every site and the applications that reference objects in the fragments. The inputs to the fragment allocation problem consists of :

- The set of n sites $S = \{s_1, s_2, \ldots, s_n\}$ with corresponding storage capacities $D = \{d_1, d_2, \ldots, d_n\}$
- The set of fragments $F = \{f_1, f_2, \ldots, f_m\}$; where $m \gg n$
- The Application Site Affinity matrix (ASA)
- The Application Fragment Affinity matrix (AFA)
- The network topology matrix (N) which holds the distance and connectivity information among sites. An entry in N, e_{ij} is 0 if $i = j$ or the shortest distance between s_i and s_j, if $i \neq j$.
- $NDiam$ = the diameter of the network[8].

---

[7] An estimate is obtained from a large number of feasible allocations by finding the storage required for each of them. The individual allocations are obtained by random placements of the fragments at different sites.

[8] Diameter of a network is the maximum distance between any two nodes of the network.

Algorithm 3.1 *Initial Allocation*

begin algorithm Initial
input: S: The set of n sites
                 D: The set of available storages at every local site
                 F: The set of class fragments
                 V: The set of fragment sizes
                 ASA: The application site affinity matrix
                 AFA: The application fragment affinity matrix
                 N: The network topology matrix
                 $NDiam$: The diameter of the network
output: $\{F_1, F_2, \ldots, F_n\}$: Initial Allocation
var: *allocated*: boolean flag indicating successful allocation of a fragment

Step(1) :
Sort the set F into F' in descending order of the fragment sizes   (1)
Step(2) :
for every fragment in the ordered set F' **do**   (2)
        allocated ← **false**   (3)
        Form sublist (S') of sites with applications referencing the fragment   (4)
        S'' is an ascending sorted sublist of S' (5)
Step(3) :
        for every site in S'' taken in order **do**   (6)
Step(4) :
                if (available storage at the site $\geq$ size of the fragment) **then** (7)
                        allocate the fragment in that site   (8)
                        decrement the available storage by the size of the fragment   (9)
                        allocated ← **true**   (10)
                        exit Step(4)   (11)
                end if
        end for
{There is no space in any of the site that make reference to the fragment so look for
space anywhere in the network to place the fragment.
We shall first try the sites that are nearest to those which refer the fragment.}
Step(5) :
        if (**not** *allocated*) **then**   (12)
                for every site in S taken in order **do**   (13)
                        for $k = 1$ to $NDiam$ **do**   (14)
                                Form sublist S''' that are a distance k from sites in S'' (15)
                                Sort S''' in ascending order of available storage   (16)
                                for every site in S''' taken in order **do**   (17)
                                        if (available storage at site $\geq$ size of the fragment)
                                        then   (18)
                                                allocate the fragment in that site   (19)
                                                decrement available storage by size of frag. (20)
                                                exit Step(5)   (21)
                                        end if
                                end for
                        end for
                end for
        end if
end for
end algorithm Initial

Fig. 2. Initial Allocation Algorithm

The output is a partition of the set F into n subsets $\{F_1, \ldots, F_n\}$, such that $\sum_{k|f_k \in F_i} v_k \leq d_i; \forall i = 1, \ldots, n;$ where v_k is the size of fragment f_k and d_i is the storage capacity of site s_i and $\bigcup_i F_i = F$.

We first try to place a fragment at a site that refers to it. If this is impossible because there is insufficient storage space at any such site, we place it at another. Since this will introduce communication overhead because of nonlocal references, we attempt to place it at a site which is at the shortest distance from any of the sites that refer to it.

The fragments are sorted in descending order of their sizes so that the allocation scheme tries to find a place for the larger fragments first. The rationale behind this is that we will have less difficulty in allocating the larger fragments at the beginning when there are sufficient large spaces to hold them at the individual sites while near the end of the initial placement process smaller fragments will be more likely to fit into any of the small "holes" in storage spaces created due to internal fragmentation. We adopt a 'best-fit' placement strategy to reduce the amount of unused storage at individual sites. To obtain better performance, the list of sites are maintained sorted in ascending order of their *remaining* storage capacities. A brief sketch of the algorithm appears in Figure 2.

3.4   Heuristic Optimization: A Graphical Approach

We adopt a heuristic approach. Our approach is based on the work done by Kernighan and Lin [KL70] on the graph partitioning problem (GPP). The GPP involves dividing a given weighted, undirected graph $G = (V, E)$, where V is the set of nodes and E is the set of edges, into n disjoint sets (called subgraphs or partitions) $V_1, V_2, \ldots, V_n$ such that the size of any subset does not exceed a given upper bound and the sum of the weights associated with the edges joining any two subgraphs is minimum. The size of any subset is the sum of the sizes of all nodes contained by the subset. The problem is NP-hard as the corresponding decision problem is shown to be NP-complete [GJ79]. Kernighan and Lin present a heuristic technique to obtain a "near optimal" partition of an undirected, weighted graph in realistic time. A two-way partitioning scheme is suggested that divides a graph of $2n$ nodes into two equal subsets of n nodes minimizing the sum of the weights corresponding to the edges that have their two endpoints in different subsets. Starting from an arbitrary initial two-way partition, an interchange process is employed iteratively that exchanges nodes between the subsets. Each interchange is performed pairwise between the nodes of either subsets. Only those nodes are selected that, when interchanged, result in a 'reduction' in the weights associated with the edges joining the nodes between the two subsets. Local optimality is reached when no further reduction is possible through the interchange process. The scheme can be generalized to obtain multiple partitions by applying the two-way partitioning scheme iteratively on every pair of subsets obtained by initial partition.

This graph partitioning problem relates closely to the problem of allocation of class fragments in a DOBS. In the graphical formulation of the object allocation problem, the set of class fragments can be considered to be the set of nodes and the inter-fragment references are analogous to edges between nodes. We can associate a reference count with every pair of fragments which gives us a measure of interactions between the object fragments. Accordingly, an *Inter Fragment Reference matrix (IFRM)*, analogous to the weight matrix W, can be constructed. The partitioning scheme suggested by Kernighan and Lin can then be applied to $IFRM$ to obtain the desired partition from an initial starting solution. In Section 3.3 we presented an algorithm that generates one such initial solution that is bounded by the constraints of storage capacities at the sites and the network topology. This initial solution becomes the input to the heuristic partitioning scheme based on the Kernighan-Lin approach.

A Heuristic Algorithm to Improve the Allocation This section describes the algorithm based on the heuristic approach adopted by Kernighan and Lin in their graph partition problem. The algorithm adjusts the initial distribution of fragments between pairs of sites by interchanging fragments to obtain a distribution that is locally optimal with respect to the nonlocal references between them. The algorithm is applied repeatedly on every pair of sites in S to obtain pairwise locally optimal solution.

A few additional definitions are required to present the algorithm.

Definition 3.10 *Fragment Reference Count $FRC(i,j)$ from fragment f_i to fragment f_j* is the total number of references made by any method in fragment f_i to any attribute or method in fragment f_j for all possible application executions.

$$FRC(i,j) = \sum_{k|q_k \in Q} \sum_{\{l|m_l \in MIS(q_k) \wedge m_l \in f_i\}} acc(q_k, m_l) * \left(\sum_{p|a_p \in f_j} MAU(m_l, a_p) \right.$$

$$\left. + \sum_{r|m_r \in f_j} MMU(m_l, m_r) \right)$$

■

To obtain *Interfragment Reference Count (IFR)* between any two class fragments we add the *fragment reference counts* in either direction between them.

Definition 3.11 *Interfragment Reference Count $IFR(i,j)$ between fragment f_i and fragment f_j* is the sum of the fragment reference count(FRC) from fragment f_i to fragment f_j and the fragment reference count from fragment f_j to fragment f_i.

$$IFR(i,j) = FRC(i,j) + FRC(j,i)$$

■

We construct the *Inter Fragment Reference Matrix, (IFRM)* whose (i,j)th element is $IFR(i,j)$.

Definition 3.12 *Site Affinity $saff(f_i, s_j)$ of a class fragment f_i at site s_j is* the sum of all references to fragment f_i made by all local applications executing at site s_j.

$$saff(f_i, s_j) = \sum_{k|q_k \in Q} ASA(q_k, s_j) * AFA(q_k, f_i)$$

■

Based on the $IFRM$, we can represent the access relationships between the fragments as an undirected, weighted graph with nodes for each fragment and the weights on the edges determined by the total reference count between the pair of fragments connected by the edge. We call this graph the *Fragment Reference Graph (FRG)*.

The fragment allocation problem can be reformulated as a partitioning problem that divides the FRG into n subgraphs, where n is the number of sites in the network, such that the number of references between the sites is minimized and the site capacities are respected. Each of the partitions can then be mapped on to a network site to obtain an allocation that minimizes the nonlocal interfragment references. Since the communication cost due to the nonlocal references dominates the overall cost of allocation, the allocation obtained can be treated as a "near optimal" solution to the allocation problem in DOBS.

The approach taken is as follows. Starting from an initial allocation (obtained by the Algorithm 2 or any other algorithm that generates an initial feasible allocation), fragments are interchanged between every pair of sites to reduce the number of nonlocal interfragment references. However, when a fragment changes site, its site affinity changes as well. A decrease in site affinity would lead to an increase in the number of remote invocations of any method, contained by the fragment, accessed by an application directly. Only those fragments are interchanged that produce a "net gain" taking interfragment references and site affinities into account.

Consider an arbitrary pair of sites (A, B), $A, B \in S$. Let the initial allocation at site A and site B be F_A and F_B respectively. For each fragment $f_a \in F_A$, we compute an *external reference count* (ERC_a) which is the sum of all nonlocal references made to f_a from all the fragments in site B and an *internal reference count* (IRC_a) which is the sum of all references made to it from within site A. Therefore,

$$IRC_a = \sum_{k|f_k \in F_A} IFR(f_a, f_k)$$

and

$$ERC_a = \sum_{k|f_k \in F_B} IFR(f_a, f_k)$$

We also calculate ERC_b and IRC_b for every fragment $f_b \in B$. The site affinities are evaluated as follows:

$$\left. \begin{array}{l} saff(f_a, A) = \sum_{k|q_k \in Q} ASA(q_k, A) * AFA(q_k, f_a) \\ saff(f_a, B) = \sum_{k|q_k \in Q} ASA(q_k, B) * AFA(q_k, f_a) \end{array} \right\} \ \forall f_a \in F_A$$

and

$$saff(f_b, B) = \sum_{k|q_k \in Q} ASA(q_k, B) * AFA(q_k, f_b) \left.\right\}$$
$$saff(f_b, A) = \sum_{k|q_k \in Q} ASA(q_k, A) * AFA(q_k, f_b) \left.\right\} \quad \forall f_b \in F_B$$

When we interchange a fragment f_a in node A and a fragment f_b in node B, all local references becomes nonlocal and vice versa for both f_a and f_b. From [KL70], for any $f_a \in F_A$ and $f_b \in F_B$, if f_a and f_b are interchanged between sites A and B, the reduction in nonlocal interfragment references is $D_a + D_b - 2r_{ab}$, where $D_a = ERC_a - IRC_a, D_b = ERC_b - IRC_b$ and $r_{ab} =$ reference count between a and b. The increase in site affinity due to the interchange for fragment f_a is $\Delta_a = saff(f_a, B) - saff(f_a, A)$, and that for fragment f_b is $\Delta_b = saff(f_b, A) - saff(f_b, B)$. Therefore the total *gain* in the interchange of fragments f_a and f_b is given by $gain = D_a + D_b - 2r_{ab} + \Delta_a + \Delta_b$.

Fragments are interchanged between pairs of sites in iterative steps. One pair of sites is selected at a time for fragment interchange. At every step, a pair of fragments, one from each site, is selected that produces a positive gain if they are interchanged between the two sites. All such fragment pairs are collected and the interchanges, permissible by the local storage constraints, are made. The process is then repeated until there are no fragment pairs producing any additional gain on interchange. The values for ERCs and IRCs are reevaluated after every interchange. When no more interchange produces any positive gain, the fragment distribution with respect to the pair of sites is optimal. The process is repeated for all pairs of sites until the global fragment distribution is optimal with respect to every pair of sites. The algorithm is formally presented in Figure 3 which can be summarized by the following logical steps:

Step One - Selection of A Site Pair

Two sites A and B are selected arbitrarily from the set S of network sites. The flag for local optimality is set to *false*. The *site affinity* measures are computed for every fragment in A and B for with respect to both the sites. Step 2 through Step 4 are repeated until a locally optimal fragment allocation between the two selected sites is obtained.

Step Two - Evaluation of D_a's and D_b's

The *Interchange Set* is initialized to null and *cumulative gain* is initialized to zero. D_a and D_b values are computed for every fragments in site A and site B.

Step Three - Selection of Subsets of Fragments from A and B for Interchange

For every possible pair of fragments in site A and site B, potential *gain* in interchanging them between the two sites is calculated. The pair that has the highest potential positive gain is selected and is added to the *Interchange set*. The selected fragment pair is removed from the set of fragments at each site. The gain corresponding to the pair selected is added to cumulative gain and the iteration is continued until either of the site runs out of fragment (for pairing with another fragment of the other site) or no pair can be formed that has a positive potential gain on interchange.

Step Four - Check for Local Optimality

If there is any pair of fragments found between the two sites that has a potential

Algorithm 3.2 *Heuristic Optimization - Graphical Approach*

begin algorithm Optimize
input: **IFRM:**The Interfragment reference matrix
      ASA: The Application Site Affinity matrix
      AFA: The Application Fragment Affinity matrix
      $\mathcal{F}_A$:The set of frags. assigned to site A by Initial Allocation
      $\mathcal{F}_B$:The set of frags. assigned to site B by Initial Allocation
      d_A:Available storage at site A
      d_B:Available storage at site B
      $\mathcal{V}_A$:Fragment sizes of the fragments in $\mathcal{F}_A$
      $\mathcal{V}_B$:Fragment sizes of the fragments in $\mathcal{F}_B$
output: $\mathcal{F}_A^{O(AB)}$:set of fragments at site A for locally optimal allocation
      $\mathcal{F}_B^{O(AB)}$:set of fragments at site B for locally optimal allocation
var: *locally optimal*: boolean variable for pairwise optimality
      $\mathcal{I}$:*Interchange Set* for collecting the fragments to be interchanged
      cumulative gain: Total gain for all selected fragment pairs

for every pair of sites (A, B), $A, B \in S$ **do**	(1)
Step(1) : *locally optimal* $\leftarrow$ **FALSE**	(2)
Compute $saff(f_a, A)$ and $saff(f_a, B) \forall f_a \in \mathcal{F}_A$ using ASA and AFA	(3)
Compute $saff(f_b, B)$ and $saff(f_b, A) \forall f_b \in \mathcal{F}_B$ using ASA and AFA	(4)
repeat step(2) through step(4) **until** (**not** *locally optimal*)	(5)
Step(2) : Initialize $\mathcal{I} \leftarrow \emptyset$.	(6)
Initialize cumulative gain $\leftarrow 0$.	(7)
Compute D_a for all $f_a \in \mathcal{F}_A$ and D_b for all $f_b \in \mathcal{F}_B$ using **IFRM**.	(8)
$\mathcal{F}'_A \leftarrow \mathcal{F}_A$	(9)
$\mathcal{F}'_B \leftarrow \mathcal{F}_B$	(10)
Initialize gain $\leftarrow 0.0001$	(11)
Step(3) : **while** $((\mathcal{F}'_A \neq \emptyset)$ **and** $(\mathcal{F}'_B \neq \emptyset)$ **and** $(gain > 0))$ **do**	(12)
Select $f_a \in \mathcal{F}'_A$ and $f_b \in \mathcal{F}'_B$ such that	
$gain = D_a + D_b - 2r_{ab} + \Delta_a + \Delta_b$ is maximum	(13)
if ($gain \leq 0$) **then** exit Step(3)	(14)
endif	
$I \leftarrow I \cup (f_a, f_b)$ /* Add the fragment pair (f_a, f_b) to $\mathcal{I}$ */	(15)
cumulative gain $\leftarrow$ cumulative gain $+$ gain	(16)
$\mathcal{F}'_A \leftarrow \mathcal{F}'_A - f_a$	(17)
$\mathcal{F}'_B \leftarrow \mathcal{F}'_B - f_b$	(18)
endwhile	
Step(4) : **if** (cumulative gain $= 0$) **then**	(19)
locally optimal $\leftarrow$ **TRUE**	(20)
exit Step(4)	(21)
end if	
for every pair of fragments $(f_a, f_b) \in I$ **do**	(22)
Interchange f_a and f_b between fragment sets $\mathcal{F}_A$ and $\mathcal{F}_B$	
subject to d_A and d_B.	(23)
end for	
end repeat	
$\mathcal{F}_A^{O(AB)} \leftarrow \mathcal{F}_A$	(24)
$\mathcal{F}_B^{O(AB)} \leftarrow \mathcal{F}_B$	(25)
end for	
end algorithm Optimize	

Fig. 3. The Optimization Algorithm

gain, the fragment pairs collected in the interchange set are interchanged between the two sites. Otherwise, there is no interchange possible that contributes any potential positive gain and hence a *local optimality* is reached. The *local optimality* flag is set. The set of fragments at the two sites are the locally optimal distribution of The algorithm is applied to every pair of sites in S. In practice, more than one pass is required. In any pass, there could be interchange of fragments between a pair of sites, of which one or both have already been considered with other sites for pairwise optimality. This will invalidate the local optimality obtained in earlier steps between any of the sites from the current pair and other sites. So all such sites pairing with other sites need to be tested again for pairwise optimality. However, application of this algorithm in graph theory shows that convergence is reasonably fast [KL70].

Algorithm 3 differs from the one presented by Kernighan and Lin [KL70] in that the latter does not consider any constraints restricting the interchange process. The two-way partitioning scheme they propose assumes the starting partitions of equal size. Our algorithm can handle initial partitions of unequal size. We also permit only those interchanges that do not violate the storage constraints for individual sites. This ensures that the final solution will be bound by the constraints. Further, Kernighan and Lin start with an *arbitrary* partition as the starting solution, though they agree that a good initial partitioning scheme helps in arriving at a near optimal solution quicker. In contrast, the starting partition we obtain from Algorithm 2 is already good with respect to storage capacities at the sites and the network topology.

4 Concluding Remarks

A heuristic method for distributing class fragments in a distributed objectbase system is presented. Our solution method builds upon previous work on DFS and DDBS. The differences that contributed additional complexity of the allocation problem in a DOBS with respect to peer problems in DFS and DDBS are highlighted. Since the general problem of allocation is NP-Hard, we have argued that an efficient optimal solution is very unlikely to be obtained in realistic time. Therefore, the solution technique we presented here is designed with the goal of obtaining a 'near optimal' allocation in realistic time to make it implementable in real-life problems.

Much work has yet to be done. The assumptions we made to simplify the allocation model need to be reviewed to make the model less restrictive and generally applicable so the solutions have an extended scope of applicability. Objectbase systems are expected to be dynamic with changes in both class libraries (schema evolution) and movement of objects themselves (to support, object migration or load balancing, for example). Further, some degree of replication is necessary for the DOBS to meet the performance requirements by the applications. Allocation schemes for nonredundant static environment need to reviewed to accommodate these changes. We are currently investigating the impact of these changes on the proposed allocation scheme.

References

[AG73]   S. R. Arora and A. Gallo. Optimization of Static Loading and Sizing of Multilevel Memory Systems. *Journal of ACM*, 20(2):307–319, 1973.

[Ape88]   P. M. G. Apers. Data Allocation in Distributed Database System. *ACM Transactions on Database Systems*, 13(4):263–304, Spetember 1988.

[BG74]   J. P. Buzen and P. S. Goldberg. Guidelines for the use of infinite source queueing models in the analysis of computer system performance. In *Proceedings of AFIPS National Computer Conference*, volume 43, pages 371–374, 1974.

[Buz71]   J. P. Buzen. *Queueing Network Models of Multiprogramming*. PhD thesis, Harvard University, Cambridge, Mass., 1971.

[CA80]   Peter Pin-Shan Chen and Jakob Akoka. Optimal Design of Distributed Information Systems. *IEEE Transactions on Computers*, C-29(12):1068–1080, December 1980.

[Cas72]   R. G. Casey. Allocation of Copies of a File in an Information Network. In *Proceedings of AFIPS Spring Joint Computer Conference*, volume 40, pages 617–625, 1972.

[CDY90]   B. Ciciani, D. M. Dias, and P. S. Yu. Analysis of replication in distributed database systems. *IEEE Transactions on Knowledge and Data Engineering*, 2(2):247–261, June 1990.

[Che73]   P. P.-S. Chen. Optimal File Allocation in Multilevel Storage Systems. In *Proceedings of AFIPS National Computer Conference*, volume 42, pages 277–282, 1973.

[Chu69]   W. W. Chu. Optimal File Allocation in Multiple Computer System. *IEEE Transactions on Computers*, C.18(10):885–889, October 1969.

[CP84]   S. Ceri and G. Pelagatti. *Distributed Databases : Principles and Systems*. McGraw Hill Inc., 1984.

[CR90]   Ge-Ming Chiu and C. S. Raghavendra. A model for optimal database allocation in distributed computer systems. *IEEE COMPUTER*, pages 827–833, May 1990.

[CY89]   D. W. Cornell and P. S. Yu. On Optimal Site Assignment for Relations in the Distributed Database Environment. *IEEE Transactions on Computers*, 15(8):1004–1009, August 1989.

[DF82]   L. W. Dowdy and D. V. Foster. Comparative Models of the File Assignment Problem. *ACM Computing Surveys*, 14:287–313, June 1982.

[EB93]   C. I. Ezeife and Ken Barker. Horizontal Class Fragmentation in Distributed Object Based Systems. Technical Report 04, University of Manitoba, October 1993.

[EB94]   C. I. Ezeife and Ken Barker. Vertical Class Fragmentation in Distributed Object Based System. Technical Report 03, University of Manitoba, April 1994.

[Esw74]   K. P. Eswaran. Placement of Records in a File and File Allocation in a Computer Network. In *Proceedings of IFIP Congress*. North-Holland, 1974.

[FH80]   M. L. Fisher and D. S. Hochbaum. Database location in computer networks. *Journal of the ACM*, 27(4):718–735, October 1980.

[Gav87]   Bezalel Gavish. Optimizing Models for Configuring Distributed Computer Systems. *IEEE Transactions on Computers*, C-36(7):773–793, July 1987.

[GJ79]   Michael R. Garey and David S. Johnson. *Computers and Intractability: A guide to the theory of NP Completeness*. W. H. Freeman and Company, 1979.

[GMM92]   D. Ghosh, I. Murthy, and A. Moffett. File allocation problem: Comparison of models with worst case and average communication delays. *Operations Research*, 40(6):1074–1085, Nov-Dec 1992.

[GP86]   Bezalel Gavish and Hasan Pirkul. Computer and Database Location in Distributed Computer Systems. *IEEE Transactions on Computers*, C-35(7):583–590, July 1986.

[GS90]   Bezalel Gavish and Olivia R. Liu Sheng. Dynamic File Migration in Distributed Computer Systems. *Communications of the ACM*, 33(2):773–793, February 1990.

[HM73]   P. H. Hughes and G. Moe. A structural approach to computer performance analysis. In *Proceedings of AFIPS Spring Joint Computer Conference*, pages 109–120, 1973.

[IK82]   Keki B. Irani and Nicholas G. Khabbaz. A Methodology for the Design of Communication Networks and the Distribution of Data in Distributed Supercomputer Systems. *IEEE Transactions on Computers*, C-31(5):419–434, May 1982.

[JFK78]   L. G. Jones, D. V. Foster, and P. D. Krolak. A goal programming formulation used in the file assignment problem. In *Proceedings of R. J. Duffin Conference*, July 1978.

[KKM92] S. B. Navathe K. Karlapalem and M. Morsi. Issues in distributed design of object-oriented databases. In *International Workshop on Distributed Object Management*, pages 47–65, August 1992.

[KL70]   B. W. Kernighan and S. Lin. An efficient heuristic procedure for partitioning graphs. *The Bell System Technical Journal*, pages 291–307, February 1970.

[Kle76]   Leonard Kleinrock. *Queueing Systems:Computer Applications*, volume 2. Wiley, 1976.

[LL83]   Laurence J. Laning and Michael S. Leonard. File Allocation in a Distributed Computer Communication Network. *IEEE Transactions on Computers*, C-32(3), March 1983.

[LM77]   K. D. Levin and H. L. Morgan. Optimal Program and Data Locations in Computer Networks. *Communications of the ACM*, 20(5):315–321, May 1977.

[MR76]   Samy Mahmoud and J. S. Riordon. Optimal Allocation of Resources in Distributed Information Networks. *ACM Transactions on Database Systems*, 1(1), March 1976.

[ÖV91]   M. T. Özsu and P. Valduriez. *Principles of Distributed Database Systems*. Prentice Hall, 1991.

[Pie75]   W. F. Piepmeier. Optimal balancing of i/o requests to disks. *Communications of ACM*, 18(9):524–527, Sept 1975.

[RC70]   C. V. Ramamoorthy and K. M. Chandy. Optimization of Memory Hierarchies in Multiprogrammed Systems. *Journal of ACM*, 17(3):426–445, 1970.

[SCW83] S. Navathe S. Ceri and S. Wiederhold. Distribution Design of Logical Database Schemas. *IEEE Transactions on Software Engineering*, SE-9, July 1983.

[SW83]   D. Sacca and G. Wiederhold. Database partitioning in a cluster of processors. In *Proceedings of 9th International Conference on Very Large Databases*, pages 242–247, October 1983.

[Wah84]   B. J. Wah. File Placement on Distributed Computer Systems. *IEEE Computer*, pages 23–32, January 1984.

[YSLC83] C. T. Yu, M. K. Siu, K. Lam, and C. H. Chen. File allocation in distributed databases with interaction between files. In *Proceedings of 9th International Conference on Very Large Databases*, pages 248–259, October 1983.

Estimating Data Accuracy in a Federated Database Environment

M. P. Reddy
Kenan Systems Corporation
One Main Street
Cambridge, MA 02142

Richard Y. Wang
Sloan School of Management
MIT
Cambridge, MA 02139

Abstract: The need for integration of data in a heterogeneous or federated database environment creates a corresponding need for estimating the accuracy of the integrated data as a function of the accuracy of the originating data sources. Even in a single database system, different base relations are frequently characterized by dissimilar levels of accuracy; however, no technique exists for defining the accuracy of this single database system in terms of the accuracy of the base relations. This need is further heightened in the case of federated environments involving multiple heterogeneous databases. To address this need, a generalized method is proposed for estimating the overall data accuracy in terms of the accuracy of relevant base relations and the actual database query. The query is examined in terms of its underlying set of base operators. A rigorous theoretical framework encompassing all these possible base operators is presented in this paper using the relational model. While the accuracy estimates are postulated on the basis of uniform distribution, the implications of non-uniform error distributions are also examined in theoretical terms. Finally, a running example is utilized to highlight the practical implications of the proposed theoretical framework.

Key Words: Data Quality, Relational Algebra, Database Management Systems

1. Introduction

In a *federated database system*, data at the federated level are derived from the data in the *component databases* of the federation. Various aspects of federated databases have been researched and reported in the database literature [Smith et al., 1981; Lander & Rosenberg, 1982; Heimbigner & McLeod, 1985; Litwin & Abdellatif, 1986; Deen et al., 1987; Templeton et al., 1987; Reddy et al., 1989; Wang & Madnick, 1989; Rajinikanth, 1990; Sheth & Larson, 1990; Pu, 1988; Sheth, 1991]. In particular, significant research has been conducted to develop mechanisms for mapping between a *federated schema* and its corresponding *export schemata* and for the retrieval and integration of data from the component databases. In general, these mechanisms assume that data associated with the component databases are accurate but not semantically compatible.

In reality, most of component databases are not free of error, and many contain a significant number of errors [Johnson et al., 1981; Liepens et al., 1982; Morey, 1982; Laudon, 1986; O'Neill & Vizine-Goetz, 1988]. Because the component databases are not error-free, the query results obtained at the federated level may be erroneous even if the *query translation* and *data integration* mechanisms are flawless. These erroneous results, in turn, may lead to erroneous decisions, resulting in significant social and economic impacts. A critical research problem that needs to be resolved, therefore, is how to develop methods that can estimate the accuracy of query results at the federated level using accuracy estimates of the component databases. Although this research problem is more obvious in a federated database system, it is also present in a single database system where different *base relations* have different accuracy levels. To the best of our knowledge, neither aspect of this problem has not been addressed in previous research efforts.

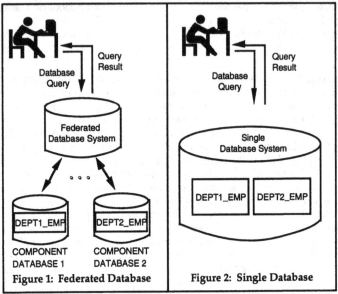

Figure 1: Federated Database   Figure 2: Single Database

In this paper, we propose a generalized method for estimating overall data accuracy. Given a database query and the accuracy of relevant base relations, this method computes the accuracy of the result. By knowing the estimated accuracy of the result, the user can make a judgment about the risk involved in using this result. Data quality analysts can also study the impact of alternative data quality control strategies. For example, by examining the accuracy estimates of different application queries, data quality analysts can identify candidate database relations for enhancing data accuracy.

This paper is organized as follows: Section 2 presents the research assumptions, definitions and notations used in this paper. Section 3 presents a *data accuracy algebra*. Section 4 provides some of the formal properties of the proposed accuracy paramaters. Section 5 presents examples to demonstrate the applicability of the algebra. Finally, concluding remarks are presented in Section 6.

2. Foundation

2.1 Assumptions

Most federated database systems adopt one of the following data models: the *relational model* [Deen et al., 1987], the *functional data model* [Smith et al., 1981], the *semantic data model* [Heimbigner & McLeod, 1985], or the *entity relationship model* [Spaccapietra et al., 1992]. We choose to extend the relational model because it is elegant and constitutes the de facto industry standard for database systems today. Accordingly, we make the following assumptions:

(A1)   Component databases in a federation are based on the relational model.

(A2)   Query processing is flawless.

(A3)   Query processing operations are based on *relational algebra*.

(A4)   Accuracy estimates of relevant base relations in each of the component databases are given at the relational (instead of tuple or cell) level.

(A5)   The accuracy of tuples in base relations can be validated.

(A6)   At the relational level, the probability of *error is uniformly distributed* across tuples and attributes.

(A7)   The cardinality of the resulting relation |R| is available.

The first three assumptions define the scope of our research problem to be one that focuses on developing a mechanism based on the relational algebra for estimating the accuracy of data derived from relational databases. We elaborate on the last two assumptions below.

In principle, the accuracy information for the relevant base relations should be provided as the input (at a certain level of granularity) for the estimation of the accuracy of data derived from these base relations. In the relational model, a base relation must be, at least, in the *first normal form,* which means each attribute value is *atomic* [Date, 1990]. Each atomic attribute value, however, can be either accurate or inaccurate and can be tagged accordingly

An *attribute-based model* can be used to store data accuracy value at the cell level [Wang & Madnick, 1990; Wang et al., 1993; Wang et al., 1995]; however, the measurement and the storage of the accuracy of each cell value is an expensive proposition. Similarly, it is also relatively expensive to measure and store accuracy information at the tuple level. Therefore, Assumption A4 states that accuracy estimates are available at the relational level, and in addition, these estimates are based on some tuple level accuracy measures.

In order to measure the accuracy of a tuple, some validation procedure should be available such as ones described by Janson [1988] and Paradice & Fuerst [1991].

In practice, | R | can be either counted <u>after</u> an operation is performed or estimated <u>before</u> an operation is performed. Algorithms for estimating |R| are described in [Ceri & Pelagatti, 1984] .

2.2 Definitions

For exposition purposes, we introduce a running example that will be expanded as the paper unfolds.

Let DEPT1_EMP (Table 1) and DEPT2_EMP (Table 2) be two employee relations for two departments, DEPT1 and DEPT2. Each employee relation consists of three attributes: EMP_ID, EMP_NAME, and EMP_SAL.

In this example, DEPT1_EMP and DEPT2_EMP could be two relations in two component databases of a federated database system (Figure 1) or two relations in a single database system (Figure 2). The ideas presented in this paper apply to both of these scenarios.

In the relational model, a relation represents a time-varying subset of a class of instances sharing the same domain values that correspond to its relational scheme. For example, the relation DEPT1_EMP is a subset or a complete set of the employees in DEPT1. Each tuple is implicitly assumed to correspond to a member of the class of instances of the same entity type as defined by a *relational scheme.* In reality, from a data quality perspective, this implicit assumption may not hold. For example, some tuples in DEPT1_EMP may not belong to DEPT1 (For example, a database administrator may, by mistake, insert into DEPT1 an employee tuple that belongs to another department, or may not have had the time to delete a tuple that corresponded to a former employee who has left DEPT1). A relation containing one or more tuples that do not meet the above implicit assumption is said to have *tuple mismembership.*

A tuple that does not satisfy the implicit assumption is referred to as a *mismember tuple*; otherwise, it is referred to as a *member tuple*.

Table 1: The Relation DEPT1_EMP

EMP_ID	EMP_NAME	EMP_SAL
1	Henry	30,000
2	Jacob	32,000
3	Marshall	34,000
4	Alina	33,000
5	Roberts	50,000
6	Ramesh	45,000
7	Patel	46,000
8	Josaph	55,000
9	John	60,000
10	Arun	50,000

Table 2: The Relation DEPT2_EMP

EMP_ID	EMP_NAME	EMP_SAL
1	Henry	30,000
2	Jacob	32,000
5	Roberts	55,000
6	Ramesh	45,000
9	John	60,000
11	Nancy	39,000
12	James	37,000
13	Peter	46,000
14	Ravi	55,000
15	Anil	45,000

In order to determine the accuracy of an attribute value in a tuple, one must know if the attribute value reflects the real world state [Kent, 1978]. We refer to an attribute value as being accurate if it reflects the real world state; otherwise, it is inaccurate. To validate that the gender, age and income data for a particular person are accurately stored in a tuple, it is necessary to know that the tuple indeed belongs to this person, and that the attribute values in the tuple reflect the real world state. The concepts of member tuple and accurate attribute values lead to Definition D1 below.

(D1) **Deterministic Tuple Accuracy Definition:** A tuple t in a relation R is accurate if and only if it is a member tuple and every attribute value in t is accurate; otherwise, it is inaccurate. Let A_t denote the accuracy of t. If t is accurate, then the value of A_t is 1; if t is inaccurate, then the value of A_t is 0.

By Definition D1, a tuple can be inaccurate either because it is a mismember tuple or because it has at least one inaccurate attribute value. For example, if an employee belongs to the department, and the employee's ID number, name, and salary are accurately stored in the tuple denoted by *emp_tuple*, then emp_tuple is said to be accurate, or $A_{emp\_tuple} = 1$. Whereas, if the employee does not belong to the department or if any of the ID number, name, or salary is inaccurate, then $A_{emp\_tuple} = 0$.

Suppose that in the relation DEPT1_EMP, the tuple for Employee 2 contains inaccurate salary, the tuple for Employee 8 contains inaccurate name, and Employee 6 no longer works for DEPT1; in DEPT2_EMP, the tuples for Employees 2 and 5 contain inaccurate salary, while the tuple for Employee 15 contains inaccurate name. These tuples are shaded in Tables 1 and 2. By Definition D1, the tuple for Employee 6 is a mismember tuple, and other tuples contain inaccurate attribute values.

(D2) **Relation Accuracy Definition:** Let A_R denote the accuracy of relation R, N the number of accurate tuples in R, and | R| the cardinality of relation R. Then

$$A_R = \frac{N}{|R|} \tag{2.2.1}$$

Let IM_R denote the inaccuracy due to mismember tuples, and IA_R denote the inaccuracy due to attribute value inaccuracy for the relation R. Let M denote the number of mismember tuples in which every attribute value is accurate. (If a mismember tuple in the relation R has an inaccurate attribute value, then it will be counted under IA_R but not under IM_R.) If P denotes the number of tuples with at least one inaccurate attribute value, then it follows that

$$IM_R = \frac{M}{|R|} \tag{2.2.2}$$

$$IA_R = \frac{P}{|R|} \tag{2.2.3}$$

$$A_R + IM_R + IA_R = 1 \tag{2.2.4}$$

In Table 1, $A_R = 0.7$, $IM_R = 0.1$ and $IA_R = 0.2$. In Table 2, $A_R = 0.7$, $IM_R = 0$ and $IA_R = 0.3$.

Suppose that DEPT1_EMP and DEPT2_EMP (Tables 1-2) are two relations in a database, then the quality profile of the database would appear as follows:

Table 3: Quality Profile

Relation Name	Size	A_R	IM_R	IA_R
DEPT1_EMP	10	0.7	0.1	0.2
DEPT2_EMP	10	0.7	0.0	0.3

(D3) **Null Relation Accuracy Definition:** In D2, if the cardinality, $|R|$, is zero, then we define that $A_R = 1$.

In order to proceed with the analysis, we make the following additional assumptions and definitions:

(D4) **Probabilistic Tuple Accuracy:** At the relational level, all the tuples in a relation have the same probability of being accurate regardless of their deterministic tuple accuracy. Let A_R denote the accuracy of relation R, then the probabilistic accuracy of each tuple in R is A_R.

In Table 1, the deterministic tuple accuracy of Tuple 2 is 1 and that of Tuple 8 is 0. Further, $A_R = 0.7$, and the probabilistic tuple accuracy for both Tuples 2 and 8 (in fact, for every tuple in DEPT1_EMP) is 0.7.

(D5) **Probabilistic Attribute Accuracy:** All attributes in a base relation are defined to have the same probability of being accurate, irrespective of their respective deterministic accuracies.

Based on the above definition, one can compute the probabilistic attribute accuracy as follows: Let A_R denote the accuracy of relation R, D the degree of the relation R, and X an attribute in relation R. Then, the probability that attribute X is accurate is given by $\sqrt[D]{A_R}$.

In Table 1, the accuracy of relation DEPT1_EMP is 0.7; and the probabilistic attribute accuracy of the attribute EMP_NAME is $\sqrt[3]{0.7}$.

Assumption A6 is relatively strong, and will not hold in general. From the theoretical viewpoint, however, it provides an analytic approach for estimating the accuracy of data derived from base relations, especially because the detailed accuracy profile of the derived data is rarely available on an a priori basis. To substantiate the analysis, we also present, in Section 3, an analysis of the best and worst cases under the scenario of non-uniform distribution of error.

3. A Data Accuracy Algebra

In this section, we compute accuracy measures for reports based on the accuracy estimates of input relations for the five algebraic operations: selection, projection, Cartesian product, union, and difference. Other traditional operators can be defined in terms of these operators [Klug, 1982].

Let A_S, A_{S_1}, and A_{S_2} denote the accuracy of Relations S, S_1, and S_2 respectively. Let R denote the resulting relation of an algebraic operation on S (if unary operation), or on S_1 and S_2 (if binary operation). Let A_R denote the accuracy of R.

3.1 Accuracy Estimation of Data Derived by Selection Operator

Selection is a unary operation denoted as $R = \sigma_c S$ where 'c' represents the selection condition. Further, $|R| \le |S|$ and A_R must satisfy the following boundary conditions:

(i) $\quad A_S = 1 \;\Rightarrow\; A_R = 1$ $\hfill$ (S1)

(ii) $\quad A_S = 0 \;\Rightarrow\; A_R = 0$ $\hfill$ (S2)

By Definition D4, the probabilistic accuracy of each tuple in the relation S is given by A_S. Since the selection operation selects a subset of tuples from S, it follows that the estimated number of accurate tuples in R is given by

$$|R| * A_S \hfill (3.1.1)$$

By Definition D2,

$$A_R = \frac{|R| * A_S}{|R|} = A_S \hfill (3.1.2)$$

Equation 3.1.2 satisfies both boundary conditions (S1) and (S2). Also, IM_R, the fraction of mismember tuples in R, and IA_R, the fraction of tuples in R with at least one inaccurate attribute value can be estimated as follows:

$$IM_R = IM_S \hfill (3.1.3)$$
$$IA_R = IA_S \hfill (3.1.4)$$

where IM_S denote the percentage of mismember tuples in S, and IA_S the percentage of tuples in S with at least one inaccurate attribute value.

Equation 3.1.2 is derived based on Assumption A6, in which errors are assumed to be uniformly distributed. When the error distribution is not uniform, the worst and best cases for the selection operation can be analyzed as shown in the following subsections.

Worst case: From Definition D2, the worst case occurs when the selection operation selects the maximum number of inaccurate tuples into the resulting relation R. Let S_i denote the number of inaccurate tuples in S. Two cases need to be considered. In the first case, $|R| \le |S_i|$. Under this scenario, the worst case occurs when all the tuples selected into R are inaccurate. Thus, $A_R = 0$.

In the second case, $|S_i| < |R|$. Under this scenario, the worst case occurs when all the inaccurate tuples in S are included in R. As such, the number of accurate tuples in R is given by $|R| - (|S| * (1 - A_S))$. By Definition D2,

$$A_R = \frac{|R| - (S * (1 - A_S))}{|R|} = 1 - \frac{|S|}{|R|} * (1 - A_S) \qquad (3.1.5)$$

Equation 3.1.5 satisfies both boundary conditions (S1) and (S2), as shown below.

In boundary condition (S1), $A_S = 1$. Substituting A_S by 1 in Equation 3.1.5, we get $A_R = 1$. Therefore, boundary condition (S1) holds for Equation 3.1.5. In boundary condition (S2), $A_S = 0$. It follows from Equation 3.1.5 that

$$A_R = 1 - \frac{|S|}{|R|} \qquad (3.1.6)$$

Since $|S|$ cannot be less than $|R|$ by the definition of the selection operation, it follows from Equation 3.1.6 that $A_R \leq 0$. By definition, $A_R \geq 0$. Therefore, $A_R = 0$. Thus, Equation 3.1.5 satisfies boundary condition (S2).

Best case: From Definition D2, the best case occurs when the maximum number of accurate tuples are selected into the resulting relation R. Let S_a denote the number of accurate tuples in S. Two cases need to be considered. In the first case, $|Sa| \geq |R|$. Under this scenario, the best case occurs when all the tuples selected into R are accurate. Therefore, $A_R = 1$.

In the second case, $|Sa| < |R|$. Under this scenario, the best case occurs when all the accurate tuples are included in R. The number of accurate tuples in R is given by $A_S * |S|$. By Definition D2,

$$A_R = A_S * \frac{|S|}{|R|} \qquad (3.1.7)$$

Equation 3.1.7 satisfies both boundary conditions (S1) and (S2).

3.2 Accuracy Estimation of Data Derived by Projection Operator

Projection selects a subset of attributes from a relation and is denoted as $R(B) = P_B S(C)$, where B and C are the set of attributes in R and S respectively, and $B \subseteq C$. By definition, $|R| \leq |S|$. In addition, the accuracy of R must satisfy the following two boundary conditions:

$$A_S = 0 \quad \Rightarrow \quad A_R = 0 \qquad (P1)$$

$$A_S = 1 \quad \Rightarrow \quad A_R = 1 \qquad (P2)$$

We consider *projection* as a two-step operation. In Step 1, the set of attributes that is not present in R will be eliminated from S to generate a relation Q, in which $|Q| = |S|$ and Q may have duplicate tuples. In Step 2, all duplicate tuples in Q will be eliminated, resulting in R.

Let $\langle v_1, v_2, ..., v_m \rangle$ be a tuple in the relation S. Let p be the probabilistic attribute accuracy for attribute v_j. Using D5, one can write $p = (A_S)^{\frac{1}{m}}$, where A_S is the accuracy of relation S. If n $(n \leq m)$ attributes are selected from S into Q, then the probability of a tuple in Q being accurate before elimination of duplicates is given by $(A_S)^{\frac{n}{m}}$. As such, the estimated total number of accurate tuples in Q before the elimination of duplicates is given by $(A_S)^{\frac{n}{m}} * |S|$.

Before elimination of duplicates from Q, we assume that Q and S represent the same class of objects as discussed earlier, except that Q provides less number of attributes than S. Therefore, the number of mismember tuples in Q is the same as that of S. As such, the inaccuracy of Q due to mismember tuples is the same as the inaccuracy of S due to mismember tuples. That is, $IM_Q = IM_S$.

By definition, the inaccuracy of Q due to inaccuracy in the attribute value is given by

$$IA_Q = 1 - (IM_Q + A_Q) = 1 - \left(IM_S + (A_S)^{\frac{n}{m}} \right)$$

Let $|R|$ be the size of the resulting relation after elimination of duplicates from Q, which implies that $|Q| - |R|$ tuples are eliminated from Q. The set of tuples eliminated may consist of both accurate and inaccurate tuples. By Definition D4, the number of accurate tuples, mismember tuples, and tuples having at least one inaccurate attribute value can be estimated as $(|Q| - |R|) * A_Q$, $(|Q| - |R|) * IM_Q$, $(|Q| - |R|) * IA_Q$ respectively.

The total number of accurate tuples retained in the relation R is given by

$$A_R = \frac{|Q| * A_Q - (|Q| - |R|) * A_Q}{|R|} = A_Q = (A_S)^{\frac{n}{m}} \tag{3.2.1}$$

Similarly, we can derive

$$IM_R = IM_S \tag{3.2.2}$$

By definition, IA_R can be determined as

$$IA_R = 1 - (A_R + IM_R) \tag{3.2.3}$$

Worst case : In the worst case, we consider two sub-cases. In the first sub-case, all tuples in Q are inaccurate, which implies that the accuracy of R is zero. In the second sub-case, some tuples in Q are accurate. In this sub-case, the worst situation occurs when all accurate tuples in Q collapse to a single tuple in R. In this situation, the accuracy is given by

$$A_R = \frac{1}{1 + \left(\left(1 - (A_S)^{\frac{n}{m}} \right) * |S| \right)} \tag{3.2.4}$$

The above accuracy estimate satisfies boundary conditions both P1 and P2 with the assumption that $|S| \gg 1$, which is a reasonable assumption in industrial-strength relational databases.

Best case: In the best case, we consider two sub-cases. In the first sub-case, all tuples in Q are accurate. In this sub-case, the accuracy of R is one. In the second sub-case, some tuples in Q are inaccurate. In this sub-case, the best situation occurs when all inaccurate tuples in Q collapse to a single tuple in R. In this situation, the accuracy is given by

$$A_R = \frac{(A_S)^{\frac{n}{m}} * |S|}{1 + \left((A_S)^{\frac{n}{m}} \right) * |S|} \tag{3.2.5}$$

The above accuracy estimate satisfies boundary condition both P1 and P2 with an assumption that $|S| \gg 1$.

3.3 Accuacy Estimation of Data Derived from Cartesian Product

Cartesian product is a binary operation; each tuple in relation S_1 is concatenated with every tuple in S_2. This operation is denoted as $R = S_1 \times S_2$ and the cardinality of R is given by $|S_1| * |S_2|$.

The number of accurate tuples in S_1 and S_2 is given by $A_{S_1} * |S_1|$ and $A_{S_2} * |S_2|$ respectively. From the definition of the Cartesian product operation and Definition D4, we know that the number of accurate tuples in R is given by $(A_{S_1} * |S_1|) * (A_{S_2} * |S_2|)$. By Definition D2,

$$A_R = \frac{(A_{S_1} * |S_1|) * (A_{S_2} * |S_2|)}{|S_1| * |S_2|} = A_{S_1} * A_{S_2} \tag{3.3.1}$$

Equation (3. 3. 1) satisfies the following boundary conditions (C1) and (C2):

$$(A_{S_1} = 0) \vee (A_{S_2} = 0) \Rightarrow A_R = 0 \tag{C1}$$

$$(A_{S_1} = 1) \wedge (A_{S_2} = 1) \Rightarrow A_R = 1 \tag{C2}$$

The inaccurate-membership of R and the inaccurate-attribute-value of the result of Cartesian Product can be estimated as follows:

$$IM_R = A_{S_1} * IM_{S_2} + IM_{S_1} * A_{S_2} + IM_{S_1} * IM_{S_2} \tag{3.3.2}$$

$$IA_R = IA_{S_1} * IA_{S_2} + IA_{S_1} * A_{S_2} + IA_{S_2} * A_{S_1} + IA_{S_1} * IM_{S_2} + IA_{S_2} * IM_{S_2} \tag{3.3.3}$$

3.4 Accuracy Estimation of Data Derived by Union Operator

Union is a binary operation that requires S_1 and S_2 to be union compatible [Codd, 1970]. Since duplicate tuples in S_1 and S_2 are eliminated during this operation,

$$|R| \leq |S_1| + |S_2|$$

In addition, A_R must satisfy the following two boundary conditions:

$$(A_{S_1} = 1) \wedge (A_{S_2} = 1) \Rightarrow A_R = 1 \tag{U1}$$

$$(A_{S_1} = 0) \wedge (A_{S_2} = 0) \Rightarrow A_R = 0 \tag{U2}$$

Let $|S_{1a}|$ and $|S_{1i}|$ denote the number of accurate and inaccurate tuples respectively in S_1; $|S_{2a}|$ and $|S_{2i}|$ represent similar values for S_2. The set S_{1i} can be decomposed into two subsets: (i) the set of mismember tuples, which is denoted by S_{1im}, and (ii) the set of tuples with at least one inaccurate attribute value, denoted by S_{1ia}. Similarly, S_{2i} is decomposed into S_{2im} and S_{2ia}. By definition,

$$|S_{1a}| = |S_1| * A_{S_1}$$

$$|S_{1i}| = |S_1| * (1 - A_{S_1}) = |S_{1im}| + |S_{1ia}|$$

$$|S_{2a}| = |S_2| * A_{S_2}$$

$$|S_{2i}| = |S_2| * (1 - A_{S_2}) = |S_{2im}| + |S_{2ia}|$$

Let a_a denote the set of accurate tuples eliminated during the union operation. By Definition D2,

$$A_R = \frac{|S_{1a}| + |S_{2a}| - |a_a|}{|R|} \tag{3.4.1}$$

In order to determine $|a_a|$, one needs to consider nine mutually exclusive and collectively exhaustive cases for a tuple t_1 in S_1, to combine with a tuple t_2 in S_2, given that each of these tuples may be (i) an accurate tuple, (ii) a mismember tuple, or

(iii) a tuple with at least one inaccurate attribute value. These nine cases are shown in Table 4, under the scenario that t_1 is identical to t_2.

Table 4: Nine cases when two tuples are identical--for Union operator

	t_2 is accurate	t_2 is mismember	t_2 contains inaccurate attribute value(s)
t_1 is accurate	Case1: t_1 or t_2 will be eliminated.	Case 2: t_2 will be eliminated.	Case 3: Invalid
t_1 is mismember	Case 4: t_1 will be eliminated.	Case 5: t_1 or t_2 will be eliminated.	Case 6: Invalid
t_1 contains inaccurate attribute value(s)	Case 7: Invalid	Case 8: Invalid	Case 9: t_1 or t_2 will be eliminated.

One can observe from Table 4, that a_a is the set of tuples eliminated due to Case 1. The set a_a is defined as follows:

$$a_a = \{t_1 \mid t_1 \in S_{1a} \text{ and } t_2 \in S_{2a} \text{ and } t_1 = t_2\}$$

Let $P(A)$, $P(B)$, and $P(C)$ denote the probability of $t_1 \in S_{1a}$, $t_2 \in S_{2a}$ and $t_1 = t_2$ respectively. It follows that

$$P(A) = A_{S_1} \tag{3.4.2}$$

$$P(B) = A_{S_2} \tag{3.4.3}$$

$$P(C) = \frac{(|S_1| + |S_2| - |R|)}{(|S_1| * |S_2|)} \tag{3.4.4}$$

Suppose that A and B are independent events. Suppose also that C is independent of A and B (this assumption will be relaxed later). Then the probability that the tuple will be eliminated due to Case 1 is as follows:

$$P(A \cap B \cap C) = P(A) * P(B) * P(C) = \frac{A_{S_1} * A_{S_2} * (|S_1| + |S_2| - |R|)}{(|S_1| * |S_2|)} \tag{3.4.5}$$

Similarly, the probabilities of a particular tuple being eliminated from S_1 due to Cases 2, Case 4, Case 5, and Case 9 are given by the following four expressions:

$$P((t_1 \in S_{1a}) \wedge (t_2 \in S_{2im}) \wedge (t_1 = t_2)) = \frac{A_{S_1} * IM_{S_2} * (|S_1| + |S_2| - |R|)}{(|S_1| * |S_2|)} \tag{3.4.6}$$

$$P((t_1 \in S_{1im}) \wedge (t_2 \in S_{2a}) \wedge (t_1 = t_2)) = \frac{IM_{S_1} * A_{S_2} * (|S_1| + |S_2| - |R|)}{(|S_1| * |S_2|)} \tag{3.4.7}$$

$$P((t_1 \in S_{1im}) \wedge (t_2 \in S_{2im}) \wedge (t_1 = t_2)) = \frac{IM_{S_1} * IM_{S_2} * (|S_1| + |S_2| - |R|)}{(|S_1| * |S_2|)} \tag{3.4.8}$$

$$P((t_1 \in S_{1ia}) \wedge (t_2 \in S_{2ia}) \wedge (t_1 = t_2)) = \frac{IA_{S_1} * IA_{S_2} * (|S_1| + |S_2| - |R|)}{(|S_1| * |S_2|)} \tag{3.4.9}$$

The total number of tuples eliminated from S_1 due to all five valid cases is given by $|S_1| + |S_2| - |R|$. The number of accurate tuples eliminated is proportional to the probability of Case 1. Since these five valid cases (Table 4) are mutually exclusive, it can be derived that

$$|a_a| = (|S_1| + |S_2| - |R|) * c_1 \tag{3.4.10}$$

where $c_1 = \dfrac{A_{S1} * A_{S2}}{A_{S1} * A_{S2} + A_{S1} * IM_{S2} + A_{S2} * IM_{S1} + IM_{S1} * IM_{S2} + IA_{S1} * IA_{S2}}$

It follows from Equation 3.4.1 that,

$$A_R = \frac{\left(|S_1|*A_{S_1} + |S_2|*A_{S_2}\right) - \left(|S_1| + |S_2| + |R|\right)*c_1}{|R|}$$

(3.4.11)

We now relax the assumption that C is not entirely independent of both A and B. In this case,

$$P(A \cap B \cap C) = P(A) * P(B) * P(C \mid A \cap B) = A_{S_1} * A_{S_2} * K_1$$

(3.4.12)

where $K_1 = P(C \mid A \cap B)$. Similarly, the probabilities of a tuple being eliminated from S_1 due to Cases 2, Case 4, Case 5, and Case 9 are given by the following four expressions respectively:

$$P\left((t_1 \in S_{1a}) \wedge (t_1 \in S_{2im}) \wedge (t_1 = t_2)\right) = K_2 * A_{S_1} * IM_{S_2}$$

(3.4.13)

$$P\left((t_1 \in S_{1im}) \wedge (t_2 \in S_{2a}) \wedge (t_1 = t_2)\right) = K_3 * IM_{S_1} * A_{S_2}$$

(3.4.14)

$$P\left((t_1 \in S_{1im}) \wedge (t_2 \in S_{2im}) \wedge (t_1 = t_2)\right) = K_4 * IM_{S_1} * IM_{S_2}$$

(3.4.15)

$$P\left((t_1 \in S_{1ia}) \wedge (t_2 \in S_{2ia}) \wedge (t_1 = t_2)\right) = K_5 * IA_{S_1} * IA_{S_2}$$

(3.4.16)

K_2, K_3, K_4, and K_5 are defined in a similar way to K_1 for their respective contexts. The total number of tuples eliminated from S_1 due to all five valid cases is given by $(|S_1| + |S_2| - |R|)$. The number of tuples eliminated due to Case 1 only is given by

$$|a_a| = \left(|S_1| + |S_2| - |R|\right) * c_2$$

(3.4.17)

where $c_2 = \dfrac{K_1 * A_{S_1} * A_{S_2}}{K_1 * A_{S_1} * A_{S_2} + K_2 * A_{S_1} * IM_{S_2} + K_3 * A_{S_2} * IM_{S_1} + K_4 * IM_{S_1} * IM_{S_2} + K_5 * IA_{S_1} * IA_{S_2}}$

It follows from Equation 3. 4. 1 that,

$$A_R = \frac{\left(|S_1|*A_{S_1} + |S_2|*A_{S_2}\right) - \left(|S_1| + |S_2| - |R|\right) * c_2}{|R|}$$

(3.4.18)

Several special cases can be derived from the above general expression. Cases 2, Case 4, and Case 5 (Table 4) arise when one or more mismember tuples exist in one or both of the relations in the union operation.

For boundary condition (U1), we substitute $A_{S_1} = 1$ and $A_{S_2} = 1$ in Equation 3.4.11. If $A_{S_1} = 1$ and $A_{S_2} = 1$ then, $IM_{S_1} = 0$, $IA_{S_1} = 0$, $IM_{S_2} = 0$, and $IA_{S_2} = 0$. It follows that

$$A_R = \frac{|S_1| + |S_2| - \left(|S_1| + |S_2| - |R|\right)}{|R|} = 1$$

For boundary condition (U2), we substitute $A_{S_1} = 0$ and $A_{S_2} = 0$ in Equation 3.4.11. One can see that $A_R = 0$. Therefore, Equation (3.4.11) satisfies both boundary conditions (U1) and (U2). One can show that Equation 3.4.18 also satisfies both boundary conditions (U1) mand (U2).

Best Case: The maximum attainable accuracy occurs when there are no duplicates among accurate tuples and as many inaccurate tuples as possible are duplicates. The maximum number of (inaccurate) tuples that could be eliminated is equal to $Min(|S_{1i}|, |S_{2i}|)$. It follows that $a_a = 0$ and $a_i = Min(|S_{1i}|, |S_{2i}|)$. By Definition D2,

$$A_R = \frac{|S_{1a}| + |S_{2a}|}{|S_1| + |S_2| - Min(S_{1i}, S_{2i})} \tag{3.4.19}$$

Worst Case: The minimum accuracy occurs when as many accurate tuples as possible are duplicates and there are no duplicates among inaccurate tuples. The maximum number of (accurate) tuples that could be eliminated is equal to $Min(|S_{1a}|, |S_{2a}|)$. It implies that $a_a = Min(|S_{1a}|, |S_{2a}|)$ and $a_i = 0$. The number of accurate tuples remaining in the result is given by $Max(|S_{1a}|, |S_{2a}|) - Min(|S_{1a}|, |S_{2a}|)$. By Definition D2,

$$A_R = \frac{Max(S_{1a}, S_{2a}) - Min(S_{1a}, S_{2a})}{|S_1| + |S_2| - Min(S_{1a}, S_{2a})} \tag{3.4.20}$$

3.5 Accuracy Estimation of Data Derived by Difference Operator

Difference is a binary operation that also requires S_1 and S_2 to be union compatible. This operation eliminates common tuples in S_1 and S_2 from S_1 and produces a new relation $R = S_1 - S_2$ From the definition of the operation, it is known that

$$|R| \leq |S_1|$$

In addition, A_R satisfies the following boundary condition:

$$(A_{S_1} = 1) \Rightarrow A_R = 1 \tag{D1}$$

$$(A_{S_1} = 0) \Rightarrow A_R = 0 \tag{D2}$$

By Definition D2,

$$A_R = \frac{|S_{1a}| - |a_a|}{|R|} \tag{3.5.1}$$

In order to determine A_R, it is necessary to compute $|a_a|$. The total number of tuples eliminated from S_1 due to the difference operation is given by the following expression:

$$|a_a| + |a_i| = |S_1| - |R| \tag{3.5.2}$$

Let t_1 be a tuple in S_1 and t_2 a tuple in S_2. Suppose t_1 is identical to t_2. Following the same rationale as presented in the previous sub-section, five cases need to be considered, as shown in Table 7 below.

Table 5: Nine cases when two tuples are identical--for Difference Operator

	t_2 is accurate	t_2 is mismember	t_2 contains inaccurate attribute value(s)
t_1 is accurate	Case1: t_1 is eliminated from S_1	Case 2: t_1 is eliminated from S_1	Case 3: Invalid
t_1 is mismember	Case 4: t_1 is eliminated from S_1.	Case 5: t_1 is eliminated from S_1.	Case 6: Invalid
t_1 contains inaccurate attribute value(s)	Case 7: Invalid	Case 8: Invalid	Case 9: t_1 is eliminated from S_1.

One can observe that a_a is the set of tuples eliminated due to Cases 1 and 2. It follows that

$$a_a = a_a^1 \cup a_a^2$$

where a_a^1 is the set of tuples eliminated due to Case 1, and a_a^2 is the set of tuples eliminated due to Case 2.

The total number of tuples eliminated from S_1 due to all five valid cases is given by $(|S_1| - |R|)$. Following the notation used in the previous subsection, the number of tuples eliminated due to Case 1 and Case 2 only is given by

$$|a_a| = (|S_1| - |R|) * c_5 \qquad (3.5.3)$$

where

$$c_5 = \frac{K_1 * A_{S_1} * A_{S_2} + K_2 * A_{S_1} * IM_{S_2}}{K_1 * A_{S_1} * A_{S_2} + K_2 * A_{S_1} * IM_{S_2} + K_3 * A_{S_2} * IM_{S_1} + K_4 * IM_{S_1} * IM_{S_2} + K_5 * IA_{S_1} * IAS}$$

By Definition D2,

$$A_R = \frac{|S_1| * A_{S_1} - (|S_1| - |R|) * c_5}{|R|} \qquad (3.5.4)$$

For boundary condition (D1), $A_{S_1} = 1$, and therefore, $IM_{S_1} = IAS_1 = 0$. Substituting these values in Equation (3.5.4), we obtain

$$A_R = \frac{|S_1| - (|S_1| - |R|)}{|R|} = 1$$

For boundary condition (D2), $A_{S_1} = 0$. Substituting $A_{S_2} = 0$ in Equation 3.5.4, one can see that $A_R = 0$. Therefore, Equation 3.5.4 satisfies both boundary conditions (D1) and (D2).

Similarly one can estimate IM_R and IA_R as follows:

$$IM_R = \frac{(|S_1| * IM_{S_1}) - (|S_1| - |R|) * c_6}{|R|} \qquad (3.5.5)$$

where

$$c_6 = \frac{(K_3 * A_{S_2} * IM_{S_1} + K_4 * IM_{S_1} * IM_{S_2})}{K_1 * A_{S_1} * A_{S_2} + K_2 * A_{S_1} * IM_{S_2} + K_3 * A_{S_2} * IM_{S_1} + K_4 * IM_{S_1} * IM_{S_2} + K_5 * IA_{S_1}}$$

and

$$IA_R = \frac{(|S_1| * IA_{S_1}) - (|S_1| - |R|) * c_7}{|R|} \qquad (3.5.6)$$

where

$$c_7 = \frac{(K_5 * IA_{S_1} * IA_{S_2})}{K_1 * A_{S_1} * A_{S_2} + K_2 * A_{S_1} * IM_{S_2} + K_3 * A_{S_2} * IM_{S_1} + K_4 * IM_{S_1} * IM_{S_2} + K_5 * IA_{S_1} * IA_{S_2}}$$

Best Case: The best case occurs when no accurate tuple from S_1 is eliminated and as many inaccurate tuples as possible are eliminated. It means $a_a = 0$ and $a_i = Min(|S_{1i}|, |S_2|)$. Therefore the maximum accuracy attainable is given by the following expression:

$$A_R = \frac{|S_1| * A_{S_1}}{|S_1| - Min(|S_{1i}|, |S_2|)} \qquad (3.5.7)$$

Worst Case: The worst case occurs when all tuples eliminated are accurate tuples and no inaccurate tuple from S_1 is eliminated. It means $a_i = 0$ and $a_a = Min(|S_{1a}|, |S_{2a}| + |S_{2im}|)$. Substituting the value of a_i in Equation (3. 5. 2), we get $a_a = |S_1| - |R|$. Substituting this result in the Equation (3. 5. 1), the accuracy of the result is:

$$A_R = \frac{|S_1| * A_{S_1} - Min(|S_{1a}|, (|S_{2a}| + |S_{2im}|))}{|S_1| - Min(|S_{1a}|, (|S_{2a}| + |S_{2im}|))} \qquad (3. 5. 8)$$

We have presented the accuracy estimates of the result of the five basic relational algebraic operations. Other relational algebraic operators can be constructed from these basic five operators. As such, the accuracy estimations for other operators can be derived from the accuracy estimations of these five operations.

4. Some Theorems and Formal Properties of Accuracy Estimations

In the previous section, we focused on the estimation of the result of each relational algebraic operation. Since complex queries involve multiple relational algebraic operations, we now concentrate on computing the accuracy estimate for such queries based on the following propositions:

⟨**Proposition 1**⟩:Accuracies of all base relations in the database are available.

This follows directly from the Relation Accuracy Assumption.

⟨**Proposition 2**⟩:Let Rel_Op $\varepsilon \{\sigma, \Pi, \times, \cup, -\}$. Given the accuracy of the operandi relations, the accuracy of the relation generated by any relational algebraic operator in Rel_Op can be computed.

The above proposition follows from the results presented in Section 3.

⟨**Proposition 3**⟩:User queries are comprised of relational operations in Rel_Op and base relations only.

A practical query involves multiple relational operators. One measure of the complexity is provided by the depth of nestedness of relational algebraic operations.

Order of a Query: We define a query to be 'nth-order-query' if the maximum depth of relational algebraic operations involved in the query is 'n'. This definition can be stated in BNF syntax as follows:

1st-order-query: Rel_Op(base_relation [,base_relation])
2nd-order-query:Rel_Op(1st-order-query[1st-order-query I base_relation])
3rd-order-query: Rel_Op(2nd-order-query [2nd-order-query I 1st-order-query I

base_relation])

In general, the operands of nth-order-query will be the results of utmost (n-1)th-order-queries.

⟨**Theorem**⟩:The accuracy of the relation generated by any nth-order-query can be estimated, where n is an integer and $n \geq 1$.

⟨**Proof**⟩: Let n=1. In other words, consider a 1st-order-query. Since operands of 1st-order-query are base relations and by Proposition 1, accuracies of all base relations are known; as such, using Proposition 2 and Proposition 3, the accuracy of the result of any 1st-order-query can be estimated. Therefore this theorem is true for n=1.

Let n=2, i.e., consider 2nd-order-queries. Since operands of 2nd-order-query are utmost the results of 1st-order-queries, the accuracy estimates of each operand can be estimated. Since accuracies of all operands can be estimated, Proposition 2 and Proposition 3 imply that, one estimate the accuracy of the result of any 2nd-order-query. Therfore the theorem is true for n=2.

Assume that the theorem is true for n=k, i.e., the accuracy of the result of kth-order-query can be estimated. Consider (k+1)th-order-query. The operands of this query must be relations resulting from utmost kth-order-queries. Since the accuracy of the result of any kth-order-queries or less can be estimated, by Proposition 2 and Proposition 3, the accuracy of (k+1)th-order-query can be estimated. Therefore, the theorem is true for (k+1)th-order-query.

By induction, the theorem is true for any value of n. In other words, the accuracy of the result of a query of any complexity can be estimated.

The above theorem shows that the accuracy estimation model presented in this paper can be used to compute the estimate the accuracy of the results for any query based on relational algebra.

Given a relational algebraic query, several equivalent *algebraic expressions* can be generated using a set of legal algebraic transformations based on the formal properties of relational algebra such as Idempotence for unary operators and associativity for binary operators. These formal properties need to be examined in the context of the data accuracy algebra in order to deterimine whether the accuracy estimates for equivalent queries will be the same or not. We have proved these properties elsewhere and not presented in this paper for the space resrictions. The following section, a few examples are provided to illustrate the utility of accuracy algebra for estimating the quality of derived data.

5. Application Examples

In this section, we demonstrate the utility of the data accuracy algebra. For exposition purposes, we use Tables 1 and 2 and the quality-profiles as shown in Section 2. Using these two base tables, the accuracy estimation process of data derived through relational algebraic operations is illustrated with two examples.

EXAMPLE-1: Generate a report consisting of names and salaries of employees who are working either in DEPT1 or DEPT2 and making less than $35,000 per year.

An algebraic expression to generate the above query can be framed as follows:

$$P_{\text{EMP-NAME, EMP-SAL}} \, \sigma_{\text{EMP-SAL} < 35,000} (\text{DEPT1\_EMP} \cup \text{DEPT2\_EMP})$$

A subset of the quality-profile of the database is selected in such a way that the quality information for all the relations referred to in the query exists. In this context, the quality-profile must consists of information regarding both Tables 1 and 2, as recapped below:

Table 6: Quality Profiles for Tables DEPT1_EMP and DEPT2_EMP

Relation Name	Size	A_R	IM_R	IA_R
DEPT1_EMP	10	0.7	0.1	0.2
DEPT2_EMP	10	0.7	0.0	0.3

The execution of the query can be decomposed into three stages. In the first stage, let the subquery be the union of DEPT1_EMP and DEPT2_EMP, and let TEMP_1 = DEPT1_EMP $\cup$ DEPT2_EMP. The result is shown in Table 7.

Table 7: Result after Stage 1-- Relation TEMP_1

EMP_ID	EMP_NAME	EMP_SAL
1	Henry	30,00
2	Jacob	32,000
3	Marshall	34,000
4	Alina	33.000
5	Roberts	50,000
5	Roberts	55,000
6	Ramesh	45,000
7	Patel	46,000
8	Josaph	55,000
9	John	60,000
10	Arun	50,000
11	Nancy	39,000
12	James	37,000
13	Peter	46,00
14	Ravi	55,000
15	Anil	45,000

Since after a union operator is involved, using Equation 3.4.10, the accuracy of Temp-1 can be estimated to be 0.6774. Similarly, other parameters of the quality-profile can be estimated. The quality profile of the results of the above subquery is shown below:

Table 8: Quality Profile of the Relation TEMP_1

Relation Name	Size	A_R	IM_R	IA_R
Temp-1	16	0.6774	0.033334	0.28831

In the second stage, the Selection operator is applied to TEMP_1, and let TEMP_2 = $\sigma_{EMP\text{-}SAL<35,00}$(TEMP_1). The result is shown in Table 9.

Table 9: Result after Stage 2---Relation TEMP_2

EMP-ID	EMP-NAME	EMP-SAL
1	Henry	30,000
2	Jacob	32,000
3	Marshall	34,000
4	Alina	33,000

Since a selection operation is involved, using Equation 3.2.1, the accuracy of TEMP_2 can be estimated. The quality profile of the query in the second stage is shown below:

Table 10: Quality Profile of Relation TEMP_2

Relation Name	Size	A_R	IM_R	IA_R
Temp-2	6	0.6774	0.033334	0.28831

In the final stage, the projection operator is applied on the relation Temp-2, and let TEMP_3 = $\Pi_{EMP\text{-}NAME, EMP\text{-}SAL}$ (TEMP_2). The result of this query is given in Table 11

Table 11: Result after Stage 3---Relation TEMP_3

EMP_ID	EMP-NAME	EMP-SAL
1	Henry	30,000
2	Jacob	32,000
3	Marshall	34,000
4	Alina	33,000

The quality profile of the query in the final stage ppears as follows:

Table 12: Quality Profile of Relation TEMP_3

Relation Name	Size	A_R	IM_R	IA_R
Temp-3	4	0.8230	0.0333	0.1437

Using Definition D2, one can see that the accuracy of the above sample report generated is 0.75 and the estimation model computes the accuracy of the report as 0.823. This discrepancy is due to the small sample size. The accuracy estimation of a more complex query is estimated in the following example.

EXAMPLE-2: Generate a report consisting of the names and salaries of employees whose salary is equal to a colleague's salary where both employees work only at one site and that being the same site.

While the Query Execution Plan (QEP) for the above query can be formulated in different ways, the following QEP involves all five fundamental relational operators.

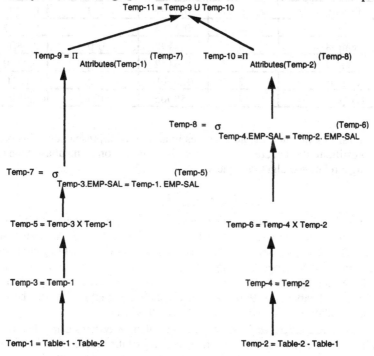

Figure 3: Query Execution Plan for EXAMPLE-2

The above diagram shows the graphical representation of the QEP. The final report generated by the query is given below, with intermediate results omitted for brevity.

Table 13: Results of Query in EXAMPLE-2

EMP_ID	EMP_NAME	EMP_SAL
5	Roberts	50,000
5	Roberts	55,000
10	Arun	50,000
14	Ravi	55,000

The quality profiles of base tables, intermediate results and the final result of the query in the example are shown in the following table.

Table 14: Quality profile of base tables, intermediate results, and the result of the query in EXAMPLE-2

Relation Name	Size	A_R	IM_R	IA_R
DEPT1_EMP	10	0.7	0.1	0.2
DEPT2_EMP	10	0.7	0.0	0.3
Temp-1	5	0.6098	0.0871	0.3031
Temp-2	6	0.565	0.0	0.435
Temp-3	5	0.6098	0.0871	0.3031
Temp-4	6	0.565	0.0	0.435
Temp-5	25	0.3785	0.1138	0.5077
Temp-6	36	0.3192	0.0	0.6808
Temp-7	4	0.3785	0.1138	0.5077
Temp-8	4	0.3192	0.0	0.6808
Temp-9	2	0.6152	0.1138	0.2710
Temp-10	2	0.5649	0.0	0.4351
Temp-11	4	0.59005	0.0569	0.35305

In a simila manner, the accuracy estimation model presented in this paper can be used to estimate the accuracy of any report derived from a database management system using a relational algebraic query.

6. Conclusion

In this paper, a theoretical framework has been presented for estimating the accuracy of derived data in a relational database environment. The proposed framework is equally valid in the context of a single database system involving multiple base relations, or multiple heterogeneous database systems in a federated environment. More specifically, a method has been developed for estimating the accuracy of any report derived from a database system through a relational query. To the best of our knowledge, no other research effort has dealt with this issue.

Since data are derived from one or more database systems using their respective data manipulation languages, and the actual accuracy of the derived data is a function of the query, the emphasis was on estimating the accuracy for the output of every operator which could be present in the relational algebra. By postulating the impact on the accuracy for each operator in theoretical terms, the accuracy profile of the output can be generated for any arbitrary query comprised of such operators. Some operators increase the degree of potential accuracy, while others tend to decrease the accuracy. This analysis, therefore, provides a mechanism for formulating queries in terms of preferred operators with the objective of attaining the optimized accuracy profile for a given set of data sources and corresponding accuracy profiles. As a corollary, when one possesses the luxury to choose from multiple data streams for the same information (such as stock prices accessed through alternative sources and channels), the cost-quality matrix can be computed to identify the optional source.

Clearly, the validity of the computed accuracy is a function of the validity of the accuracy figures for the base data. The latter figures obtained via sampling and other techniques are themselves prone to error. In the absence of more stringent information for each base relation, the data quality profile has been assumed to be uniform; however, the impact of non-uniform accuracy profiles has been studied in this paper. The issue of defining rigorous techniques for estimating the idiosyncrasies of individual data parameters is an area requiring further research.

Finally, in this paper, the focus has been exclusively on data accuracy, which is only one of the constituencies of the broader term - data quality. The latter also encompasses dimensions such as completeness and timeliness. The incorporation of such dimensions into the theoretical framework proposed in this paper is an area that merits further investigation.

7. References

[1] Ceri, S. & Pelagatti, G. (1984). *Distributed Databases Principles & Systems.* McGraw-Hill.

[2] Codd, E. F. (1970). A relational model of data for large shared data banks. *Communications of the ACM,* 13(6), 377-387.

[3] Date, C. J. (1990). *An Introduction to Database Systems.* Reading: Addison-Wesley.

[4] Deen, S. M., et al. (1987). Implementation of a prototype for PRECI*. *Computer Journal,* 30(2), 157-162.

[5] Heimbigner, D. & McLeod, D. (1985). A federated architecture for information management. *ACM Transactions on Office Information Systems,* 3, 253-278.

[6] Janson, M. (1988). Data Quality: The Achilles Heel of End-User Computing. *Omega Journal of Management Science,* 16(5), 491-502.

[7] Johnson, J. R., et al. (1981). Characteristics of Errors in Accounts Receivable and Inventory Audits. *Accounting Review,* 56(2), 270-293.

[8] Kent, W. (1978). *Data and Reality.* New York: North Holland.

[9] Klug, A. (1982). Equivalence of relational algebra and relational calculus query languages having aggregate functions. *The Journal of ACM,* 29, 699-717

[10] Lander, T. & Rosenberg, R. (1982). An Overview of Multibase. In proceedings of Second Symposiam on Distributed Databases, Sept. 1982.

[11] Laudon, K. C. (1986). Data Quality and Due Process in Large Interorganizational Record Systems. *Communications of the ACM,* 29(1), 4-11.

[12]   Liepens, G. E., et al. (1982). Error localization for erroneous data: A survey. *TIMS/Studies in the Management Science*, 19, 205-219.

[13]   Litwin, W. & Abdellatif, A. (1986). Multidatabase interoperability. *IEEE Computer*, 10-18.

[14]   Morey, R. C. (1982). Estimating and Improving the Quality of Information in the MIS. *Communications of the ACM*, 25(5), 337-342.

[15]   O'Neill, E. T. & Vizine-Goetz, D. (1988). Quality Control in Online Databases. In *Annual Review of Information, Science, and Technology*. (pp. 125-156). : Elsevier Publishing Company.

[16]   Paradice, D. B. & Fuerst, W. L. (1991). An MIS data quality methodology based on optimal error detection. *Journal of Information Systems*, 5(1), 48-66.

[17]   Pu, C. (1988). Superdatabases for Composition of Heterogeneous Databases. J. Carlis (Ed.), In *IEEE 1988 Data Engineering Conference*, Los Angeles, 548-555.

[18]   Rajinikanth, M. (1990). Multiple Database Integration in CALIDA: Design and Implementation. In *First International Conference on Systems Integration*, inproceedings of first international conference on systems integration, (April).

[19]   Reddy, M. P., et al. (1989). *Query Processing in Heterogeneous Distributed Database Management Systems*. (Ed.) Amar Gupta, IEEE Press, New York.

[20]   Sheth, A. (1991). Special Issue: Semantic Issues in Multidatabase Systems. *SIGMOD Record*, 20(4), (December).

[21]   Sheth, A. & Larson, J. (1990). Federated Database Systems for Managing Distributed, Heterogeneous, and Autonomous Databases. *ACM Computing Surveys*, 22(3).

[22]   Smith, J. M., et al. (1981). Multibase - Integrating Heterogeneous Distributed Database Systems. In *Proceedings of AFIPS*, 50, 487-499.

[23]   Spaccapietra, S., et al. (1992). Model Independent Assertions for Integration of Heterogeneous Schemas. *The VLDB Journal*, 1(1), 81-126.

[24]   Templeton, M., et al. (1987). MERMAID - A Front-end to Distributed Hetergeneous Databases. *Proceedings of the IEEE*, 1(5), (May), 695-708.

[25]   Wang, Y. R., et al. (1993). Data Quality Requirements Analysis and Modeling. In *the Proceedings of the 9th International Conference on Data Engineering*, Vienna: IEEE Computer Society Press, 670-677.

[26]   Wang, Y. R., et al. (1995). Toward Quality Data: An Attribute-based Approach. *Journal of Decision Support Systems* (March).

[27]   Wang, Y. R. & Madnick, S. E. (1989). Facilitating connectivity in composite information systems. *ACM Data Base*, 20(3), 38-46.

[28]   Wang, Y. R. & Madnick, S. E. (1990). A Polygen Model for Heterogeneous Database Systems: The Source Tagging Perspective. In *the Proceedings of the 16th International Conference on Very Large Data bases (VLDB)*, Brisbane, Australia, 519-538.

A Study of Distributed Transaction Processing in an Internetwork *

Bharat Bhargava, Yongguang Zhang **, Shalab Goel ***
Raid Laboratory
Department of Computer Sciences
Purdue University
West Lafayette, IN 47907, USA
{bb,ygz,sgoel}@cs.purdue.edu
Fax: +1-317-494-0739

Abstract. The rapid growth of interconnected computer networks has
generated a lot of interest in migrating the distributed systems to a wide
area network (WAN) environment, such as the Internet. We have developed mechanisms for studying the performance of the distributed transaction processing on the Internet, without *actually* having to move the
database sites to remote Internet hosts. We have conducted experimental
studies to analyze and understand the behavior of this transition on various transaction processing algorithms, such as concurrency control and
atomicity control algorithms. The throughput, per-transaction response
time, and abort rate of an industrial standard benchmark transactions
have been measured and evaluated.

We conclude that the optimization of the number of messages for atomicity control is the single most important criteria for effectiveness of the
algorithm in the Internet environment. We also suggest some directions
for improvement of both the communication facilities and the transaction
processing algorithms.

Keywords: distributed databases, performance, internetwork, communication.

1 Introduction

There are two dimensions for scaling a distributed system: **horizontal** scaling
and **vertical** scaling. Horizontal scaling refers to an increase in the number
of sites participating in the distributed computation. Vertical scaling relates
to the geographical dispersion of sites interconnected by long-haul networks.
Most previous research efforts in the distributed systems design have assumed
(either implicitly or explicitly) an environment with a small number of sites

* This research is in part supported by Army Research Lab (Software Technology
Branch) and AT&T.
** Yongguang Zhang's current address is Hughes Research Laboratories, Malibu, CA
90265, USA.
*** Please send future correspondence to Shalab Goel.

interconnected by a local area network (LAN). It remains incomprehensible as to how the applications and theory developed in such an environment apply to the horizontally or vertically scaled distributed system. The understanding of the implications of scaling the system, in a possibly unreliable environment, will have a significant impact on the development of existing as well as future applications for distributed systems [ÖV91, SSU91].

One particular application of distributed systems that interests us is the Distributed Transaction Processing (DTP). The communication software is a crucial component of DTP system [Spe86]. All the other software components in a DTP system rely on the efficient communications for their performance and reliability. For example, the two-phase commit (2PC) protocol, an atomicity control protocol for ensuring the unilateral commit/abort of a transaction at all participating sites, is not resilient to network partitioning. Some distributed deadlock detection algorithms require that the information to detect and resolve the deadlock in a distributed "waits-for" graph be exchanged periodically. Replication control protocols require the transmission of both the data items and the control (e.g., votes) messages for maintaining the consistency of replicated copies. The control messages eventhough small (in tens of bytes usually), are exchanged frequently. The frequency of the message exchanges will determine the effectiveness of these and many other distributed algorithms. For example, in the deadlock detection algorithm, the periodicity of the message exchange will dictate the efficiency of detecting a deadlock.

The performance of the communication software is largely dependent on the underlying communication media. Networks with varied technologies and different characteristics have emerged in the recent past, and have been connected together by internetwork connections. Thus, a DTP system spanning over a large number of geographically dispersed sites will vary in speed, bandwidth, reliability, and processing capability. With the availability of newer network technology, this variance in these parameters will become even more pronounced [Com88, Wit91]. For example, a system spanning over ATM and ethernet networks will have a bandwidth ranging between 145 Mb/s and 10 Mb/s. To summarize, the requirement of a high performance and reliable communication facility by the DTP software becomes more crucial and difficult to achieve in a WAN[4] environment.

From the preceding discussion, the need for designing efficient and reliable communication facility and scalable distributed algorithms for DTP is apparent. But before embarking on this goal, we think that understanding the behavior of the existing DTP algorithms in a WAN environment will give us an insight into the design of algorithms that will be suited to this new environment. In this paper, the mechanisms developed at Purdue University for understanding the implications of vertically scaling the distributed transaction processing are used to measure the performance of various concurrency and atomicity control protocols in the WAN environment. We utilize the experimental results to analyze the change in the behavior of these protocols due to the transition to the WAN

---

[4] The word *WAN* and *internetwork* are used interchangeably in this document.

environment. A novel technique called *emulation*, via WANCE tool [ZB93], is used to realize these experiments over the geographically dispersed sites. We discuss the problems we encountered and the solutions proposed by the other researchers in moving the transaction processing from LAN to WAN.

2 Communication in the Internet and its Impact on DTP

To study the performance of DTP in a WAN environment, the need for evaluating the communication latencies and packet losses in an internetwork is apparent. In our previous research [BZ94], we have conducted experimental studies for performance of message delivery in the Internet. We present here the results of those studies and discuss their implication on the performance of DTP.

2.1 Background: from LAN to WAN

Issues	LAN	WAN
Scale	small(10-100 sites)	large (100s of sites)
Geographic span	within a mile	over 100 miles
Topology	bus, ring	hierarchical, irregular mesh, interconnected
Routing	simple, static direct link or 1–2 hops	complex, dynamic multiple hops
Speed	high (10–1 Gbps)	low (1 Kbps–10 Mbps)
Error rate	low	high
Ownership	private	public
Access	under the same authority	shared & public network user has no control

Table 1. Difference in LAN and WAN environments

WAN differs from LAN in several ways (see Table 2.1). The performance characteristics are significantly different:

1. The communication delay and failure rates are small in LAN. In the Internet, error rates and packet latencies are large and non-uniform [Wit91]. For example, the bandwidth of a link between two sites in the Internet can range from 56KB/sec (lease line) to 45MB/sec (T3 link).
2. LAN routing is simple and static. Source routing in the Internet is dynamic and time variant. This further complicates the task of defining a useful relationship between the communication parameters and the distributed application performance.

3. LAN consists of a fewer number of hosts under the same administrative unit. Internet connects millions of hosts spanning over several organizational boundaries. The experimentation in the Internet environment will be cumbersome due to the bureaucratic administrative policies.

These dissimilarities make many communication solutions, applicable in LAN environment, unfounded in the WAN environment. For example, physical multicasting, widely accepted for distributed transaction processing in a LAN environment, is inadequate in a WAN environment.

2.2 Communication Performance

We have conducted measurements on the Internet in three dimensions: the *time* dimension by periodically repeating the experiments, the *site* dimension by repeating experiments with different sites, and the *size* dimension by varying the sizes of a message. We have been particularly interested in the following two communication performance measures: the message round-trip time and the message loss rate. In our DTP model, round trip time is the time for a site to send a request message to another site and receive a reply message. A message is "lost" if the transport service layer of the Internet fails to deliver the message on time. This is a important parameter for the DTP application, since a lost message not only blocks or aborts the transaction but also increases the contention for the shared data, e.g. the relational indices.

Our experiments involved over 2000 sites and 500 networks in the United States. We probed the Internet with ICMP and UDP messages periodically and collected the data [BZ94]. Based on these measurements, we can make the following observations.

- We observed that the time of day has strong influence on the message delivery. For example, the message loss rate is much higher in the noon working hours than in early mornings. The round-trip time, however, is immune to the time of the day, except for the hourly peeks which can be attributed to the jobs scheduled to run periodically on the gateways.
- We observed that the message delivery is non-uniform on the Internet. Most of the communicating hosts, however, reported a round-trip time of within 400ms. The *"clustering"* effect on the Internet, i.e. the communication between a host and many different hosts on another local network has similar performance, is also observed.
- We noticed that small messages which can fit in an IP datagram (i.e. without fragmentation), demonstrated a close to linear correlation between the transit time and the size of a message. However, the performance of message delivery remained unaffected by this size.
- And last but not the least, as anticipated we noticed a large variation in these performance parameters for different hosts across different networks in the Internet.

2.3 Impact on Distributed Transaction Processing

The performance of communication in the Internet will have a significant impact on the DTP performance. The time to deliver a transaction message in the WAN is a number of magnitude longer than in a LAN. While it takes only a few milliseconds to deliver a message in a LAN [AK91], on the Internet it is several hundreds of milliseconds to send a message across the continent [GL91]. This means that a transaction stays longer in the system, implying the larger lock holding time for data items, if two phase locking (2PL) is used for concurrency control. This will lead to increased contention in the database, affecting the throughput adversely.

The already difficult problem of finding a "good" timeout value in a LAN environment is further aggravated in Internet. A timeout is required by a DTP system to trigger special treatment for the transactions that are unable to complete in time. In a LAN environment, its value is usually determined as a constant multiplied by the number of read/write operations in a transaction. This flat timeout is not adequate in the Internet because both the physical distance (number of hops) and location of the host affects the message delays and loss rates. Furthermore, with the continual improvement in the CPU and I/O technology, it can be discerned that more time will be spent by the transaction waiting in the message queues than performing actual computations. Thus, the timeout value must be carefully computed, may be even dependent on the number and destination of remote messages.

Autonomous control over LAN allows the modification of the communication software to provide physical multicasting, light weight protocols, etc. [BZM91]. Unless dedicated links or special networks are adopted, not much can be altered in the shared public WAN such as the Internet. The performance of message delivery is determined by traffic and various other factors beyond the designer's control. Therefore, the focus of improving the DTP performance must shift towards reducing the number of messages exchanged in its various components.

DTP systems seldom adopt a transport service that guarantees reliable delivery because of the associated overheads [Spe86, BZM91]. This approach works fine because of the high reliability in the LAN environment. We have observed message losses as high as 30% in the Internet [BZ94]. This will lead to an increased transaction aborts resulting in a degraded DTP system performance.

The DTP algorithms have to adapt to large variations in message latencies and losses in WAN. For example, the time dependent and dynamic site-to-site performance data, and not just static data specified as a function of geographic location of the site, must be available to the quorum consensus replication control and surveillance control protocols [HZB92]. However, the detailed cost matrix to reflect the network dynamics grows as $O(n^2)$ and will become unmanageable, for large number of sites n. This problem can be tackled by taking advantage of the **clustering** effect in an internetwork, i.e. the latency between two networks can be used to estimate the communication delay between two hosts in these networks respectively.

3 Experiments on DTP in the Internet

In the past two years, we have conducted a series of experiments to study distributed transaction processing in the Internet environment. The ongoing experimental research in the Raid laboratory will pave the path for better understanding of the problems in migrating the DTP to the WAN environment and suggest some concrete directions for tackling them. In this section, we first present the apparatus for our experimentation and then the preliminary result of our experimental study.

Raid System: Raid is a reliable and adaptable distributed database system developed for conducting experimental research in distributed transaction management [BR89]. It provides an experimental infrastructure for our research in communication issues in DTP [BZM91].

In Raid system, the components of DTP are implemented as separate server processes. The servers can be arbitrarily distributed over the network and interact with each other via a high-level communication subsystem. The Raid system has the following seven servers: the Access Manager (AM), the Concurrency Controller (CC), the Atomicity Controller (AC), the Replication Controller (RC), the Surveillance Controller (SC), the Action Driver (AD), and the User Interface (UI). The UI is a front-end that allows a user to invoke SQL-type queries on the distributed relational database. The AD translates the parsed queries into a sequence of low-level read and write actions. The AM is responsible for the storage, indexing, and retrieval of data on the physical device. The CC assures that the concurrently executing transactions are conflict serializable. The AC ensures that a transaction is either atomically committed or aborted at all the sites that participated in its execution. The RC maintains the mutual consistency of the replicated data items and provides system reliability. The SC collects availability information about the Raid sites and advertises the view changes to the RC for quorum control. The details of the Raid system can be found in [BR89].

Benchmark: The benchmark we used is called the Raid benchmark. It uses a small banking database with banking transactions. It is based on TPC-A and TPC-B benchmarks and their ascendant the DebitCredit benchmark [Gra91]. All of them are widely accepted industry standards and TPC-A/B are published by TPC (the Transaction Processing Council).

Similar to DebitCredit, TPC-A, and TPC-B, the Raid benchmark uses a small banking database, which consists of three relations: the teller relation, the branch relation, and the account relation. The tuples in these relations are 100 bytes long and contain an integer key and a fixed-point dollar value. In addition, there is a sequential history relation, which records one tuple per transaction. Its tuples are 50 bytes long and contain a teller identifier, a branch identifier, an account number, and the relative dollar value specified in the transaction. Each transaction updates one tuple from each of the three relations and appends a logging tuple to the sequential history relation.

The Raid benchmark uses a multiple-sites distributed database. Each data item in the database is replicated on all sites. ROWA (Read One Write All) replication control protocol is used in our experiments, i.e. the read access to a data item can be processed locally, but the data item update has to be simultaneously propagated to all sites to maintain the consistency of the database. The transactions are randomly generated and satisfy the following specifications: 20% of the data items are marked "hot spot"; 80% of all transactions access some items in the "hot spot"; the average number of read/write actions per transaction is 4.0 with a variance of 2.0; the average percentage of writes actions per transaction is 50% (and 10% in some experiments if so indicated). The size of the workload (number of transactions in the batch) ranges between 50 and 250 depending upon the communication latency and the load on the hosts participating in the experiment. The measured attributes are the per transaction response time, the system throughput, and the transaction abort rate (percentage of the transaction failed).

We have conducted experiments with transactions distributed over more than ten sites and found that the DTP performance is usually bound by the site pair connected by the worst performance link. Thus, we report the results of experiments involving only two sites in this paper. We believe that these results will furnish a good understanding of how the wide area networking influences the performance of existing distributed transaction processing.

Experimental environment: Our research is conducted in the Raid laboratory with access to ten Sun SparcStations running SunOS 4.1 and a Sun Sparcserver 10 running SunOS 5.3. The Raid Ethernet is a 10Mbps individual subnet of the Purdue CS departmental network. The connection from the Purdue–CS net to the Internet backbone is first through a NSC hyperchannel in Purdue campus, and then via a T1 line to the NSFNET T3 backbone. All Raid machines have local disks and are also served by departmental file servers.

We use the Internet as our wide area network testbed. Internet connects over two million computers in over 50 countries around the world. The United States portion of the Internet is organized bottom up by linking many regional networks by high speed NSFNET backbones. It provides a good representative interconnection network for our study.

WANCE tool: We have developed a Wide Area Network Communication Emulator (WANCE) tool [ZB93] to *emulate* Internet communication in a LAN environment. This is clearly helpful because it can save the researchers from the bureaucracy in different organizations/countries for conducting an experiment on actual geographically separated sites. Even if we assume that some such sites are available at one's disposal, the repetition of the experiment in the *site* dimension, i.e. with different configurations, will be limited. For example, changing a database site in an experiment may need to install all the Raid software at the new site. The basic idea behind emulation is to force communication between two local hosts to go through the real Internet. We chose the emulation approach because it provides the real WAN communication in a LAN environment, capturing the dynamics of the Internet. As shown in figure 1, to emulate a three-site system linking A, X, Y on the Internet, we only need to find two hosts B, C in the same LAN as A and *comparable* in performance to X, Y respectively. We run the transaction managers on the hosts A, B, C instead of hosts A, X, Y. The user of the WANCE tool can specify it to route all transaction packets from A to B through X, and A to C through Y. Thus, eventhough our experiments are running on the local hosts A, B, C (three SparcStations in the Raid laboratory), the same results as running them on the remote hosts X and Y are obtained. The justification for the emulation approach is based on the observation that the difference between the behavior of a distributed system running on a LAN and that on a WAN is primarily due to the communication performance. The validation of the WANCE tool has been studied and is found to report a deviation of $\pm 3\%$ compared to the *real* experiments on the Internet.

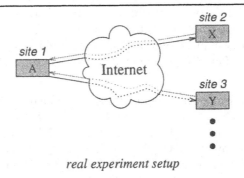

real experiment setup

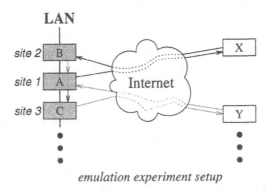

emulation experiment setup

Fig. 1. Configuration of emulation experiments

3.1 Experiment I: Raid Transaction Processing in an Internetwork

Statement of the Problem: In section 2, we summarized that large variations in the communication performance metrics is a reality on the Internet. The time of the day and the location of the host are two of the main contributors to this variance. Since the DTP software components rely on the underlying communication software, we can safely claim that the performance of transaction processing will demonstrate a similar behavior. This experiment strengthen our assertion. It studies the effect of the time of day on the DTP performance, and also how this performance is dependent on the geographical location of the site.

Procedure: The workload was a periodic batch of 20 transactions every 10 minutes over a 24 hours time span. The 10 minutes interval between two batch submissions was small enough to capture the network dynamics. We discerned that a smaller interval would unnecessarily clobber the Internet without contributing to our study. The batch of 20 transactions was large enough to generate meaningful average response time, throughput, and the abort pattern; but small enough for their execution (plus experimental setup and bookkeeping) to fit in the 10 minute interval.

We used the WANCE tool to conduct emulation experiments between Purdue and hosts in Germany, Finland, Norway, Israel, India, Japan, Hong Kong, Thailand, Australia, Brazil, Zambia, etc.

Data: We present the results of experiments between the hosts at Raid Laboratory and Helsinki (Finland); and between the hosts at Raid Laboratory and Hong Kong. The choice of these two remote sites can be explained as follows. First, the results from these two remote sites are representative of all the other sites we have experimented with. And secondly, because they cover much of the global span and many different time zones, they provide a better understanding of how network metrics behave at different hours of the day. We measured the response time, throughput, and transaction abort rate for each transaction batch.

The results are graphically represented in Figure 2. X−axis represents the time of the day. The experiment was repeated over a period of 40 days. We also average the data for each day and list the mean, variance, and the 95% confidence intervals for these averages in this figure.

Discussion: The data shows that the performance of DTP on the Internet can change drastically at different times of the day. For example, the transaction processing between Raid Laboratory and Helsinki (Finland) has much better performance at night than during the day. This can be attributed to the fact that the minimum time zone difference between North America and Europe connected via a few trans−atlantic links is only about three hours. This suggests that the network traffic patterns at different times of the day are similar on at least these "shared" links. This is consistent with our previous studies that shows the message delivery is slower and less reliable during the working hours [BZ94]

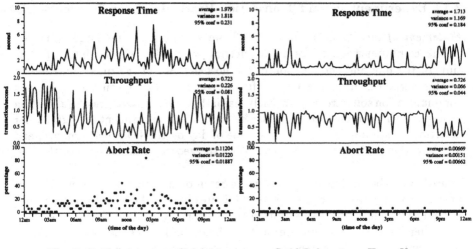

Fig. 2. Raid Laboratory–Helsinki   Raid Laboratory–Hong Kong

in this region. The throughput was almost 2 transaction/second at night, but dropped to below 0.5 transaction/second during the day time.

The DTP performance between Raid Laboratory and Hong Kong is better than the previous case. We attribute this to the following two reasons: First, the time difference between these sites is 13 hours. Thus, when it is day in USA, it is night in Hong Kong. Second, the US–Hong Kong Internet link is more reliable as has been evaluated in [BZ94]. Unlike the US–Europe Internet communication, where so many countries share a few cross-Atlantic links, the US–Hong Kong cross-Pacific link serves only Hong Kong, an area smaller than New York city.

3.2 Experiment II: Multi-Programming Level and Concurrency Control

Statement of Problem: The number of concurrent transactions in a system at any time is called its multi-programming level (MPL). This experiment investigates the relationship between the MPL and the DTP performance on the Internet using different combination of concurrency and atomicity protocols. We have selected the two popular concurrency control and atomicity control protocols: two-phase locking (2PL) and timestamp ordering (TO); and two-phase commit (2PC) and three-phase commit (3PC) respectively.

Procedure: We have compared the performance of DTP in three different site configurations: within Raid Laboratory (LAN); Raid Laboratory and University of Texas, Arlington (USA); and Raid Laboratory and University of Melbourne (Australia). A workload of 250 transactions with 50% average updates is used. We varied the MPL as 1, 2, 3, 4, 5, 10, 15, and 20 (MPL=1 is the normal batch mode). The WANCE tool is used to emulate the Internet communication for distributed transactions. We measured the performance of four different protocol configurations: 2PL with 2PC, TO with 2PC, 2PL with 3PC, and TO with 3PC. For each value of MPL, we repeated the experiment 50 times and averaged the measured data.

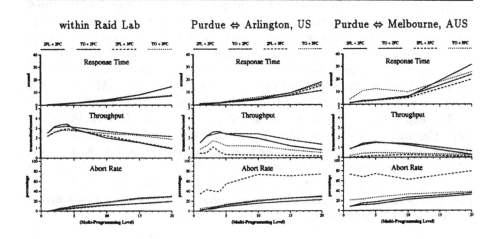

Fig. 3. MPL experiments with LAN, UTA, and AUS site respectively

Data: Figure 3 shows the response time, throughput, and abort rate of the benchmark transactions using different combinations of concurrency and atomicity control protocols, as the MPL is varied. The results for DTP in LAN (leftmost graph in fig 3) environment are included for the comparing the performance.

Discussion: As the MPL increases, the response time increases monotonically in all three configurations, which is as expected. The increase in MPL of the system implies more data item conflicts and hence the blocking/aborting of a transaction, delaying its completion.

The abort rate also rises with the increase in MPL in all the three configurations. In 2PL based combinations, the higher contention for data items increases the probability of deadlocks which is resolved by aborting the transactions. In TO based combinations, the probability of "out of order" arrival of read/write actions on a data item increases as the number of transactions in the system (MPL) increase. TO scheduler rejects each such action, aborting the transaction. The $2PL + 3PC$ case shows the highest abort rate. We suspect that this is due to the extra round of message required in the Internet for committing a transaction. This extra message round increases the lock holding time for a transaction which increases the probability of deadlocks. We conjecture that this might indeed be the case because a similar behavior is absent in the local LAN case.

The throughput increases initially for small MPL, but starts to dip for larger values of MPL. The increased throughput is due to the "non blocking" concurrent execution of the transactions. As the MPL increases beyond a certain value, the transaction aborts increases due to increased number of conflicts and prevents any useful work to be done at the local site, thus reducing the system utilization.

The "best" MPL value is located in the range of 4 to 7 for each of the three configurations. This is the MPL at which the DTP system can provide acceptable response times without too much compromising on the concurrency, and hence the throughput. However, the throughput achieved at the "best" MPL value is lower and response times higher in comparison to LAN case. Furthermore, for MPL equal to one, the difference in the throughput between LAN and WAN cases can be factored out as entirely due to the message delays and losses.

The experiment suggests no significant impact of the MPL on the abort rate. This means that the unreliability of the Internet is the main contributor for the transaction aborts. In the LAN environment, the performance of TO scheme is inferior to the locking scheme. However, improved throughput with lower response times are observed in WAN environment with TO schemes. This preliminary result has motivated us to look into the utility of locking schemes for the WAN environment more thoroughly.

3.3 Experiment III: Atomicity Control and Commitment

Statement of Problem: This experiment explores the cost of different commit protocols in an internetwork. It will reaffirm our intuition that additional round of message in 3PC protocol may actually create a bottle neck in the performance of transaction processing.

Procedure: We used the same two-site fully replicated benchmark database. A workload of 250 transactions with 50% update rate was used as before. The experiment was repeated with the following sites configurations: within the Raid Laboratory; Raid Laboratory and Arlington (USA); Raid Laboratory and Helsinki (Finland); and Raid Laboratory and Hong Kong.

Data: Figure 4 displays the total execution time of a transaction and the portion of this time spent in AC (Atomicity Controller) in Raid system for both 2PC and 3PC protocols. To correlate the results, we have also measured the communication delay between the two database sites at the time of experiment. These are listed at the bottom of the figure, below the corresponding configurations.

Discussion: We observe that when the communication delay increases, both the response time and the time to commit increase. However, the time to commit shows a sharper rise in case of 3PC protocol. Although the contribution of 3PC to the actual transaction response time is not significantly larger than that of 2PC in a LAN because of very small communication delays (i.e. 0.001 sec), it results

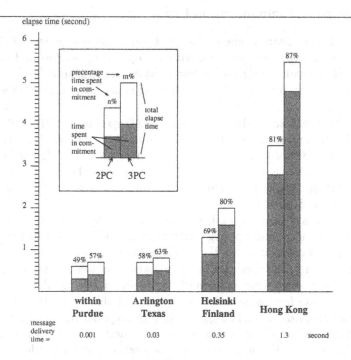

elapse time (second)

Fig. 4. Breakdown of the total time for transaction in a system

in a much larger response time as compared to 2PC in the WAN environment because of large communications delay (e.g. 1.3 sec for Hong Kong).

This experiment also indicates that the commit phase of DTP will create a bottleneck when we move our transaction processing software to WAN environment. This is due to the rounds of messages required to form global consensus. Efforts to reduce the overhead can be directed into the following directions: streamlined commit protocol, optimal replica placement, and perhaps relaxed consistency requirements, etc.

4 Issues in Transition of DTP from LAN to WAN

The communication experiments in the Internet have embarked the rethinking of our transaction processing software. We have recognized some potential problems in the transition of distributed transaction processing from LAN to WAN. We believe that the following two directions are important for accommodating the DTP in a WAN environment: improving the communication software and tuning/modifying/rewriting the DTP algorithms and protocols.

4.1   Improving Communication Facility

Communication software is one of the most important component of the DTP
system. However, the ones provided with the operating systems and computer
networks are not tailored for the on-line transaction processing. To improve the
communication support for DTP, the following issues have to be addressed.

Reliability vs. Efficiency: Most commercial DTP systems use UDP/IP protocol
from the Internet protocol suite for their communication facility. UDP is a "best-
effort delivery" protocol and does not guarantee message delivery. Since the
error rate in a LAN is less than 0.001% [AK91], most DTP systems can use
UDP/IP protocol to meet the performance and real-time requirements of the
application [Spe86, BZM91].

Large volumes of network traffic may be generated by a transaction executing
indexed queries or queries on large data objects, which may lead to the over-
flow of receiver buffers at either the destination or at an intermediate gateway.
Furthermore, the packet loss rate in some portions of the internetwork may be
so high that even the retransmission of the messages with UDP/IP may be inadequate. A brute force approach is to use the reliable TCP/IP protocol from
the Internet protocol suite. The guaranteed message delivery comes at the expense of increased overheads in setting up a connection for data transmission.
A compromise is to design a communication scheme in which the amount of
unreliability that is acceptable to an application is a *tunable* parameter and the
mechanism to achieve it is not opaque to the application designer.

Scalability: The performance of communication component of the DTP system
in the WAN environment must not degrade with the horizontal scaling of the
system. There are two communication models often used in DTP: the `virtual
circuit model` and the `stateless message passing model`. The communica-
tion between two processes is via a *pre-established* channel (e.g. a TCP connection) in a virtual circuit model. In the stateless message passing model, each
process has a "message port" (e.g. a UDP port in Unix socket interface) to
forward a self-addressed message packet.

The virtual circuit model is not scalable because the maximum number of
connections per process is usually restricted by the operating system. Furthermore, the buffer space for maintaining the state information and un-acknowledged
packets on either side of connection is limited. The message port model poses
no such problems for scalable communications.

We propose a *"two-level"* communication scheme in the WAN environment.
A set of local sites connected "reliably enough" is a *cluster*. A link is "reliable
enough" if the packet loss rate is so low that its effect on the performance of
DTP will be marginal. Such occasional packet loss will be remedied by a timeout
of a DTP server and roll back of the transaction to a previous checkpoint. In
this scheme, *intra-cluster* communications utilize the efficient datagram message
passing interfaces [BZ94]. For *inter-cluster* communication, the DTP server sends
the message to a designated communication server for its cluster which will

forward the message using a more reliable protocol such as TCP, based on the condition of the links.

4.2 Distributed TP Algorithms

There have been several performance studies on the transaction processing algorithms in local area networks. However, their performance in a wide area network environment can only be conjectured at this time.

Concurrency Control: In transition from LAN to WAN, the communication latency has changed from several milliseconds to several hundred milliseconds. Given a two site database in a LAN, a typical benchmark transaction takes a few hundred milliseconds to finish on Sun SPARCstations. It will take a couple of seconds to complete in a WAN environment, depending upon the distance and the traffic on the network [BZ94]. A transaction stays longer in the system, increasing the data contention. A classic locking concurrency control would block transactions for a longer period of time and therefore degrades the throughput of the system. The chances of transactions getting in deadlocks also increase, leading to higher abort rates. A possible solution may be to relax the concurrency correctness constraints. For example, epsilon-serializability concurrency control allows inconsistency for an *epsilon* time period.

Atomicity: The two-phase commit protocol (2PC) is not resilient to site failures while the three-phase commit protocol (3PC) needs one extra round of messages. The throughput of a DTP system in LAN environment was found to be similar with both 2PC and 3PC in [BFHR90]. The response time was slightly higher in case of 3PC protocol. Thence, it was suggested that 3PC be used for atomic control since it provided higher availability at a little additional cost. It has been shown in [SJR91] that the time to commit accounts for one-third of the transaction duration in a general purpose database. Experiment III shows that this ratio can be as high as 80% of the total transaction time in the Internet. This explains the transaction throughput degradation in experiments II when 3PC protocol is used. These results, contrary to our advocating the use of 3PC protocol in the LAN environment in the previous studies, suggest that 3PC protocol must be used with caution in WAN environment and only when availability can not be compromised. Several possible optimizations to minimize the number of messages for atomicity control have been proposed in [SBCM93]. We will conduct more experiments to study their feasibility and performance.

Availability & Replication: Darrel Long, in [LCP91] has studied the availability of an Internet host and has found it to be below 86%. This low availability has been attributed to their periodic maintenance, software/hardware crashes, and high local work load. Furthermore, the end-point hosts may be up and functional, but the in-route gateways may drop the message packets because of the localized network congestion. In either case, the low availability will result in degraded DTP performance. One proposal is to pair each database site with a backup site.

The notion is similar to *disk mirroring* for improved disk availability. The backup site is placed in a different geographic location such that it uses a different link to the internetwork backbone. However, this increased resilience to failure comes with the cost of maintaining the consistency of the two sites. A detailed analysis of the availability of a replicated database can be found in [BHF92].

4.3 Relevant Work and Related Results

To understand the implication of moving a distributed system to an internetwork environment, the communication behavior of the internetwork has been studied by many researchers. The studies include the analysis of the Internet host availability [LCP91], packet arrival behavior in NSFNET backbones [Hei90], network dynamics [AS92].

Communication in an internetwork While our study focuses on estimating the time for a transaction message to reach a remote site in the Internet, Darrel Long et al [LCP91] have analyzed the reliability and availability of a remote site. They have measured the mean-time-to-failure for each host connected to the Internet and derived a mean-time-to-repair estimation.

Golding and Long [GL91] have assessed the reliability of message delivery, which validates our measured packet loss rates in the Internet. They have also observed that retrying a message beyond two or three times does not add to the system performance.

Packet arrival behavior study [Hei90] and network dynamics [AS92] have focused on the behavior of the packet delivery, while we have concentrated on the end-to-end performance of sending a transaction message.

Transaction processing in WANs Pu et al [PKL91] measured the performance such as response-time and failure rate in Camelot and Webster dictionary lookup over the wide area networks. The "Layered Refinement" measurement method, simultaneously collecting data at various communication software layers (network, transport, and application layers), was utilized. Inspired by their work, we started on the analysis of the different transaction processing protocols performance in the Internet.

5 Conclusion and Future Work

We have studied the transition of distributed transaction processing systems from a local area network to a wide area internetwork environment. We have analyzed the effects of this migration on the concurrency control and atomicity control protocols via an experimental study. Concomitant with the intuition, we come to the conclusion that isolating the impact of different transaction control protocols on the system performance is not easy in a WAN environment. Furthermore, many of the solutions that were pronounced adequate in the LAN environment, fall apart when applied to the WAN environment. For example, the timestamp ordering concurrency control scheme is better than locking scheme in the WAN environment. The experiments have commenced our rethinking of the DTP software, to adapt to uncontrollable and dynamically changing WAN environment. We continue our research in two directions: to develop a communication facility that scales horizontally and vertically in WAN environment; and to demonstrate experimentally the utility of the techniques such as clustering, two-level communication, and surveillance for dampening the effects of uncontrolled and dynamic behavior of the internetwork on the DTP.

References

[AK91]   Bandula W. Abeysundara and Ahmed E. Kamal. High-speed local area networks and their performance: A survey. *ACM Computing Surveys*, 23(2):221–264, June 1991.

[AS92]   Ashok K. Agrawala and Dheeraj Sanghi. Network dynamics: An experimental study of the Internet. In *Proceedings of GLOBECOM'92*, Orlando, FL, December 1992.

[BFHR90]   Bharat Bhargava, Karl Friesen, Abdelsalam Helal, and John Riedl. Adaptability experiments in the Raid distributed database system. In *Proc of the 9th IEEE Symposium on Reliability in Distributed Systems*, Huntsville, Alabama, October 1990.

[BHF92]   Bharat Bhargava, Abdelsalam Helal, and Karl Friesen. Analyzing availability of replicated database systems. *International Journal of Computer Simulation*, 1:393–418, 1992.

[BR89]   Bharat Bhargava and John Riedl. The Raid distributed database system. *IEEE Transactions on Software Engineering*, 15(6), June 1989.

[BZ94]   Bharat Bhargava and Yongguang Zhang. A study of distributed transaction processing in wide area networks. Technical Report CS-94-016, Purdue University, March 1994.

[BZM91]   Bharat Bhargava, Yongguang Zhang, and Enrique Mafla. Evolution of communication system for distributed transaction processing in Raid. *Computing Systems*, 4(3):277–313, Summer 1991.

[Com88]   Douglas E. Comer. *Internetworking with TCP/IP*. Prentice-Hall, Englewood Cliffs, NJ, 1988.

[GL91]   Richard Golding and Darrel D. E. Long. Accessing replicated data in an internetwork. *International Journal of Computer Simulation*, 1(4):347–372, December 1991.

[Gra91]   Jim Gray, editor. *The Benchmark Handbook for Database and Transaction Processing Systems*. Morgan Kaufmann, San Mateo, CA, 1991.

[Hei90]   Steven A. Heimlich. Traffic characterization of the NSFNET national backbone. In *Proceedings of the 1990 ACM SIGMETRICS Conference on Measurement and Modeling of Computer Systems*, pages 257–258, Boulder, CO, May 1990.

[HZB92]   Abdelsalam Helal, Yongguang Zhang, and Bharat Bhargava. Surveillance for controlled performance degradation during failure. In *Proc of the 25th Hawaii Intl Conf on System Sciences*, pages 202–210, January 1992.

[LCP91]   D. D. E. Long, J. L. Carroll, and C. J. Park. A study of the reliability of Internet sites. In *Proceedings of the 10th Symposium on Reliable Distributed Systems*, pages 177–186, Pisa, Italy, September 1991. IEEE.

[ÖV91]   M. Tamer Özsu and Patrick Valduriez.   Distributed database systems: Where are we now? *IEEE Computer*, 24(8):68–78, August 1991.

[PKL91]   Calton Pu, Frederick Korz, and Robert C. Lehman.   An experiment on measuring application performance over the Internet. In *Proceedings of the 1991 ACM SIGMETRICS Conference on Measurement and Modeling of Computer Systems*, San Diego, CA, May 1991.

[SBCM93]   George Samaras, Kathryn Britton, Andrew Citron, and C. Mohan. Two-phase commit optimizations and tradeoffs in the commercial environment. In *Proceedings of 9th IEEE International Conference on Data Engineering*, pages 520–529, Vienna, Austria, April 1993.

[SJR91]   P. Spiro, A. Joshi, and T. K. Rengarajan. Designing an optimized transaction commit protocol. *Digital Technical Journal*, 3(1), Winter 1991.

[Spe86]   Alfred Z. Spector. Communication support in operating systems for distributed transactions. In *Networking in Open Systems*, pages 313–324. Springer Verlag, August 1986.

[SSU91]   A. Silberschatz, M. Stonebraker, and J. Ullman.   Database systems: Achievements and opportunities. *Communications of the ACM*, 34(10):110–120, October 1991.

[Wit91]   Larry D. Wittie. Computer networks and distributed systems. *IEEE Computer*, 24(9):67–76, September 1991.

[ZB93]   Yongguang Zhang and Bharat Bhargava. Wance: A wide area network communication emulation system. In *Proc. of IEEE workshop on Advances in Parallel and Distributed Systems*, pages 40–45, Princeton, NJ, October 1993.

The Use of an Object Repository in the Configuration of Control Systems at CERN

Wayne Harris, Richard McClatchey and Nigel Baker

Dept. of Computing, Univ. West of England, Frenchay, Bristol BS16 1QY UK
Phone: (44) 1179 656261 ext. 3163 FAX: (44) 1179 763860
email: Richard.McClatchey@csm.uwe.ac.uk

Abstract: The CERN-based CICERO project is creating a general purpose Control Information System (named Cortex) which will enable physicists and technicians in integrating computer control systems across computer networks using standard protocols and interfaces. Cortex has two major functionalities: a configuration based tool for describing the components to be connected and their interfaces, and an on-line message passing system. An Object Oriented Database is seen as an essential feature of such a system, both for the control system configuration process and to provide information for the on-line system. This paper reports on an investigation into the requirements for the use of an Object repository in providing these features. The requirements are compared to the functionality provided by a set of current, stable commercially available Object Oriented Databases, to demonstrate areas where features need to be developed, either by the application programmer or the database developers.
Keywords: Distributed Systems, Object Database, Control Systems, Object-Oriented Design

1. Introduction

Modern experiments into the fundamental nature of particles at the High Energy Physics (HEP) laboratory at CERN utilise complex equipment for the detection of particles in experiments. The operation of these experiments is managed by sophisticated control software. In essence the experiments carried out at CERN use control software to gather data, control large and complex devices and to monitor and alert the operators to various alarm conditions.

CERN has recently approved the next generation of experiments and is now preparing to build the Large Hadron Collider (LHC), which will require further equipment and networking capabilities. In order to meet the demands of the future, the CERN project CICERO, (Control Information system Concepts based on Encapsulated Real-time Objects) [CIC93] is implementing a system to assist users in configuring the control information systems without requiring detailed knowledge of any details of the control system other than the interfaces to the data distribution system.

In the past, HEP control systems have been developed as a response to the needs of the experiments on an ad-hoc basis, with a variety of different short term solutions to similar problems [BARI91]. A general solution to these problems is being developed as a central part of CICERO, and this necessitates a novel approach incorporating a wide range of new problems, many of which are now being addressed in database research. In particular, the CICERO research provides an opportunity to investigate the requirements for Object Oriented Databases (OODB) in a distributed systems environment.

The development of the CICERO project follows a Spiral Model [BOE88] or evolutionary approach which allows for the planned delivery of multiple releases of the software. In order to understand the full requirements for CICERO, a prototype system [GOV95] has been designed and built, using GemStone [GEM94] as the selected OODB. This Prototype has helped to refine the requirements for the control system and for the use of OODBs.

This paper reflects on the outcomes of using GemStone as the OODB for the CICERO Prototype and from this identifies the main database requirements for Control Systems at CERN and compares these requirements to the features supplied by typical commercial OODBs available today.

The requirements for the use of an Object Repository in CICERO are examined in greater detail in the following section. In Section 3, a short summary of the main features of OODBs is provided as a background for the following two sections, which discuss the database oriented features of the Cortex [BARI94] element of the CICERO project. Section 4 demonstrates the use of the off-line repository for configuring and modelling the control information system (called Cortex) and section 5 describes the on-line requirements of the database. Section 6 outlines the use of GemStone as a repository in the Cortex Prototype and section 7 reviews in a qualitative manner the Cortex design requirements and compares them to the facilities available in a set of commercial OODBs. The final section presents conclusions.

2. Repositories in a Control Environment

The HEP community at CERN has considerable experience in the use of databases for capturing control system design. The present generation of experiments has relied on the use of Relational Database Management Systems (RDBMS) such as Oracle to act as Control System repositories [BAI94], [DON94] & [MAT94]. Due to the performance limitations of RDBMSs, however, these repositories have only been used in an off-line environment at the start-up of the Control System, rather than on-line while the system is operative and when response is required at the level of a few seconds. The next generation of HEP experiments will not only be an order of magnitude more complex than the present experiments with the consequent increase in complexity of the Control System software, but will also require on-line access to large configuration repositories. One of the main aims of CICERO is to investigate the feasibility of using OODBs to address these requirements.

Individual control elements (see figure 1) of a HEP experiment hierarchy are normally developed by different institutes in their laboratories and integrated later following a

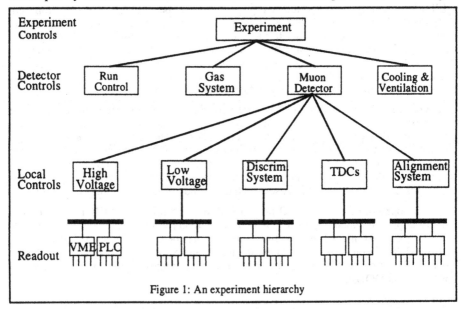

Figure 1: An experiment hierarchy

programme of testing. Any software framework aimed at integrating these disparate experiment control elements must support a life cycle for the collaborative distributed control systems of these experiments. In particular, it should be possible to develop control (sub-) systems and later to integrate them in a distributed control system without major code modifications. Users should only have to configure the existing collaborative control system, by introducing their control (sub-) system and specifying the information and commands to be exchanged between it and the global control system.

The architecture of any distributed control system will change periodically during the lifetime of the experiment it is operating. The proposed software integrating framework in CICERO, Cortex, must therefore support a mechanism to allow a new version of the distributed control system to be in preparation while an older one is operated on-line. Cortex also needs to support the backup and the restoration of a given version of the distributed control system i.e. it must support multiple versions of the control system. Furthermore, tests and validation (and possibly simulations) of a new configuration will be needed before it is applied to the operating on-line system.

The above constraints have led to a so-called dual-face approach (see figure 2.) being adopted in the design of Cortex:

• an off-line Cortex representation is implemented to handle the logical descriptions of the architecture of the distributed control system and to describe the various information and commands to be exchanged between the different control elements. This so-called *Repository* also holds the description of the hardware model from which the on-line distributed control system is constructed, and

• an on-line Cortex representation is implemented through which the control elements can exchange information and commands in a pseudo- 'plug-and-play' fashion. Control elements operating within the Cortex *Infrastructure* can access the Cortex Repository through this so-called Infrastructure. A generation and validation mechanism is provided to facilitate updates of the on-line Infrastructure according to the Repository contents.

The responsibility of the Cortex integrating framework is twofold: on the one hand it has to support the description of the architecture of the distributed control system and the definition of the information and commands to be exchanged between the control elements. On the other hand, Cortex must also transport and distribute these data and commands to the appropriate control elements when part or all of the distributed control system is in operation. The integrating framework must be flexible enough to sup-

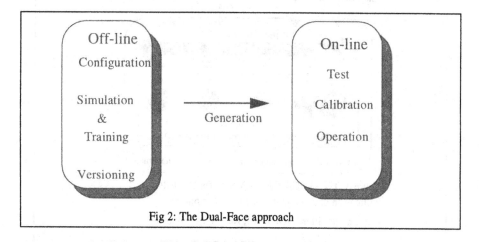

Fig 2: The Dual-Face approach

port the addition or removal of control elements, without deteriorating the operation of the rest of the distributed control system.

Much of the complexity for the type of control systems proposed here, is a direct result of the distribution of the control elements across heterogeneous systems which gives rise to problems of location, naming, protection, access, data consistency, data presentation, synchronisation and other control problems. Object encapsulation is an ideal way of hiding and managing this complexity. Objects can be combined in arbitrary ways to provide new information processing and controlling capabilities. The object based approach also allows encapsulation of existing non-object elements to provide interoperability with other control (sub-) systems. Encapsulating all elements has the disadvantage that internal details critical for management and monitoring of the system will be hidden. Consequently objects will require a separate interface on which they can publish and subscribe management events and items of interest and on which their integration behaviour can be controlled.

Given these considerations and the fact that the development of control systems is generally carried out in a multi-user environment, OODBs are the natural candidates for the Cortex Repository. Also, the use of standards is essential to enable large numbers of physicists to use the Cortex system, and also to protect the development from the variability in future technological developments. The OMG [OMG92] standard, embodying the CORBA model (see figure 3) and the ODMG [ODMG93] standards, (including the OQL query language) are therefore seen as central to the development of Cortex and provide a specific constraint on any database chosen for use in CICERO.

3. OODB Features Required in Cortex

Hughes [HUG91], Kim [KIM95] and others have identified the characteristic features of the family of OODBs. In essence these features can be considered in three categories: object orientation (OO) features, database (DBMS) features and additional features.

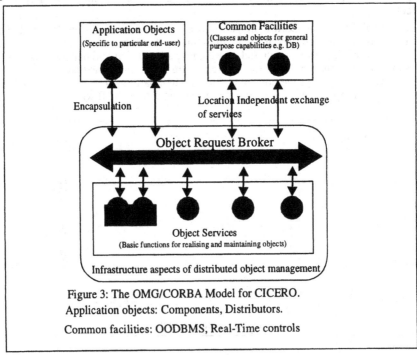

Figure 3: The OMG/CORBA Model for CICERO.
Application objects: Components, Distributors.
Common facilities: OODBMS, Real-Time controls

The additional features are often domain-specific: they are needed as a result of the application areas in which OODBs are applied such as CAD/CAM systems, Geographic Information Systems and Computer Aided Software Engineering.

The essential object oriented features that any OODB used in CICERO must provide include the standard OO features of Object Identity, inheritance, complex object handling and secure message passing. In addition the minimum DBMS features that CICERO requires of an OODB include persistence of objects, controlled access to objects, effective querying and browsing, a multi-user capability and full recovery facilities. Additional OODB requirements which are particularly important in the domain of CICERO include:

i) Controlled data distribution which should be heterogeneous, thereby enabling the OODB to exist on many machines and to access existing ('legacy') databases eg RDBMSs.

ii) Schema Evolution. As the control architecture matures so the supporting Object Model must evolve. Kim & Chou [KIM88] and Agrawal & Jagadish[AGR89] in particular emphasise the importance of the controlled evolution of object schema.

iii) Long transactions may be required in Cortex since the Repository will act as a *design database* with multiple concurrent designers. Techniques for handling long transactions are becoming almost essential in OODBs due to the application areas they are aimed at. Examples include CAD/CAM and CASE environments ([BAR91], [SCI94]) Several strategies have been suggested to enable long term shared developments such as in [DAY90] and [BEE89].

iv) Rollback to a point in time (such as that proposed in [JEN92]) is a further feature that will be required in CICERO. Rapid access to earlier versions of the design database can be facilitated using this feature.

v) Since CICERO wishes to reduce development costs by incorporating commercial products, use of industry standards e.g. ODMG [ODMG93] is a prerequisite. This standard enables the database to access foreign systems, including legacy systems and objects distributed on other machines across a network.

vi) Development tools. These must include as a minimum a schema design tool and a tool for generating user interaction programs to allow users to access and populate the database.

vii) Given the complexity, cost and sensitive nature of Control Systems software, security is a major concern in the development of Cortex. This is particularly true when the control software is designed and implemented by geographically separated groups of developers.

Several commercial databases are now available that meet the essential OO and DBMS requirements in CICERO and which, in addition, address many of the additional desirable features. The CICERO project has used GemStone as its basis for its initial prototype. Section 6 will show that this Prototype revealed, amongst other things, that the C++ mechanisms for GemStone are insufficient to satisfy the requirements of CICERO [GOV95]. Consequently in the second phase of CICERO development, a comprehensive review of several other OODBs has been carried out. The additional features of OODBs listed above are used for a comparative study of the following OODBs in Section 7: O2, ObjectStore, Matisse and Illustra.

4. Cortex Off-line Database Requirements

The Cortex off-line Repository is used to define the communications paths between *components*. It is based on a graphical tool that represents components, which may be *composite* in nature and the interfaces between these components. Any component can contain a set of *items* which are directly related to data produced or consumed in the

Control System. Items are complex structures with a format, based on CORBA types, and various descriptions and constraints on the item. Any item can be *published* and hence made available for another component to *subscribe* to. Components publishing items become producers on-line and components subscribing to item become consumers.

The components communicate through collaborative *groups* which share *interfaces*. The off-line configuration process declares the components, their sub components and their items. It also declares the groups in which components are *engaged* (ie publish or subscribe) and the items that are actually published and subscribed within the groups. Figure 4 demonstrates an example of a control (sub-) system (such as that of fihgure 1) as it might appear in the configuration tool. A validation process ensures that publications and subscriptions match as well as performing various other cross-checks.

The validation procedure is central to the Cortex environment. Since the resulting system to be built must be coherent and efficient, some combinations of interactions between the components must be excluded. However, at configuration time, the users need to be free to create and delete objects and references on a temporary basis. To meet these requirements, each configuration must be validated via a set of rules before it can be used to generate an on-line system. Such validation can be used as a final check on the development process, since users can concurrently make changes to the stored model that might ultimately lead to an inconsistent on-line system being generated.

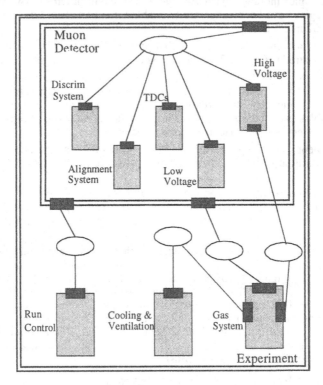

Fig 4: (Composite) Components & Grouping

Key: Groups are represented by ellipses, Components by rectangles and Component interfaces by shaded boxes. Composite Components are shown as double rectangles.

This continuous operation implies that massive changes to the on-line operation need to be minimised thereby reducing system downtime. A mechanism is required for the system to identify the differences between new versions of objects and the existing on-line objects. These differences can then be used to create a set of changes (or deltas) to the on-line system. Such changes can also be used to implement a rollback database at the application level, so the OODB need not necessarily provide access to past versions of the system. There is no identified need for past versions of temporary configuration objects to be available.

Since large teams develop control system software in a collaborative manner, there must be some measure of security in the design process to minimise accidental or deliberate corruption of items. These security needs to be addressed by the OODB kernel. A major consequence of the number of developers involved is that the resulting system must enable multiple users to share the design and development of the information transportation system. Such sharing is similar to the features of any design database, where users will be working on design components for extended periods of time, possibly sharing the work. Similarly, it is essential that the users can refer back to old versions of their data model at any time in the future. This implies the need for a rollback capability in the database, although, this could be implemented at the application level.

Finally, a standardised object interface is needed for the off-line development procedure (as well as the on-line operation) to provide upward compatibility with future technical developments in OODBs and to allow access to the products of a variety of vendors. The ODMG standard is the only comprehensive standard to be suggested to date, and therefore ODMG compliance is seen as a requirement of the CICERO off-line database.

The following section describes the on-line requirements for tools to monitor and control the data exchange and control process levels i.e. the on-line Cortex environment.

5. Cortex On-line Database Requirements

The characteristics of on-line operation of a Control System for a HEP experiment are that it must be reliable, fail-safe, provide 24-hour availability and provide response to operators to command and control messages in the order of a few seconds. Control Systems do not address the sub-second response required for the fast acquisition of data from the detectors rather the so-called *Slow Controls* required to maintain equipment in a quiescent state. The safety of the experiments is handled by a combination of authorisation procedures to ensure that no unauthorised commands are executed and the use of special independent alarm systems which monitor the experiments and alert the users. Alarms are the only area in which Control Systems have to exhibit quasi- real-time (i.e. at the level of tenths of second) response.

Under normal operation control elements (such as operator interfaces, archivers, loggers, producers and consumers of data) will access the Cortex Repository through the Cortex Infrastructure. In order to provide this desired level of access to objects the OODB must be able to translate the Object Identifiers (OIDs) of the (description held in the) OODB to actual machine and disk addresses and then to *swizzle* to memory addresses within the time limits required. This is a difficult feature to quantify and test, since the objects will be distributed and the OIDs require translation to disk addresses. This table lookup from OID to object location must be as efficient as possible and provide the limiting quasi real-time performance required by the Control System.

A set of tools is required to be available for monitoring the on-line processes, both the data as they are transferred throughout the distributed control system and the status of the individual control elements. To ensure that users can operate the control system effectively at all times, whether the experiments are executing in a normal, stable manner or in an alarm condition, as much information as possible must be available in a concise and easily accessible manner. This includes rapid access to a help system, with intuitive navigation to points of interest, and also multi-media features such as annotated pictures and videos of items of equipment, their locations and their modus operandi.

These tools require fast (of the order of one second) access to component descriptions held in the database, and therefore the database should provide adequate response as it references and swizzles the objects in its internal tables. Response time is particularly important when concurrent processes of reconfiguration conflict. Since these actions may attempt to lock objects for extended periods of time it is essential that they do not interfere with the normal on-line access to the database.

6. GemStone as the Prototype Cortex Repository

Cortex exhibits many requirements similar to other application areas in which OODBs have been utilised. One purpose of the Prototype, produced as the first phase (or 'Spiral') of Cortex development, was to facilitate an investigation of where OODB techniques from other application domains (e.g. CAD/CAM and CASE) could be applied in Cortex. This Prototype has also helped to indicate areas where further requirements are needed in Cortex as well as identifying those requirements which are of lesser importance or which are not addressed by available OODB technology.

No ODMG compliant OODB was available when the architectural design of Cortex started. It was identified that the GemStone ODBMS was the commercial product which best suited the needs of the Cortex Repository. It provided robustness, easy prototyping capabilities and the possibility to implement the construction rules for the Cortex architecture as active objects inside the Repository. The GemStone object language, Opal, is very similar to SmallTalk80. The Cortex Repository has been written in Opal for all the persistent objects and in C++ for the objects interfacing with the on-line Infrastructure. The design concept of *Compositeness* identified in section 4 has been used to isolate a high level (operator console) subsystem from a low level (actuators & sensors) subsystem in the Cortex Prototype. Grouping has been implemented as collections of collaborating control components, and is modelled as such in the GemStone OODB, acting as the off-line Repository.

As the version of GemStone used for the Prototype was not ODMG compliant, no standard object interface for CORBA objects (components) could be implemented to access the Cortex Repository. For the Prototype access to the Repository was supported by the use of a so-called *Query Server* which acted as the interface between the Cortex on-line Infrastructure and the Cortex Repository. GemStone offered very limited capabilities for C++ objects to become persistent in the Repository. No C++ method for registration was supported by GemStone. In addition SmallTalk objects inside the Cortex Repository could only access C++ objects through static member functions in C. These two important limitations made the full integration between the Repository and user-specified components impractical.

Concurrent access, versioning and configuration management have been included in the Prototype, but not fully implemented since multi-user access to the Repository was given a low priority in the Prototype. Neither was the OODB optimised for performance since the Prototype was acting as 'Proof-of-Concept' rather than a final demonstrator.The second phase of the CICERO project, presently underway, is aimed at investigating, amongst other aspects, multi-user access and performance issues in Cortex operation..

7. Cortex Requirements Evaluated Against Commercial OODBs

As all OODBs can handle both the essential Object Oriented and Database features listed in Section 3,the only significant areas to consider for evaluation are the additional features supplied by the database vendors, against the actual requirements of Cortex. A prerequisite for consideration in this study has been ODMG compliance and the OOD-Bs considered are GemStone (once compliant), O2, Matisse, ObjectStore and Illustra.

Taking each of the additional features supplied by OODBs:

• Heterogeneous distribution is crucial for Cortex, especially if it is to be implemented in existing experiments which use RDBMSs. The only OODB that has been considered which provides full heterogeneous distribution is GemStone. It is able to do so since its underlying language is interpreted and hence able to be executed by any machine. O2, Matisse, Illustra and ObjectStore are all based on compiled C++ or C and so the code is not directly portable from one machine to another. This problem can be overcome through the use of a Client-server architecture operating with CORBA objects.

• Cortex users need to be able to concurrently access and update objects over extended periods of time if they are to use the OODB as a repository of a design or configuration. This requirement is normally met in commercial OODBs through the use of long trans-actions. As yet, it is not clear whether Cortex will need to use the OODB long transaction features since it is based on a *change-oriented system*. As Cortex needs to generate and validate changes, and since no long transaction system handles the merging of versions of the object model satisfactorily, mechanisms for allowing users to share development environments are required to be written with any of the OODbs considered. Provided the OODB can provide optimistic locking, concurrent access to objects from the configuration tools and the on-line system will be possible.

• Schema evolution. Ideally, Cortex requires a schema evolution strategy which enables the user to select when objects are converted to the updated schema. The OODBs investigated can be divided into two groups: those that implement schema evolution partially by requiring all objects to be copied out before conversion to the updated schema; and those with full historical capability, such as Matisse and Illustra. The use of linked C or C++ code to implement the methods affects the ability for a database schema to be rolled back to a past date, so that integrated systems such as GemStone and Illustra are preferred for their ability to integrate the schema with the database. Illustra, with its dual ability for rollback access and a schema included with the database is the preferred OODB for these features.

• The requirement for development tools is central to the use of the OODB as a design database in Cortex since the Cortex system is being developed in a rapid prototyping environment. At present, of the OODBs considered, GemStone and O2 provide the best set of development tools. It is clear, however, that all OODB vendors are aware of the importance of such tools and are attempting to improve the range and power of their toolsets.

• All the OODB vendors are presently developing query languages, usually based on SQL. In Cortex, objects use navigation to search for related objects, either by referencing sub-objects or by holding specific relationships, hence the Cortex on-line system only requires to use object identifiers (OIDs). Off-line, a search facility, based potentially on a query language, is required to locate components in the control system configuration. The O2 query langauge has been identified as providing the functionality required in Cortex.

• The separation of the physical structure from the logical structure in the database does not have the same importance in an OODB that it has in a RDBMS. This is due to the encapsulation of the objects, enabling them to be restructured easily to improve efficiency. Also, since the query language is less important in an OODB than in a RDBMS, optimisation becomes a minor consideration in Cortex. No differentiation can be made between the OODBs considered on this basis.

• Security. Cortex requires high security for the on-line and off-line systems, due to the large number of developers and the possibilities for corruption on-line. No single OODB in the sample at present provides the level of security required in Cortex, however O2, ObjectStore and GemStone continue to develop strong security strategies.

• The speed of access of each OODB in the sample (using OIDs and swizzling) is appropriate for the response required in the Cortex Prototype. Further investigation is required in the second phase of Cortex development, when the complete database has been implemented and a multi-user environment has been established, to quantify the on-line response to queries. Each OODB vendor provides direct access to objects using OIDs and each release of their software exhibits significant performance improvements.

To summarise, while OODBs provide all the required features for Object Orientation and Database Management, there do seem to be some areas where there are specific weaknesses, such as in heterogeneous distribution and in schema evolution. The conclusions in the next section consider these weaknesses to identify their sources and whether they can be overcome. For the second phase of Cortex development, it is likely that a C++-based OODB will be finally used and the favoured product is O2.

8. Conclusions

OODBs provide features that are necessary for the Cortex application, over and above the standard object oriented and database features. There are, however, problem areas where they do not meet all of the Cortex requirements. Many of these problems areas are not new or technically difficult. Some are standard problems solved in RDBMSs but now requiring new solutions for OODBs.

For the Cortex application the problem areas can be divided into three groups:

•required features that are technically difficult to implement, that involve difficult trade-offs, and are being addressed by the OODB developers. These areas are: heterogeneous distribution; schema evolution; separation of the physical model from the logical model; fast access and swizzling of objects using their OIDs.

•required features that should be straightforward to adapt from RDBMS solutions but are either poorly handled by current OODBs or not handled at all. In many cases, this is because OODB developers have been concentrating on the technical problems in the other two groups and are only now starting to consider these area (in particular development tools and security systems).

•features that are not specifically required in the Cortex application, but are supplied by OODBs. In some respects, these are the problems that developers have been concentrating on at the expense of the above groups. They are: long transactions or versioning; historical access; and an SQL based query language.

To summarise these conclusions, no commercial standard tool can be expected to provide the ideal solution to a single application but it does appear that OODBs need to continue to improve to meet some of the essential requirements in application areas such as Cortex. Is is appropriate, however, for the CICERO project to implement the Cortex Repository on OODB rather than RDBMS technology, since RDBMSs cannot address the time constraints in Cortex, since the project is largely object-based, the majority of its requirements are satisfied by that technology and there are good prospects of that technology continuing to improve.

OODBs are required for recent new developments in network based control systems. With a few improvements they will be able to provide an ideal solution to the requirements for this fast growing area and thus meet the requirements of ongoing developments like Cortex in the near future.

9. Acknowledgments

The CICERO project involves the following groups to which the authors extend thanks both for financial support and for technical assistance: BARC (Bombay), CERN (Geneva), CIEMAT (Madrid) IVO International (Helsinki), KFKI (Budapest), OBLOG (Lisbon), SEFT (Helsinki), SPACEBEL (Brussels), UID AB (Linkoping), USDATA (Dallas), UWE (Bristol), VALMET Automation (Tampere) and VTT (Oulu).

References

[AGR89] Agrawal, R., Jagadish, H.V., "On Correctly Configuring Versioned Objects" in Proceedings of the 15th VLDB, Amsterdam, 1989.

[BAI94] Bailey, R et al., "Development of the LEP High-Level Control-System Using Oracle as an On-Line Database" in Nuclear Instruments & Methods A Vol. 352 No. 1-2 pp 430-433 (1994).

[BAR91] Barghouti, N.S., Kaiser, G.E., "Concurrency Control in Advanced Database Applications" in ACM Computing Surveys, Vol. 23, No.3, September, 1991.

[BARI91] Barillere, R et al., "Ideas on a Generic Control System Based on the Experience of the Four LEP Experiments Control Systems". Presented to the ICALEPCS'91 Conference, Tsukuba, Japan, Nov. 11-15 1991, pp 246-253.

[BARI94] Barillere, R et al., "The Cortex Project: A Quasi- Real-Time Information System to Build Control Systems for High Energy Physics Experiments". Nuclear Instruments and Methods A 352 pp 492-496 1994.

[BEE89] Beeri, C., Bernstein, P.A., Goodman, N., "A Model for Concurrency in Nested Transaction Systems" in JACM, Vol. 36, No. 2, April 1989.

[BOE88] Boehm, B., "A Spiral Model of Software Devbelopment and Enhancement", in IEEE Computer Vol. 21 No. 5 pp 61-72 (1988).

[CIC93] CICERO: "Control Information system Concepts based on Encapsulated Real-time Objects". CERN/DRDC/93-50.

[DAY90] Dayal, U., Hsu, M., Ladin, R., "Organizing Long-Running Activities with Triggers and Transactions" in Proceedings of the 1990 ACM SIGMOD International Conference on Management of Data, May, 1990.

[DON94] Donszelmann, M & Gaspar, C., "The DELPHI Distributed Information-System for Exchanging LEP Machine Related Information" in Nuclear Instruments & Methods A Vol. 352, No.1-2 pp 280-282 (1994)

[GEM94] GemStone.Object-oriented database produced by Servio, USA.

[GOV95] Govindarajan G et al. CICERO: 1994 Status Report. CERN RD-38/LHCC/95-15.

[HUG91] Hughes, J. Object-Oriented Databases. Prentice Hall 1991.

[JEN92] Jensen, C.S., Clifford, J., Gadia, S.K., Segev, A., Snodgrass, R.T., "A Glossary of Temporal Database Concepts", in SIGMOD Record, Vol. 21,No. 3, September, 1992.

[KIM88] Kim, W., Chou, H.T., "Versions of Schema for Object-Oriented Databases" in proceedings of the 14th VLDB Conference, Los Angeles, California, 1988.

[KIM95] Kim, W. Modern Database Systems: The Object Model, Interoperability & Beyond. Addison Wesley 1995.

[LHC93] LHC: "The Large Hadron Collider Accelerator Project". CERN AC 93-03 1993.

[MAT94] Mato, P et al., "The New Slow Control-System for the ALEPH Experiment at LEP" in Nuclear Instruments & Methods A Vol. 352 No.1-2 pp 247-249 (1994)

[OMG92] The Object Management Architecture Guide, version 2.1, OMG Pubs 1992. The Common Object Request Broker: Architecture and Specifications, OMG Pubs 1992.

[ODMG93] The Object Database Standard ODMG-93, Atwood T, Duhl J, Ferran G, Loomis M, Wade D & Cattell R 1993.

[SCI94] Sciore, E., "Versioning and Configuration Management in an Object Oriented Data Model", in VLDB Journal 3, 1994.

Implementation and Performance Evaluation of Compressed Bit-Sliced Signature Files

Kazutaka Furuse, Kazushige Asada, and Atsushi Iizawa

Software Research Center, RICOH Co., Ltd.
Tomin-Nissei-Kasugacho Bldg., 1-1-17 Koishikawa, Bunkyo-ku, Tokyo, Japan

Abstract. This paper describes design and implementation techniques of the signature file access method. It has been implemented and embedded in RICOHBASE, a commercial DBMS product based on an extended relational data model. This is our first experience of practical implementation of this access method, although some experimental implementations exist. We discuss some design topics which turn into issues in practical environment.

We also describe the results of experimental simulations which were performed upon the implementation of the signature file access method in RICOHBASE. From these results, it is possible to examine characteristics of the signature files. The performance of the signature files is largely dependent on the tuning of parameters. The results of the simulations indicate changes in performance changes in the values of the parameters. Furthermore, the results give values of the parameters that achieve optimal performance in text database applications.

1 Introduction

Text retrieval is increasingly becoming an important feature of DBMSs for various application fields such as electric office information systems and hypertext database systems.

Signature file is one of the most popular access methods designed for improving performance of the full text retrieval[2][6][7]. For this access method, there are some discussions on performance analysis and experimental implementations in the literature[3][7]. However, there are few implementations embedded into DBMS for practical use.

We have implemented a signature file access method and embedded it into the extended relational DBMS RICOHBASE. To the best of our knowledge, this is the first experience of implementation of this access method embedded in a commercial DBMS product. This paper describes design and implementation technique of the signature file access method. It also describes the results of experimental simulations. So far, a few research studies have been done for analyzing the nature of the signature files. However, there are very few performance evaluations which are based on the simulations with the real implementations in the DBMS like the one described in this paper.

The reminder of this paper is organized as follows. Section 2 presents an overview of our implementation. In Section 3, we discuss on several design issues.

Section 4 describes experimental simulations and the results. Section 5 gives a summary and concludes this paper.

2 Implementation of a Signature File Access Method

2.1 Summary of Signature File Mechanism

The signature file access method is a mechanism designed for improving performance of text retrieval[2][6][7]. In this access method, data objects stored in databases are indexed by signature files, which are composed of pairs of signatures and object identifiers of the data objects.

words	word signatures
electric	0001000100000000
power	0000000001000100
supply	0000000100100000

target signature: 0001000101100100

Fig. 1. Superimposed Coding

Signatures are fixed length sequences of bits which are generated for data objects. Among various approaches to generate signatures proposed so far, the *superimposed coding* is one of the most popular. In this approach, signatures for data objects are constructed as follows. First, data objects are divided into *logical blocks*. Then, for each word in a logical block, *word signatures* are generated by hashing. Finally, a target signature for the logical block is obtained by bit-wise OR-ing (i.e., superimposing) all the word signatures of the logical block (Figure 1).

Figure 2 illustrates the physical file organization of a signature file. As shown in this figure, pairs of the target signatures and data objects' identifiers are stored. When size of a data object is large, it is divided into logical blocks, and the target signatures are generated for each logical block. This division helps in improving efficiency. Thus, in general, there may be more than one entries in the signature file for each data object.

The procedure of the text retrieval using signature files is composed of following steps. First of all, for the served query, a *query signature* is generated. Query signature is a signature that is generated by superimposing signatures of words that appear in the query. Next, the signature file is scanned. During the scan, the query signature is compared in a bit-wise manner with each target signature in the signature file to find out *drops*. A drop is an entry in the signature file, and is a candidate (but not necessarily an element of the result) that may satisfy the query condition. In superimposed coding approach, an entry is a drop, iff all

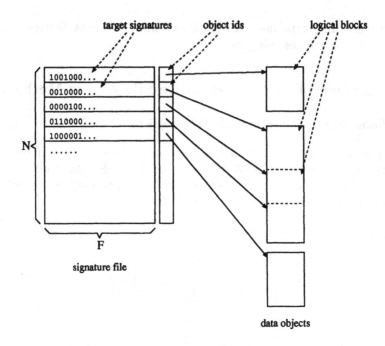

Fig. 2. Signature File

"1"s in the query signature are set in the entry's target signature. That is, let Q be the query signature and T be the target signature of the entry, the entry is a drop, if

$$(Q \& T) = Q,$$

where $\&$ is a bit-wise AND operator.

Next step of the retrieval procedure is *false drop resolution*. Unlike the other common access methods, in the signature file access method, it is not sufficient to find out drops, because there may be some entries that do not satisfy the query condition. Thus, to construct the final result of the query, there is a need to filter out unsatisfactory drops by checking whether the contents of the data objects really satisfy the query condition or not. This process of qualification is called false drop resolution. In this way, drops are divided into two categories; satisfactory drops and unsatisfactory drops. They are called *actual drops* and *false drops*, respectively.

In general, data objects are apt to be very large in text retrieval applications. Therefore, the false drop resolution tends to be a considerably heavy and time-consuming process. Thus, it is quite desirable that the number of false drops is as small as possible, as compared with the total number of data objects. This ratio is called *false drop probability* P_{fd}[5], and is defined as

$$P_{fd} = \frac{D_{fd}}{N - D_{ad}},$$

where N is the number of entries in the signature file, and D_{fd} and D_{ad} are the number of false drops and the number of actual drops, respectively. As shown in the results of the experimental simulations described later, the false drop probability can be a parameter for performance measurement of signature files. In general, the lower the false drop probability of a signature file is, the faster the retrieval procedure is.

2.2 RICOHBASE

Recently, the authors have implemented a signature file access method and embedded it into the extended relational DBMS RICOHBASE. RICOHBASE is a commercial DBMS product which is designed and implemented by RICOH Software Research Center. RICOHBASE is an extended relational DBMS and some distinctive features including the graph data model and a rich set of data types.

The graph data model is a variant of the extended relational data models[10][11]. In the graph data model, the notion of links has been introduced. Links are used to connect instances of entities. With use of links, in the graph data model, it is possible to represent the entity-relationship model[1] directly.

The rich set of data types is one of the advantages of RICOHBASE. It includes unlimited variable length string data type and bulk data type. The former can be used to build full text databases. The latter is absolutely invaluable for multimedia database applications.

Figure 3 illustrates the architecture of RICOHBASE. RICOHBASE is based on the client/server architecture. The server is called data manager. It accesses databases and serves DBMS services to clients. Applications (i.e., clients) send requests to the data manager and receive resultants through the database library. The data manager is composed of some modules including query optimizer, DBMS kernel, access methods, schema manager, and page buffer manager.

2.3 Implementing Signature File Access Method in RICOHBASE

In RICOHBASE, access methods are implemented as submodules as shown in Figure 3. It can be said that mechanism of the access method submodules in RICOHBASE is just like the mechanism of device drivers in operating systems. There is a specification of a set of functions which implements an access method mechanism (i.e., creation of a file, insertion of a record, scanning of a file, etc.). Each submodule includes definitions of such functions, and exports them to the DBMS kernel.

We have implemented a set of functions of the signature file access method, and embedded it into the data manager.

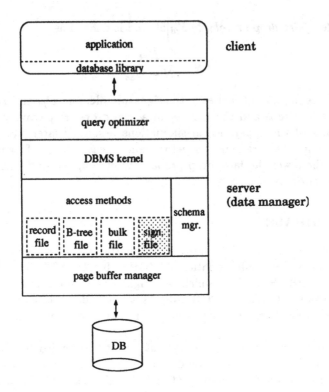

Fig. 3. The Architecture of RICOHBASE

3 Design Issues for Improving Performance

This section describes some topics based on the decisions made in the course of design of the access method.

3.1 Compressed Bit-Sliced Signature Files

There are a few variants of physical storage organization for signature files[5]. Among the candidates, we have selected *bit-sliced signature files.*

Bit-sliced signature file is one of the most popular file organizations. Figure 4 illustrates the structure of a bit-sliced signature file. As shown in this figure, signatures are *sliced* in a column-wise manner, and columns are stored in distinct files (bit-slice files, or *slices*). In this file organization, to find out drops in a signature file, it is not necessary to access all slices. If there are n bits of "1"s in a query signature, only n slices are opened and scanned. Thus, as compared with normal (non-sliced) signature files, the number of disk page accesses during retrieval process is considerably small.

One of the disadvantages of the bit-sliced signature files is that it takes much more time to insert records. To insert a record, all slices have to be opened

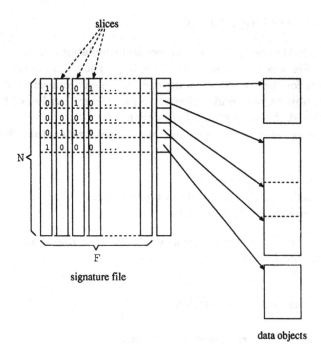

Fig. 4. Bit-Sliced Signature File

and updated. However, in the application such as text databases, performance of retrieval is much more important than that of insertion. Thus, we concluded that bit-sliced signature files are preferable.

Next decision we have made is choice of a compression algorithm. It is well known that it is possible to decrease the number of disk accesses by compressing signature files. Generally, contents of signature files are very sparse. That is, number of "1"s in signature files is much smaller than number of "0"s. Thus, any compression algorithms can reduce the size of signature files considerably. In other words, the compression ratio is not a criterion for selecting a compression algorithm for signature files. Instead of that, we settled the following criteria.

– It should be possible to decompress during the scan of slices.
– It should not take much time to decompress slices.

As we stated earlier, in text database applications, the performance of retrieval is the most important. Therefore, it is desirable that the (de)compression algorithm doesn't reduce the speed for scanning slices.

In conclusion, we picked up MH (modified Huffman) coding algorithm. This is a common compression algorithm that is adopted in the G3 FAX protocol, and qualifies the abovementioned criteria.

3.2 Parameters of Signature Files

As we can see in the results of the experimental simulations, the performance of the signature file access method varies according to some parameters. Since the optimal values for these parameters depend on the environment and characteristics of data objects, we decided that some parameters can be set by users for each databases at the stage of database design.

The followings are the user-definable parameters for signature files in RICO-HBASE.

- *length*; size of a signature in bits
- *weight*; number of bits set to "1" in a word signature
- *unit*; size of a logical block in bytes

In the next section, we will show how the setting of these parameters has its impacts on the performance of text retrieval.

4 Experimental Simulation

4.1 Simulation Environment and Parameters

In order to evaluate the performance of the proposed implementation, we performed several simulations. We prepared simple test databases in which up to 50000 records were stored. The schema of the test databases is the following:

```
record article (
    id              integer(4),
    title           variable character(64),
    author          variable character(64),
    body            variable character(4096),
    registered      timestamp,
    last_modified   timestamp on update
) key(id);
```

index article(body) signature *F m U*;

Each record has a field body, in which text data is stored and is indexed by signature files. The italic symbols F, m, and U in the schema are parameters which were varied in the experiments. These parameters are listed in Table 1. We made various signature files by changing values of the parameters in this list. We served a set of queries for each signature file, and measured average elapsed time to get results (including time of the process of the false drop resolution) and false drop probability.

Experiments were done on Sun microsystems' SPARCstation 10 workstation running under SunOS 4.1.3. The amount of main memory mounted on the machine is 32MB.

Symbol	Definition	Default	Range
N	total number of data objects	30000	10000–50000
F	length; size of a signature in bits	2048	1024–4096
m	weight; number of bits set to "1" in a word signature	1	1–4
U	unit; size of a logical block in bytes	512	256–2048

Table 1. Parameters of Simulations

4.2   Experimental Results

With and Without Signature Files

In the first experiment, we study how much the introduction of the signature file access method contributes to the improvement of performance. Figure 5 shows the average elapsed time of served queries, varying the number of records from 10000 to 50000, for the databases with and without signature files (other parameters are set to the defaults).

We see that the performance of text retrieval has improved by significant orders of magnitude by introducing the signature file access method. From this result, it can be said that RICOHBASE has a clear advantage over other commercial DBMS products which do not use any special access methods for text retrieval.

Figure 6 shows the sizes of the databases. Although the size of a signature file changes according to its parameters, the introduction of a signature file has no impact on the total size of databases.

Length of Signatures

Figure 7 shows how elapsed time changes by varying the length of signatures F. In general, increasing the length of signatures decreases false drop probability, and hence improves the retrieval performance. Figure 8 shows the variations in the false drop probability. From this figure, we see that doubling the length of signatures causes considerable reduction of the false drop probability.

Note that further decrease of the elapsed time has stopped at point $F = 2048$. This is because the size of the logical block of the text data stored in the test databases is 512. Therefore, further increase of the size does not make any effect. The false drop probability does not improved after the point $F = 2048$ as shown in Figure 8.

It is interesting to note that Figure 7 and Figure 8 are very similar. This means that the false drop probability can be used as a measure of performance in our implementation.

Weight of Signatures

Next experiment examines the effect of the weight (m). Figure 9 shows the result. Clearly, the smallest weight (i.e., $m = 1$) wins in all cases.

In superimposed coding, it is known that the performance will be optimal when the value of weight to be set to

$$m_{\text{opt}} = \frac{F \ln 2}{D},$$

where D is the number of words superimposed in a signature[6]. However, as we can see in Figure 9, it is obvious that this rule does not hold in our implementation. We conjecture that the following may be the reason: the increase of weight means the increase of number of occurrences of "1"'s in signatures, and hence it makes the compression ratio worse. Consequently, the number of disk accesses grows, and the performance goes down.

It is future work to figure out this nature by analysis.

Length versus Unit

Further, we examine the effect of unit (U), the size of logical blocks. Figure 10 shows the average elapsed time, varying the unit of signatures from 256 to 2048. In general, as opposed to the length of signatures, the performance has improved as the unit decreases. When the value of unit gets smaller, the number of words in a logical block gets fewer. As this causes the decrease of the false drop probability, the performance improves. However, in our experiments, this is not true for the case $F = 4096$. In this case, the performance is getting a bit worse as the value of unit gets smaller. This is because the total size of the signature file is too big to fit in cache memory of the data manager.

Consequently, here is an interesting question: what makes the performance better, to increase length or to decrease unit? Figure 11 shows the answer. In this figure, we compare two configuration: one is $F = 4096, U = 512$, and the other is $F = 2048, U = 256$. From this figure, we can study that it is more effective to increase length than to decrease unit.

5 Summary and Conclusion

In this paper, we have described design and implementation techniques of the signature file access method embedded in RICOHBASE. With this implementation, we performed experimental simulations and studied characteristics of the signature file access method from various aspects.

Our study from the simulations shows:

- our implementation of the signature file access method improves the performance of text retrieval;
- the false drop probability can be used as a performance measurement in such implementations;

- the bigger the length of signatures, the better the performance, although there are upper bounds on this;
- the performance is optimal when the weight of signatures is 1; and
- to improve the performance, it is better to increase the value of the length than to decrease the value of the unit.

As a suggestion for future work, we propose to build a cost model and analyze it, and verify the analysis by further experimental simulations.

Acknowledgments

The authors are grateful to Dr. H. S. Kunii for her encouragement to this research. The authors would like to thank members of Software Research Center, RICOH Co., Ltd.

References

1. Chen, P. P.: The entity-relationship model: Toward a unified view of data. *ACM Trans. on Database Systems*, 1(1):9–36, Mar. 1976.
2. Faloutsos, C.: Access methods for text. *ACM Computing Surveys*, 17(1):49–74, Mar. 1985.
3. Faloutsos, C.: Signature files: Design and performance comparison of some signature extraction methods. *Proc. of ACM-SIGMOD 1985 Int'l. Conf. on Management of Data* , Austin, Texas, 14(4):63–82, May. 1985.
4. Faloutsos, C.: Signature-based text retrieval methods: A survey. *IEEE Data Engineering*, 13(1):25–32, March 1990.
5. Faloutsos, C., Chen, R.: Fast text access methods for optical and large magnetic disks: Designs and performance comparison. In *Proc. of the 14th Int'l. Conf. on Very Large Data Bases*, Los Angeles, USA, pages 280–293, Aug. 1988.
6. Faloutsos, C., Christodoulakis, S.: Signature files: An access method for documents and its analytical performance evaluation. *ACM Trans. on Office Information Systems*, 2(4):267–288, Oct. 1984.
7. Faloutsos, C., Christodoulakis, S.: Description and performance ayalysis of signature file methods for office filing. *ACM Trans. on Office Information Systems*, 5(3):237–257, July 1987.
8. Furuse, K.: Implementation fo signature file access method in a dbms. Technical Report of IEICE DE94-58, The Inst. of Elec., Info. and Comm. Eng., Sep. 1994. (in Japanese).
9. Kanasaki, K.: Database management system g-base. Ricoh Technical Report 15, Ricoh Co., Ltd., Apr. 1986. (in Japanese).
10. Kunii, H. S.: *Graph Data Language: A High Level Access-Path Oriented Language*. PhD thesis, The University of Texas at Austin, May 1983.
11. Kunii, H. S.: *Graph Data Model and Its Data Language*. Springer-Verlag, 1990.
12. Lee, D. K., Kim, Y. M., Patel, G.: Efficient signature file methods for text retrieval. *IEEE Trans. on Knowledge and Data Engineering*, 7(3):423–435, June 1995.

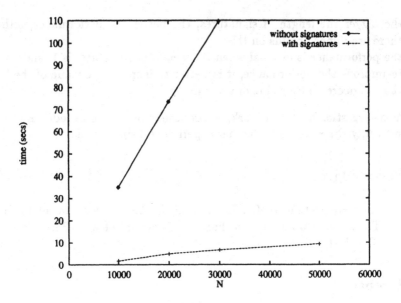

Fig. 5. With and Without Signature Files

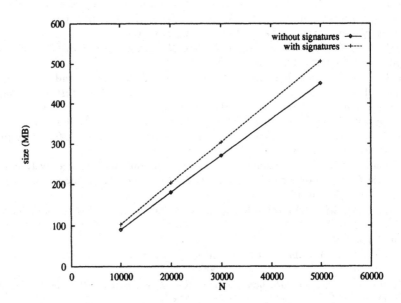

Fig. 6. Size of Databases

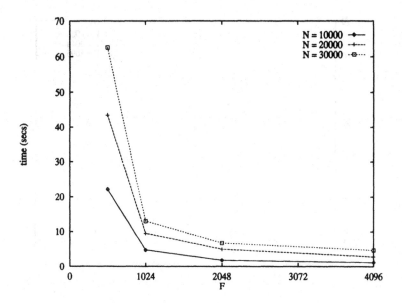

Fig. 7. Length of Signatures and Performance ($m = 1, U = 512$)

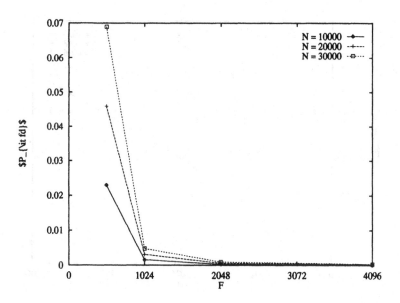

Fig. 8. False Drop Probability ($m = 1, U = 512$)

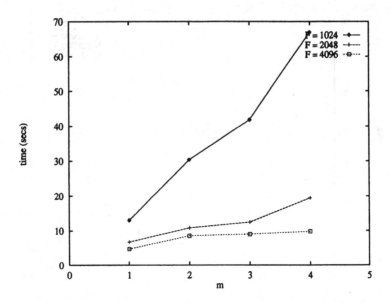

Fig. 9. Weight of Signatures ($N = 30000, U = 512$)

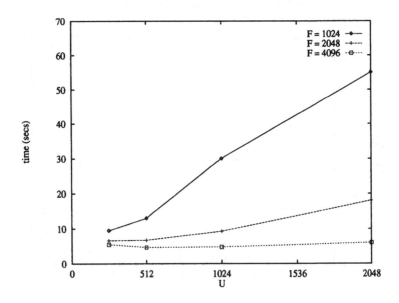

Fig. 10. Unit of Signatures ($N = 30000, m = 1$)

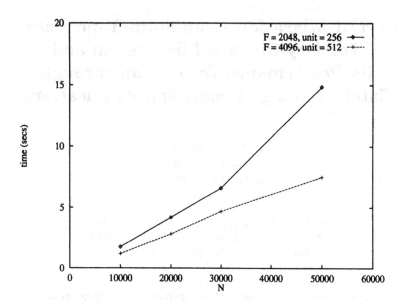

Fig. 11. Length versus Unit ($m = 1$)

Partial Migration in an 8mm Tape Based Tertiary Storage File System and its Performance Evaluation through Satellite Image Processing Applications

Kazuhiko Sako*
Toshihiro Nemoto
Masaru Kitsuregawa
Mikio Takagi

Institute of Industrial Science, The University of Tokyo
7-22-1, Roppongi, Minato-ku, Tokyo, JAPAN
{nemoto,kitsure,takagi}@tkl.iis.u-tokyo.ac.jp

Abstract. Recent attention on global environmental changes has stimulated the development of large scale global information systems. Satellite images play a very important role for understanding these global changes. However, the data size is very large and current commercial hierarchical file management systems are not efficient enough to handle such huge data files. Migration of a whole file from tape to disk takes a very long time. Usually users are not interested in a whole image but in only a small portion of it. Thus, there is no need for full migration.

We designed and implemented a partially migratable file system based on 8mm tape robotics. The file system migrates only the necessary portion of a file onto the disk. Two real application programs: radiometric/geometric correction and NDVI (Normalized Difference Vegetation Index) generation, were chosen and executed using our experimental file system. Large performance improvements were achieved when compared to the conventional file level migration scheme.

1 Introduction

Recently, global information systems have attracted strong attention since global changes such as the green house effects and enlargement of ozone hole are becoming non-negligible. At our institute, we have so far received more than 20,000 scenes from the NOAA satellite over the last ten years. Each scene comes to around 100 MBytes. Since the storage technology was immature when we started to receive the satellite imagery, we could not have the database on-line, thus we stored all of the imagery on AMPEX analog tapes. However recently, very inexpensive high density 8mm tapes have become available. So we initiated a project with the objective that all the imagery acquired so far will be available on-line

---

* This work was done while he was a student at Univ. of Tokyo. Currently he belongs to Hitachi, Ltd.

through the internet so that any scientists can easily retrieve and use the imagery. We designed a 200 cartridge tape robotics library, which has a larger capacity than the one provided by Exabyte Corp. and in addition has a cassette migration mechanism. At present four tape libraries are installed in our institute. The total capacity of the system comes to almost 6 TBytes.

In this paper, we report on the effectiveness of the partial migration mechanism employed in our system. Usually, users are interested in some small portion of an image rather than the full-size image, such as a particular area centered around a specific city. If the file system can migrate only the necessary portion of the file, I/O efficiency would be much improved. On the other hand, currently available hierarchical file management systems can support only file level migration. The unit of migration is a whole file. Even if the application is interested in just a small portion of the file, all of the data has to be migrated from the tape to the disk system. Migration of the whole file usually takes a very long time. Partial Migration helps to reduce the time to migrate significantly.

In addition to performance improvements, space efficiency also will increase, since only the necessary data is loaded onto the disk. For example, users are usually not interested in the areas covered by clouds. The unnecessary areas covered by clouds do not occupy disk space in our scheme.

In this paper, we present the organization and performance of PFS (Partially Migratable File System) which we designed and implemented for our satellite imagery database system. Two major applications, radiometric/geometric correction and NDVI (Normalized Difference Vegetation Index) generation, were chosen and executed on our experimental file system. Large performance improvements were achieved compared with the conventional file level migration scheme.

Although global information systems have recently attracted strong attention, little research has been done on the design of the file system for huge image repositories. One of the major efforts is the SEQUIOA 2000 under development at UC Berkeley in the U.S.[1][2]. NASA is also one of the major sites archiving huge amounts of information[3]. Both are attempting to build file systems to support the EOS (Earth Observing System) project, where it is expected that 1 TBytes of satellite data per day will be received by the year 2000. But none of them are considering a partial migration scheme. In this paper, we concentrate on the partial migration issues of our system, even though there are other interesting features of our system, such as cassette migration and intelligent communication through an ATM network.

The next section identifies the problems of current hierarchical file management systems for building huge image file databanks based on tertiary storage devices. We also explain our design principles and some features of our system. Section 3 describes the organization of our experimental file system named PFS:Partially Migratable File System. For analysis, we chose two typical application programs in earth engineering. The performance of PFS is evaluated using these two applications in addition to fundamental measurements. Section 4 clarifies the effectiveness of our system through performance evaluation. Section 5 gives the conclusion and current status of our project.

2 System Design

In this section, several design consideration will be given, which we faced in the design of the tertiary storage file system for our satellite imagery database system.

2.1 User Transparency

From the user's point of view, the interface should not be changed at all. Namely, the same access interface as the UFS (Unix File System) file system should be provided. Transparency to the different devices should also be kept. Users are not interested in the internal behavior of the file system. That is, migration should be completely hidden from the users. The portions of the file being accessed appears to reside on disk to a user program. The organization of the file system to achieve this transparency will be given in Section 3.

2.2 From a Hierarchical File Management System to a Partially Migratable File System

At present, database management systems (DBMS) treat tertiary storage devices as back-up devices. DBMS cannot directly manipulate the records stored on the tapes. On the other hand there are several hierarchical *file* management software packages commercially available, these systems do not understand the internal structure of a file. Since their unit of migration is a file, application programs can be invoked after the whole file is staged onto disk. For applications such as satellite image processing, the size of a file is usually very large. Migration of a file from the tape to the disk takes a very long time. This holds true for not only satellite applications but also for all applications which handle large files. Applications usually do not necessarily need to access the whole file, but only some small portion of it. Currently available commercial hierarchical file management systems have to wait a long time for the whole file migration to compete, which results in low efficiency.

We could resolve this problem through several approaches. One extreme approach is to build a DBMS which can directly manipulate the data on the tertiary storage device. Although this is sound and is not impossible, implementation would be very expensive. Also, the tuple/attribute manipulation routines would be very complex. However, if we focus on remote sensed image applications, the structure of the data is rather simple. Thus, powerful record manipulation capabilities are not required. Thus we could choose another alternative, a partially migratable file system.

For remote sensed imagery, the file consists of an arbitrary number of lines. A line has a clear semantics, one scan of the earth by a sensor. The size of the line is fixed. The line is a unit of processing and a unit of calibration. Thus it is relatively easy to locate a specific line in a file. If the file system can selectively transfer the lines from/to the tape to/from disk, its migration efficiency significantly improves. In other words, if we have a file system that can migrate a portion of a file which can be specified by a relative byte address from the start of the file, we could run our existing application program much faster than using currently available commercial hierarchical file systems.

We call the mechanism where the file system can migrate a portion of a file from/to the tape to/from disk the partial migration scheme. We call a file system which utilizes the partial migration schema a partially migratable file system.

2.3 Consideration for Further Performance Improvements

(a) Prefetch

Many application programs for remote sensed imagery tend to process the images sequentially. Thus, block prefetch usually works very effectively. While a user program processes the current block, the file system prefetches the next block from the tape to disk. This overlaps application program execution with I/O accesses against the tertiary storage devices. Block prefetch is introduced in our file system. As you will see in Section 4, the prefetch mechanism improves performance.

(b) Dedicated Device Driver for Quick Positioning

The mt tape driver in Sun 4.1.3 OS which we used for the pilot system is not very sophisticated. To seek the tape using mt usually takes a very long time, which is because mt checks every file mark for all files before the target file. It cannot skip files to the destination file on the tape. In order to resolve this problem, we developed our own driver for the 8mm tape recorder. The Exabyte 8500 recorder provides a 75 times faster skip read operation. Our drive fully utilize this facility and can locate the specified file position very quickly.

(c) Striping

In addition, each jukebox can be equipped with two recorders. To improve the data transfer bandwidth, two way striping is employed. Since only one robot arm is available, two cassettes are loaded into the recorders sequentially. But data transfer can be performed in parallel.

3 Partially Migratable File System PFS

3.1 Organization of PFS

Fig. 1 shows the components of the software in PFS. The three major components include the device driver, the I/O library, and the migrator. Drivers provide the

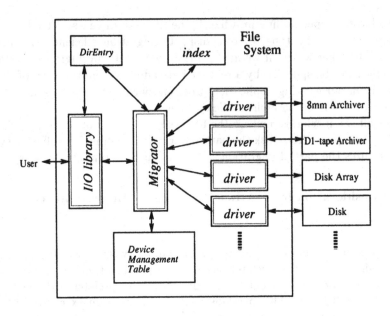

Fig. 1. The components of the software in PFS

access methods for each of the specific storage devices. User can access these devices with much higher abstraction through I/O libraries. Most of the file management is done by the migrator. The migrator keeps track of the amount of free storage on each device and controls the file migration between devices. The migrator maintains the access frequencies for each block of each file as well as the storage hierarchy. That is, frequently accessed images are on the disks and infrequently accessed ones remain on the tapes.

To achieve file location transparency, that is, to handle the file as if on a disk under UFS is useful, but its implementation is expensive. So, PFS adopts a simpler and more practical way. A user can see a PFS file as if it were on disk when the user employs our own I/O libraries.

The migrator is running as a server process. It manages all of the devices of PFS and provides a hierarchical file system to the client processes. Client processes, which usually use special I/O libraries for PFS, and the migrator communicate with each other using a socket. To operate the hierarchical file system and to perform partial migration, the migrator uses a directory entry, block index and device management table.

In PFS, the same tree structure as in UFS, that is, hierarchical directories and files, is offered. Therefore users can deal with a file on PFS using easily understandable directory and file names abstractions just as in UFS. To realize a tree structured directory and a file name on PFS, directory entries on PFS are stored as files on UFS with the same file name as that of PFS. File name, user ID and group ID of owner, file mode (permission), access time, modify time etc.

of the directory entry file are stored for the UFS file. The directory of each file in PFS corresponds to the directory of the directory entry file in UFS. Accordingly, the tree structured directory of UFS is also maintained in PFS. In each directory entry file, a magic number, a file index number and the file size are kept. The magic number is used to indicate that it is a directory entry file. The file index number, which intuitively corresponds to an i-node in UFS, represents a pointer to the file in PFS. The file size is the size of a file in PFS in bytes.

In PFS, all files are divided into several blocks on each device. The partial migration is performed in units of blocks. The index has information about the location of all file blocks on PFS. At the time of partial migration, it is used by the migrator in order to determine the device where the requested block is.

Device management tables hold the status of each device. A device management table exists for each PFS device. The migrator determines the free space of each device by examining the device management table and uses this information to determine when to migrate files from the secondary storage file system to the tertiary one.

The I/O library provides the access methods for PFS to user programs. The I/O library provides functions similar to the UFS system calls for operations like open, read, write and so on. The I/O library also provides communication with the migrator.

The interface of the functions in the I/O library for PFS are much the same as the corresponding system calls in UFS. Accordingly, the user can access files on PFS without worrying about whether it is on PFS.

3.2 Example of read/write Operation on PFS

Here, examples of the read/write operation are shown. Read/write on PFS is performed as follows.

Read

1. I/O library linked to the client process sends the migrator a request message to read the file blocks the user is requesting.
2. Migrator gets the index number of the file block from the directory entry, and refers to the index for the location where the file block exists.
3. When the requested file block is on secondary storage, the migrator sends information about the file block to the I/O library.
4. When the requested file block is not in secondary storages, the migrator transfers the requested file block from tertiary storage to secondary storage and returns the information about the file block in secondary storage to the I/O library.
5. User reads the data through the I/O library. Simultaneously, the migrator moves the next file block from tertiary storage to secondary.

Write

1. I/O library linked to the client process sends the migrator a request message to write the file block the user specified.
2. Migrator gets the index number of the file block from the directory entry, and refers to the index for the location where the file block exists.
3. When the file block already exists on PFS, that is, when the user requests overwrite, the migrator migrates the block if necessary and sends the information about the file block to the I/O library.
4. When the file block is not in PFS, the migrator looks for a free area in the secondary storage referring to the device management table and returns information about the file block when it is found.
5. User writes the data through the I/O library.

4 Performance Evaluation of the PFS Experimental System

After the organization of the pilot system is given, the performance evaluation results are presented. First, fundamental performance characteristics such as tape seek speed and naive sequential read speed are measured. Then real applications such as the geometric and radiometric correction program and the vegetation index generation program are used to clarify the effectiveness of the partial migration mechanism.

4.1 Experimental Environments

Fig. 2 shows a block diagram of the experimental pilot system. Four robotic arm tape jukeboxes are used; each one has two Exabyte recorders and holds up to 200 8mm tapes. They are connected through a SCSI interface to a SUN SPARCstation10/40 with 64 MBytes of main memory. The capacity of a single 8mm tape is at present around 5 GBytes. The size of one NOAA satellite image amounts to about 100 MBytes. Thus a tape can store 50 scenes. One jukebox can store around 1 TByte. Current Exabyte recorders support on-the-fly compression/decompression mechanisms. Since we found that the compression ratio is not high enough for NOAA images, in this experiment, we did not use the compression capabilities.

4.2 Performance Characteristics

(a) Tape Seeking

Fig. 3 shows the tape seek times for the SCSI driver we developed and the standard "mt" command. In this experiment, several 50 MBytes files are stored on a tape one after the other. The size, 50 MBytes was chosen because we employ two-way tape striping and one image is around 100 MBytes. The horizontal axis

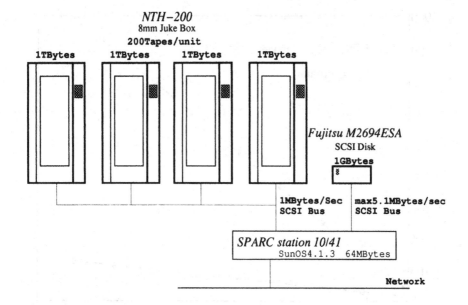

Fig. 2. The block diagram of the experimental pilot system

depicts the file number to be located. The "mt" command from SUN OS 4.1.3 appears to stop the tape at each file mark, which is why the mt seek command is very slow. Whereas the developed driver skips all the file marks before the target address and can locate the target file very quickly.

Actually, recent operating systems such as Solaris 2.3 employ similar mechanisms to ours. However, even Solaris 2.3 cannot skip to an arbitrary position within a file. Our driver directly seeks to any location of any file. Tape seek speed is 75 times faster than the normal read speed. By using this driver, the access time can be significantly improved.

(b) Tape Striping

Our 8mm tape jukeboxes have two Exabyte tape recorders in each unit. In order to further improve the throughput of the file transfer, the file is stored in a two-way striped fashion. Table 1 shows the time to read out a 100 MBytes file from the tape for striped and non-striped file systems. You can see that the transfer time decreases almost by half. For applications which handle large files, striping works very effectively.

4.3 Performance Evaluation of the PFS through Remote Sensing Application

In order to understand the effectiveness of the partial migration mechanism, an experimental file system was evaluated using real application programs for

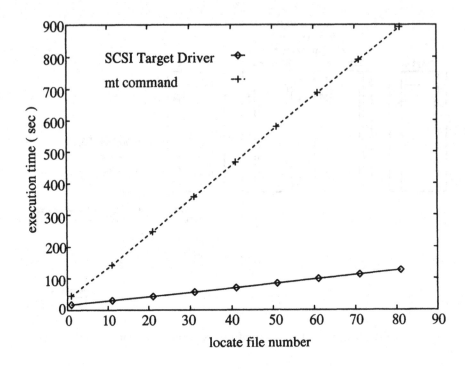

Fig. 3. Comparison of tape seek time between mt command and our own driver

Table 1. The time to read out a 100 MBytes file from the tape

File system	time (sec)	transfer speed
single device	220	465 KBytes/sec
2-way striping	116	883 KBytes/sec

remote sensed image processing. We used two kinds of programs. One extracts a specified area and applies radiometric and geometric corrections to it. The other extracts a specified area and generates a vegetation index (NDVI). The reason we picked these two applications is that the former is computation intensive, while the latter is I/O intensive. Thus, these two applications have completely different behaviors. It should be noted that these two are very frequently used application programs for earth engineering.

Table 2 shows the execution parameters. One line of the NOAA satellite image is 2048 pixels. Additional information for calibration and the 2048 pixels amounts to 22.18 KBytes per line. The number of lines for the images depends on the case. In our experiments, 4297 line images were used.

Table 2. The execution parameters

NOAA satellite image

Total file size	95307460 bytes
Size of a line	22180 bytes
Number of lines	4297 lines

File system

Image block size	1 MBytes
Disk block size	256 KBytes
8mm tape block size	256 KBytes

(a) Radiometric and Geometric Correction

This program first converts the 10 bit sensor count into a temperature. This is done pixel by pixel. Then the geometric correction routine converts the distorted image into a specified map image such as Polar-stereo or Mercator. Coast line information is used for fine fitting of the final calibration. Fig. 4 shows the execution time for the program. The horizontal axis denotes the size of the extracted area in degrees. For example, 5 degrees means the extracted area is a rectangle centered at Tokyo with ± 5 degrees longitude and latitude. The size of the output image is always set to 512 × 512 pixels independent of the size of the extracted area. The vertical axis shows the program execution time.

For ordinary file systems, the whole file on the tape must first be migrated onto the disk system. Thereafter the application program begins to process the data on the disk, which is denoted by the uppermost line of the graph, "file on disk + file migration time". On the other hand if the file is already on the disk, no migration is necessary. The line, "file on disk" denotes this situation. Between these two lines, you can see another two lines. The line named "file on 8mm with prefetch" shows the execution time for the radiometric and geometric correction program against the file on the 8mm tape using PFS. In this case, the prefetch mechanism is invoked. It can be seen that these results are very close to the results for when the file is on disk. Since the application program is CPU intensive, block prefetch works very effectively. Partial migration with prefetch behaves almost ideally as if the file were on the disk. Partial migration without prefetch is also shown in this figure. By comparing prefetch and non-prefetch, we can see that prefetch works very well.

Fig. 5 shows the execution time relative to the full migration scheme. If the size of the extracted area is relatively small compared with the whole file, the figures verify that execution time could be substantially reduced through the partial migration scheme.

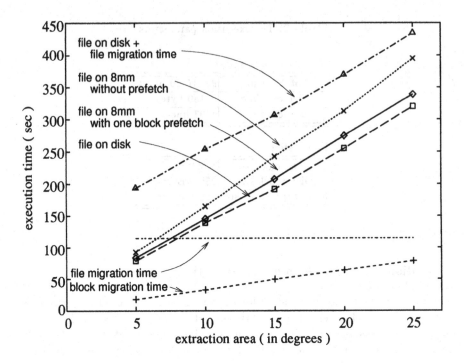

Fig. 4. The execution time of the extraction and correction program

(b) NDVI Generation

Normalized Difference Vegetation Index (NDVI) is being interpreted as an indicator of the degree of photosynthetic activity and is widely used to observe the changes in land covered with vegetation. NDVI is calculated from visible and near-infrared channels as follows.

$$NDVI = \frac{NIR - VIS}{NIR + VIS}$$

VIS : digital count of visible channel (channel 1)
NIR : digital count of near-infrared channel (channel 2)

Fig. 6 shows the execution time of the NDVI generation program. The horizontal axis denotes the number of lines from which the NDVI is generated. The vertical axis denotes the execution time. The uppermost line, "file on disk + file migration time", denotes the ordinary full file migration scheme. The horizontal line at just below 120 seconds shows the migration time for the full file transfer from 8mm tape to disk. The difference between these two lines corresponds to the NDVI generation time. The line, "file on disk" shows the execution time for the NDVI application program against the file on the disk. The line "read file from the disk" shows the time required to read just the specified lines of the file

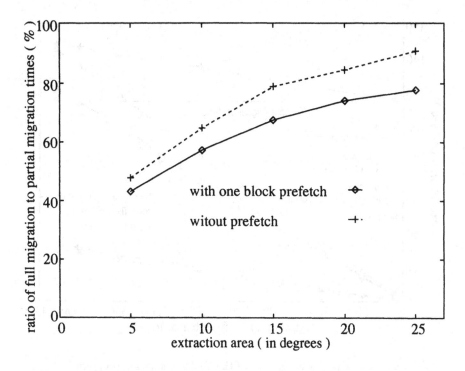

Fig. 5. The relative execution time relative to the full migration scheme

from disk. The fact that these two lines are very close to each other indicates that this NDVI program is I/O bound. In the middle you can see the line, "file on 8mm with prefetch", which shows the execution time under the partial migration scheme using PFS. In addition, "read file on 8mm tape" is also plotted. The difference between these two lines is very small, which means that NDVI generation processing can be done at nearly the same rate as the I/O rate from the 8mm tape system. The one block prefetch mechanism also works very effectively in this application. Fig. 7 shows that the execution time of the partial migration scheme relative to the full file migration scheme. As can be seen from this figure, the execution time is almost linearly dependent on the number of extracted lines. Thus by employing PFS, performance is significantly improved.

5 Conclusion

We designed the PFS file system based on 8mm tape robotics tertiary storage devices for remote sensed image processing applications. The size of satellite images is usually very large,thus the efficient handling of tertiary storage devices is an important file system factor for such applications.

We introduced a partial migration mechanism and implemented it under an

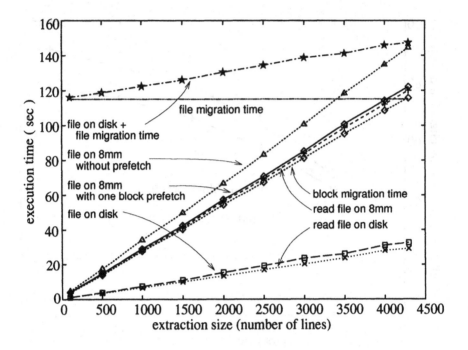

Fig. 6. The execution time of the NDVI generation program

experimental pilot system. We ran two representative applications and examined their performance. Both the radiometric/geometric correction program and the NDVI generation program should significant reductions in execution time. The performance of PFS was examined in detail.

In addition, the block prefetch mechanism, the two-way striping scheme, and the quick tape positioning driver contribute to the overall performance improvement.

From the user's point of view, the only change is to use a special I/O library. Users do not care about the underlying mechanism.

After finishing the experiments, we are now porting all the software to a SPARCcenter 2000E server running Solaris 2.4. More than 20,000 satellite images which we acquired over the last 10 years are being placed into the PSF system.

In addition to the 8mm tape robotics and disks, we plan to integrate high density and high speed D1 digital tape archivers. One cartridge stores 90 GBytes and its data transfer rate is 15 MBytes/sec. This will play the role of intermediate storage between two layers of the hierarchy. Since this device has higher bandwidth than current disk drives, we also plan to integrate RAID-3 disk arrays. At present we are now extensively doing experiments using these two new devices.

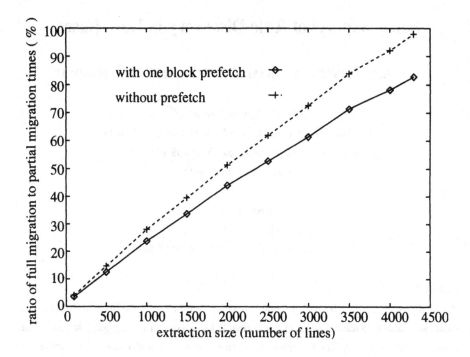

Fig. 7. The relative execution time relative to the full migration scheme

Acknowledgements

Mr. Kazuo Takahashi developed the original version of the driver for the 8mm tape recorder. This research is partially supported by the Ministry of Education, Japan. NCL Communication Inc. helped us to develop the 8mm tape based file system.

References

1. Stonebraker, M., Frew, J., Dozier, J.,"The Sequoia 2000 Architecture and Implementation Strategy", Sequoia 2000 Technical Report 93/23, University of California, Berkeley, CA, April 1993.
2. Stonebraker, M., and Olson, M., "Large Object Support in POSTGRES, Proc. 9th Int'l Conf. on Data Engineering, Vienna, Austria, April 1993.
3. Thomas S. Woodrow, "Hierarchical Storage Management System Evaluation," Third NASA Goddard Conference on Mass Storage Systems and Technologies, NASA Ames Research Center, 1993.

Some Aspects of Rule Discovery in Data Bases

L. FLEURY[(1 et 2)]   C DJERABA[(1)]   H BRIAND[(1)]   J PHILIPPE[(1 et 2)]

IRESTE, Nantes University[(1)]
La Chantrerie, CP 3003, 44087 Nantes cedex 03, France
tel : 33 40 68 30 00 fax : 33 40 68 30 66
e-mail : cdjeraba@ireste.fr

PerformanSe[(2)]
3, rue Racine, 44000 Nantes
tel 33 40 73 18 81, fax : 33 40 69 59 51

Abstract

Rule Discovery in Databases integrates machine learning, probabilistic techniques and database concepts to learn a range comprehensible knowledge in sparse, noisy and redundant data. The discovery enables the learning of rules from data and extract their underlying structure. In this paper, we present the probabilistic index and the notion of minimal set of discovered rules which enhance runtime performance, improve discovery accuracy, resist noise, converges with the size of the sample, and eliminates coarse and redundant rules. This index can be used within the framework of an incremental discovery system.

In other words, in this paper, we describe the rule intensity measurement which is an index that answers the question `What is the probability of having a rule of the form `IF premise THEN Conclusion`; the premise and conclusion are conjunctions of propositions ?`.

Keywords

Knowledge Discovery in Databases, discovered rules, rule intensity measurement, noisy, sparsity, redundancy.

1. Introduction

Together with networks, databases have become one of the most important components in information technology environment. We know, of course, why we need large databases, their main purpose is to provide facility in storing, retrieving, and manipulating large amount of information. We know too, that Databases have been designed to support lot of challenging applications. In this paper, we will focus on Knowledge Discovery in Databases (KDD) which is one of these challenge applications. Knowledge Dicovery in Databases is the search for relationships hidden in a large datatabase [Mat 93]. These relationships represent valuable, interesting and useful knowledge with respect to the end user's objectives [Fra 91], such as relationships between demographics and market factors, or between diseases and diagnoses, etc.

KDD is a form of machine learning using a large database as a training set. But the database characteristics differ from the training sets used in machine learning:

1) A database is generally very large, it contains thousands of attributes. The number of objects in databases is large [Mat 93], so querying the entire database is expensive. the large number of possible relationships prohibits the search for the correct ones by simply validating each of them [Klo 91].

2) The database is frequently updated, data is added, modified or removed. Any knowledge that was previously extracted from this database can therefore become inconsistent. So, the KDD system should consider these changes. For example, the most recent data should be valued more than older data, we can, for example, assign higher weights to recent discoveries in the discovery process [Gra 93].

3) In databases the attribute values of objects may be overlooked. The KDD system must construct descriptions even when some of the attribute values are missing.

4) Finally, the database, invariably supports, noisy (uncertain), sparse and redundant data [Pia 92] [Mat 93].

a) The data supplied in a set of objects is noisy (uncertain) because it is not entirely correct. The objects may include attributes based on measurement of subjective judgements, which induce errors in the value of attributes. Some of the objects may even be misclassified. Noise is the term used to describe non-systematic errors of this kind in the values of attributes. The database is often contaminated by errors, so the KDD system has to cope with noisy and uncertain data [Mat 93] [Pia 92]. Hence, probabilistic techniques and intelligent knowledge discovery systems should be applied to search and evaluate the reliability of discovered relationships.

b) Examples should represent a large object space. Unfortunately, some examples in a database are often sparse because they represent only a small subset of all possible objects in an instance space. So, the discovery class cannot be determined exactly, and seems vague and incorrect.

c) Finally, knowledge often reoccurs in a database. Our work is concerned with two forms of redundancy, the first one is functional dependencies, e.g A ---> B, in which a set of attributes is defined as a function of another attributes. The problem with redundant knowledge is that many implications, i.g a ---> b, a is a suprest of A and b is a suprest of B, can be mistakenly discovered, even though they are usually uninteresting to the end user because they can either be deduced from implications which belong to minimal set of implications, or correspond to functional dependencies that already exist in the database. The elimination of redundancies is essential for the gradual and explanatory characters of the knowledge.

To deals with the sparse and noisy data, our work present the intensity of implication that measures the quality of a rule extracted by the KDD system. So, the KDD system may eliminate the implications whose intensities are greater than a certain value fixed by the user.

In this paper, we will first present how to position our work in a KDD system (section 2), so we show how useful the intensity of implication is when dealing with the noisy (uncertain) and sparse data (sections 3, 4). These points are important characteristics of databases, and a knowledge discovery system has to take them into account.

2. Knowledge Discovery system

We would situate the evaluation of the implication intensity in a KDD system to handle the specific requirements for discovery in databases. An ideal model of KDD [Mat 93] may be seen as a collection of parts that together can efficiently extract significant and new knowledge from data stored in databases. This ideal model is composed of a controller, a database interface, a focus module, an extraction module, and an evaluation component. The controller controls the other components. This control is based on expert knowledge and data provided by the database and the user input. The database Interface is able to access a database, and generates database queries, such as retrieving, modifying or removing objects according to specified constraints. The focus component selects data that would be retrieved from the Database, and used by the extraction component. The extraction component is composed of a set of algorithms used to extract relationships (i.e. rules) from data stored in database. Finally, the evaluation component evaluates the usefulness of extracted relationships. Generally, it is not always possible to cleanly separate the parts of the KDD model, for example, in most KDD systems the evaluation is a part of the extraction part. That is why we found this ideal model of KDD useful in situating of our approach of intensity implication evaluations.

In other words, data is selected from the database and then computed by the extraction algorithms that returns candidate relationships. These relationships are evaluated and some are identified as significative discoveries. Our approach of the implication intensity measurement of a rule contributes to the evaluation of discoveries in a KDD system.

After extraction, a database contains many rules, we assume that discoveries are rules, many of them are not interesting, accurate and useful enough with respect to the end user's objectives [Pia 92]. The quality of each generated rule has to be verified. So, we need probabilistic tests to check if the rule actually describes some regularity in the data. So, the evaluation has to determine the usefulness of extracted rules, and decide which to save in the database. We propose a probabilistic measurement, based on the measurement of rules in a noisy and sparse database, to determine the usefulness of a rule. If a rule is not valid, by using the intensity of the implication, then it is not interesting because the intensity of the rule is lower than a certain value given by the user, and will not be saved in the database.

3. Implication intensity measurement

Before presenting how to use the implication intensity to evaluate a rule, it is important to define some concepts, that will be frequently used.

Let $E = \{e_1, e_2, \ldots, e_n\}$ be a set of objects, and $D = \{d_1, d_2, \ldots, d_k\}$ be the set of propositions which may characterise an object of E, d_i is a value of an attribute .

We look for all the conjunctions between the propositions in the form of an implication such as *Premise⇒Conclusion* in which *Premise* is a conjunction of propositions from *D*, and *Conclusion* a proposition from *D*. A rule is considered logical if it does not allow any counter-examples. When set *E* is large and/or noisy, it may be interesting to generate rules which allow a certain number of counter-examples. We then refer to probabilistic rules. In that case, we have to ask ourselves a question.

Can we associate a measurement with the *Premise⇒Conclusion* "quasi-implication", whose value would be an index of the quality of this "quasi-implication" ? The intensity of implication, an index resulting from the work of R.Gras and A.Larher [Gra 93] based on simple probability concepts, provides an answer to this question. In this paper, we use R. Gras works in the context of KDD.

3.1 Principle of the implication intensity

The cardinals of the subsets A and B are determined by the objects of the database belonging to A and B. Let us define two random sub-sets X and Y which have the same cardinals as A and B. The implication $a{\Rightarrow}b$ is denied in the sub-set $A \cap \overline{B}$. We compare the cardinal of $A \cap \overline{B}$ (given by the database) with that of $X \cap \overline{Y}$ (random variable), supposing there is no link between X and Y. If the cardinal of the set $A \cap \overline{B}$ is unusually small compared with the one we could expect from the distribution of the cardinals of $X \cap \overline{Y}$, we will accept a quasi-implication, $a{\Rightarrow}b$. The probability of $a{\Rightarrow}b$ is measured by the complement on one of the measurement of the probability inferred from the comparison between the aleatory cardinal of $X \cap \overline{Y}$ and that of $A \cap \overline{B}$.

☞ Intuitively, the value of the intensity of implication measures the degree of statistical astonishment of the size of $A \cap \overline{B}$, considering the sizes of *A*, *B* and *E*, and assuming there is no a priori link between *A* and *B* [Gra 93].

3.2 Definition of the implication intensity

Card(X $\cap \overline{Y}$) may obey a hypergeometric distribution, but when we consider certain conditions (very large databse), this distribution should be approximated by Binomial distribution.

✱ The random variable $Card(X \cap \bar{Y})$ obeys a Binomial distribution because :

    (i) the random variable can occupy only two complementary states

        • to be described with the conjunct of propositions $x \wedge \bar{y}$

        • not to be described with the conjunct $x \wedge \bar{y}$

    (ii) the random variable has a constant probability of occupying one of the two states.

✱ With other conditions, it is proved that the Binomial distribution may be approximated by Poisson distribution. In other words, the random variable $Card(X \cap \bar{Y})$ obeys the Poisson rule with the parameter λ. Its formula is given below : $\Pr\left[Card(X \cap \bar{Y}) = k\right] = \dfrac{\lambda^k}{k!} . e^{-\lambda}$, and $\lambda = n.p(a \wedge \bar{b})$, with : $n = Card(E)$ and $p(a \wedge \bar{b})$, the probability for the random variable to be described with the conjunctions of propositions $a \wedge \bar{b}$.

Let λ be : $\lambda = Card(E).p(a \wedge \bar{b}) = Card(E).p(a).p(\bar{b})$ where $p(a) = \dfrac{Card(A)}{Card(E)}$, and $p(\bar{b}) = \dfrac{Card(\bar{B})}{Card(E)}$.

then $\lambda = \dfrac{Card(A).Card(\bar{B})}{Card(E)}$ (1)

✱ In short, the random variable $Card(X \cap \bar{Y})$ obeys the Poisson distribution with the parameter λ : $\lambda = \dfrac{Card(A).Card(\bar{B})}{Card(E)}$.

☞ The quality of implication is even better if the number of counter-examples is smaller than the expected number, in other terms if the quantity $\Pr(Card(X \cap \bar{Y}) \leq Card(A \cap \bar{B}))$ is small. It is the observation "smallness" of $Card(A \cap \bar{B})$ compared with $Card(X \cap \bar{Y})$ which is taken as a basic pointer of the quasi-implication $a \Rightarrow b$. The implication intensity is then defined with the function φ_1 : $\varphi_1(a,\bar{b}) = 1 - \Pr\left[Card(X \cap \bar{Y}) \leq Card(A \cap \bar{B})\right] =$, with

$\Pr\left[Card(X \cap \bar{Y}) \leq Card(A \cap \bar{B})\right] = \sum_{i=0}^{Card(A \cap \bar{B})} \dfrac{\lambda^i}{i!} . e^{-\lambda}$ (2)

In the cases where the cardinalities of sets A, B and E are so that $\lambda \geq 3$, the probability calculation of the random variable $Card(X \cap \bar{Y})$ according to the

Poisson distribution with parameter λ can be made from the variable $Q\ (a,\bar{b})$

observation of $Q\ (a,\bar{b}) = \dfrac{Card(X \cap \bar{Y}) - \lambda}{\sqrt{\lambda}}$ obeying the normal distribution with 0 and

variance 1 ($N(0,1)$).

As a conclusion : We can say that the implication $a \Rightarrow b$ can be admitted at a level of

confidence 1-α if and only if :   • If $\lambda \leq 3$, then $\varphi(a,\bar{b}) = \varphi_1(a,\bar{b}) = \left[1 - \sum\limits_{i=0}^{Card(A \cap \bar{B})} \dfrac{\chi}{i!}.e^{-\lambda}\right] \geq 1 - \alpha$

$$(3)$$

• If $\lambda \geq 3$, then $\varphi(a,\bar{b}) = \varphi_2(a,\bar{b}) = \left[1 - Pr\left[Q(a,\bar{b}) \leq q(a,\bar{b})\right]\right] \geq 1 - \alpha$ $\qquad(4)$

To appreciate this index we will compare it with others which are widely represented in the literature of learning machine.

3.3 Comparative study of indices
The four indices that we will compare are :

• **the intensity of implication** : $1 - \sum\limits_{i=1}^{card(A \cap \bar{b})} \dfrac{\chi}{i!}.e^{-\lambda}$ with $\lambda = \dfrac{Card(A) \times Card(\bar{B})}{Card(E)}$,

• **ratio of the number of examples to the number of counter-examples** : $\dfrac{Card(A \cap B) - Card(A \cap \bar{B})}{Card(A \cap B)}$; We can find this index or a similar one in [Did 91], [Bao 91].

• **the conditional probability** : $P(b|a) = \dfrac{Card(A \cap B)}{Card(A)}$; The conditional probability is

employed in [Gan 88], [Cen 87].

•   the   **J-Measure**   [Goodman   and   Smyth   89]   :

$J(B;A) = p(a) \times \left[p(b|a) \times \log_2\left(\dfrac{p(b|a)}{P(b)}\right) + p(\bar{b}|a) \times \log_2\left(\dfrac{p(\bar{b}|a)}{p(\bar{b})}\right) \right]$; This index is built from entropy

that is also used in [Qui 87], [Clar 88].

The study, we are going to carry out, will be done by simulated comparative study. Those indices depend on six cardinals which are: $Card\,(A), Card(\bar{B}), Card(B), Card(A \cap B)$, $Card\,(A \cap \bar{B})$ and $Card(E)$. To study the influence of cardinals on indices, we will make one of them vary and consider the others constant. To illustrate the behaviour of the intensity of implication we are going to plot its graph in accordance with the variation of the cardinal of the set $A \cap B$. But as a result of $Card\,(A \cap B) + Card(A \cap \bar{B}) = Card(A)$, it is not possible to make $Card(A \cap \overline{B})$ constant. We make the parameters vary according to the following schematic representation (Figure n°1).

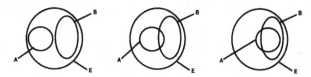

- Figure n°1 : Evolution of the cardinals of sets $Card\ (A \cap B)$ and $Card(A \cap \overline{B})$ -

The intensity of implication is slow to react initially and then changes fast whereas the conditional probability and the ratio of examples to counter-examples vary in a linear or hyperbolic way. As for the J-Measure, its variation can be broken down into two distinct periods. The first one, in which the Right-Part prevails (evaluation of the robustness of $a \Rightarrow \overline{b}$) decreases until it becomes null in $Card(A \cap B) = p(b).Card(A)$. The second one, in which the Left Part prevails, increases according to the robustness of the implication $a \Rightarrow b$.

Example : "Ferraris are usually red" can be expressed by the quasi-implication : *if Ferrari=true then Red=true*. If we see a green Ferrari then we would rather believe that we are wrong than question our first knowledge. We would probably have to live through the same experience several times before we started having doubts about our first knowledge. From the moment when doubt appears, the refutation of our previous rule will not need as many counter-examples as those necessary the first time.

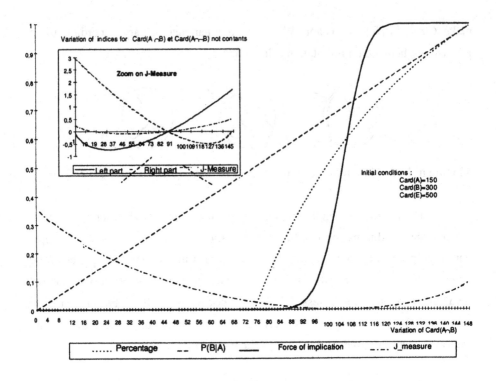

- Figure n°2 : Graph n°3 -

The evolution of our certainties is modeled by the intensity of implication according to the curve of graph n°3. Conversely the curve representing the ratio of examples to counter-examples is linear which means that there would be an equal loss of trust with each counter-example.

3.4 Behaviour of the implication intensity for a population widely described with a proposition

We deal with the kind of situation where an attribute is almost always identical for a given set of examples. *Example* : The walker makes the experiment again in another forest (Figure n°3).

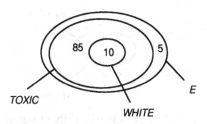

- Figure n°3 : *TOXIC* is nearly equal to *E* -

✖ Let us calculate the conditional probability.

$$P(Toxic|White) = \frac{Card(TOXIC \cap WHITE)}{Card(WHITE)} = \frac{10}{10} = 1, \quad \varphi(White, \overline{Toxic}) = 1 - \sum_{i=0}^{0} \frac{\lambda^i}{i!}.e^{-\lambda} = 1 - \left[e^{-4.5}\right] = 0,40$$

The cardinal of the set *TOXIC* (*Card(TOXIC)*=95) is great compared to the one of *E* (*Card(E)*=100), so a mushroom picked at random among *E* set is likely to belong to the set *TOXIC* (Pr(*Toxic*)=0,95). The probability that all the elements of the set *WHITE* (*Card(White)*=10) belong to the set *TOXIC* is not negligible ($Pr(Card(WHITE \cap \overline{TOXIC}) = 0) = 0,60$). So we cannot reasonably conclude in favour of the existence of an implicative link between the propositions *White* and *Toxic*. That is why the intensity of implication associated to the rule *White* ⇒ *Toxic* is low. If sets *A* and *B* are fixed, the intensity of implication decreases when set *B* is directed towards set *E*. This does not apply to the conditional probability which remains unchanged. In the case when *B=E*, the following property has been proved : For *Card (E)* and *Card (A)* fixed, we have : $\lim_{Card(B) \to Card(E)} \varphi(a, \overline{b}) = 0$ (5)

4. Intensity of implication and sparse data

The intensity of implication increases with the representativeness of the population samples which are studied. This notion particularly interesting in the case of sparse data, will be studied by comparing the intensity of implication to conditional probabilities. Let us take notice of the remarks resulting from this comparison would have been strictly identical with the ratio of the number of examples to the number of counter-examples. *Example* : Let us look for the regularities in physical features of 10 persons. We will be more particularly interested in the implicative link between the two propositions *Red-Haired* and *Blue-Eyed*.

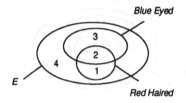

- Figure n°4 : One sample not really representative -

✖ Let us calculate the conditional probability :

$$P(BlueEyed|RedHaired) = \frac{Card(RedHaired \cap BlueEyed)}{Card(RedHaired)} = \frac{2}{3} = 0,67$$

$$\varphi(RedHeared, \overline{BlueEyed}) = 1 - \sum_{i=0}^{1} \frac{\lambda^i}{i!}.e^{-\lambda} = 1 - [e^{-1,5}.(1+1,5)] = 0,44 \cdot$$ The reduced size of the sample does not allow us to conclude that there is an implication from *Red-Haired* towards *Blue-Eyed*. Let us multiply the sample size by 10 (Figure n°5) and keep the proportions in E set.

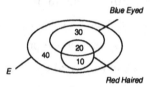

- Figure n°5 : Sets *E, Red-Haired, Blue-Eyed, Raid-Haired∩Blue-Eyed* -

✖ One more time, we calculate the intensity of implication.

$$P(BlueEyed|RedHaired) = \frac{Card(RedHaired \cap BlueEyed)}{Card(RedHaired)} = \frac{20}{30} = 0,67$$

$$\varphi(RedHeared, \overline{BlueEyed}) = 1 - \sum_{i=0}^{1} \frac{15^i}{i!}.e^{-15} = 0,90$$

This time, the sample is large enough to conclude that there is an implication *Red-Haired* to *Blue-Eyed*. We can notice that the conditional probability of *Blue-Eyed* given that *Red-Haired* remains unchanged.

☞ The intensity of implication seems to be particularly adapted to algorithms which deal with disparate volumes of data.

5. Intensity of implication and noisy data

The measure of the quasi implication is well adapted to the noisy data because the counter-examples do not necessarily delete the implication. Moreover, using this measure allows us to know whether the implication remains valid when the number of counter-examples increases. Noise is a prominent problem that must imperatively be taken into account in a KDD system. We are going to study the variation of $q_{(a,\bar{b})}$ with that of $Card(A \cap \bar{B})$. Let us consider a real situation for which we know $Card(E)$, $Card(A)$, $Card(\bar{B})$ and $Card(A \cap \bar{B})$.

We have : $q_{(a,\bar{b})} = \dfrac{Card(A \cap \bar{B}) - \dfrac{Card(A).Card(\bar{B})}{Card(E)}}{\sqrt{\dfrac{Card(A).Card(\bar{B})}{Card(E)}}} \leq \beta$ and $\varphi(a,\bar{b}) \geq 1 - \alpha$

What error of measurement on $Card(A \cap \bar{B})$ does the implication still allow (with 1-α, level of confidence), even though $Card(E)$, $Card(A)$, and $Card(\bar{B})$ remain constant. In other words, what is the sensitiveness of the implication to a variation of $Card(A \cap \bar{B})$? It has been proved [Gra 93] that the implication is still valid ($\varphi(a,\bar{b}) \geq 1 - \alpha$) for any variation k of the size, such as :

$$Card\ (A \cap \bar{B}) \leq k \leq \beta \sqrt{\frac{Card(A).Card(\bar{B})}{Card(E)}} + \frac{Card(A).Card(\bar{B})}{Card(E)} - Card(A \cap \bar{B}) \qquad (6)$$

☞ According to the learning parameters linked with the field, it is possible to measure the resistance to noise of extracted implications. Concerning our example, the generated rules are very noise-proof.

6. Conclusion

Knowledge Discovery in Databases (KDD) is the search for relationships, such as rules, in large databases. These relationships should be interesting and useful with respect to the end user's objectives. One of the main problems, that a KDD system has to cope with, is the noisy (uncertain), sparse and redundant data. In this paper, we have found a way for measuring the sensitiveness of implication to size variations. We have shown how important the notion of intensity of implication is in the measurement of the resistance to noise of extracted rules, and in the case of sparse data. If, for every rule, we combine the maximum limit of counter-examples, the addition of an example which transgresses one rule of the system or more, does not necessarily imply the redesign of a new rule system. Thus, the cost of generating a new system is spared. This approach is used in the domain of human resources management by PerformanSe company [Gra 93]. Furthermore, we have found a way of eliminating redundant rules discovered in the database. To achieve this goal, we have adapted an algorithm for finding a minimal set of dependencies to proposition rules. Eliminating rule redundancies make rules discovered less voluminous, more evolving, and better adapted to explanation.

References

[Anw 92]   T. M. Anwar, H. W. Beck, S. B. Navathe. "Knowledge Mining by Imprecise Querying: A Classification-Based Approach, "IEEE 8 th Int. Conf. on Data Eng. Phoenix, Arizona, Feb. 1992.

[Bri 86]   Briand H., Crampes J. B., Hebrail Y., Herin-Aime D., Kouloumdjan, Sabatier R., "Les systemes d'Information", DUNOD edition, 1986.

[Cla 88]   Clark P., Niblett T. "The NC2 induction algorithm. Machine Learning", 3. p 261-283 1988.

[Cen 87]   Cendrowska J. "An Algorithm for inducing modular rules", Int J. Man-Machine Studies, p 349-370, 1987.

[Did 91]   Diday E. Mennessier M.O. "Analyse symbolique pour la prévision de séries chronologiques pseudo-périodiques", Induction symbolique et numérique à partir de données, p 179-192,Cépaduès-éditions, 91.

[Fra 91]   Frawley W. J., Piatetsky-Shapiro, C. J. Matheus, "Knowledge discovery in databases: an overview", in Knowledge Dicovery in Databases. Cambridge, MA: AAAI/MIT, 1991, pages 1-27.

[Gan 88]   Ganascia J. G., "CHARADE : A rule system learning intelligence", IJCAI, Milan, Italy, Août 1987.

[Goo 89]   R. M. Goodman, P. Smyth, "The induction of probabilistic rules set - the rule algorithm", Proceedings of the sixth international workshop on machine learning, Splatz B. ed., p 129-132, San Mateo, CA Morgan Kaufmann 1989.

[Gra 93]   Gras R., Larher A. "L'implication statistique, une nouvelle méthode d'analyse de données", Mathématiques, Informatique et Sciences Humaines n°120.

[Bao 91]   Ho Tu Bao, Tong Thi Thanh Huyen, "A method for generating rules from examples and its application", Symbolic Numeric Data Analysis And Learning, p 493-504, Nova Sciences Publishers 1991.

[Kod 93]   Kodrattof Y., Tecuci G., "Techniques of design and DISCIPLE learning apprentice", Knowledge acquisition and learning, Kaufmann edition, pages 655-668, 1993.

[Klo 91]   W. Klosgen, "Visualization and adaptivity in the statics interpreter EXPLORA", in Workshop Notes from the 9th Nat. Conf. Art. Intell.: Knowledge Discovery in Databases. American Association for Artificial Intelligence, Anaheim, CA, July 1991, pages 25-34.

[Mat 93]   C. J. Matheus, P. K. Chan, G. Piatetsky-Shapiro, "Systems for Knowledge Discovery in Databases", IEEE Trans. Knowl. Data Eng., vol 5, n 6, 1993.

[Qui 87]   J. R. Quinlan,"Generating Production Rules from Decisions Trees". The 10 th International Conference on Artificial Intelligence, p 304-307, 1987.

[Seb 91]   M. Sebag, M. Schoenauer, "Un réseau de règles d'apprentissage", Induction symbolique et numérique à partir de données, p 241-255, Cépaduès-éditions, 91.

Telesensation
Fusion of Multi-Media Information
and Information Highways

Nobuyoshi Terashima

president

ATR Communication Systems Research Laboratories

E-mail terasima@atr-sw.atr.co.jp

Abstract

Telesensation is a concept of combination of virtual reality and telecommunications. Through it, an image of a scene from nature or an image of a museum exhibit from a remote place is instantly transmitted over high speed transmission network to a viewer.

Displaying such an image stereoscopically, the viewer can enter the scene which is a virtual world and walk through it. They can not only walk through but also touch the leaves on a tree and a wall in the museum as if they were in a same place.

Therefore telesensation can smash the bonds of time and space.

In this paper, the concept and technologies of telesensation are introduced.

Introduction

In many countries, especially advanced nations, information superhighway projects are going on.

Through information highways, multi media information - voice, image, video and so forth are instantly transmitted from remote place. Telesensation is a concept of combination of virtual reality and telecommunications.

Through it, viewers can go to any place where they would like to go as if they went there actually.

Telesensation can provide a variety of services such as television type services, interactive services and teleconferencing services.

Therefore it will have many possibilities of creating a new industry.

1 Telesensation concept

Fig.1 shows an image of telesensation concept. The right picture shows an scene of a street in Stockholm. The scene is taken by a camera and recognized by computer vision technology.

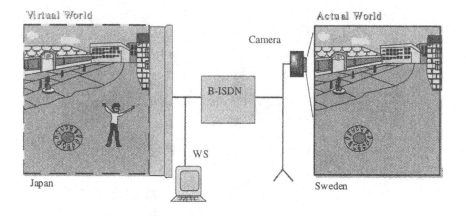

Fig. 1 Telesensation

The information is transmitted over an information highway to Japan.

In Japan, the scene is reproduced and displayed on the screen. The viewer in Japan can eater the street and walk through.

He can go behind the building and see what it looks like.

He go to the entrance of the building, open it and go inside just as he does in Stockholm. In this way, telesensation can smash the bonds of time and space.

2 Telesensation Service

Telesensation services are classified into following field: television type services, interactive services and teleconferencing services.

3 Television Type Services

Video pictures of a remote place are transmitted over information highways to a viewer. Displaying such images stereoscopically, the viewer can enter the scene, enjoy watching and walk through.

These services include 3-D video theaters, video murals and fashion shows.

Fig.2 shows an image of video murals. The scenes of a mountain, a sea and a city area are transmitted over high speed transmission networks to this room.

The scenes are reproduced and displayed stereoscopically to viewers. Viewers can enter the scenes and walk through.

Fig. 2 Video Murals

4 Interactive Services

Video pictures of a remote place are instantly transmitted over information highways to viewers.

Displaying such images stereoscopically. Viewers can enter the scenes and walk through. Not only walking through, but they can handle the objects in the scenes.

These services include teleshopping, simulated experiences of golf lessons.

Fig.3 shows an image of teleshopping service. They can have a front view, a side

Fig. 3 Teleshopping

view and a back view of a new cloth. She can try to wear a new cloth using her human form and see how it works and what it looks like.

5 Teleconferencing Service

In this system, participants from remote places, in reality their images, are brought together into a virtual conference room where they can hold a meeting as well as a cooperative work, as if they were gathered in a same place.

Fig.4 shows an image of virtual space teleconferencing environment.

Fig. 4 Virtual Space Teleconferencing system

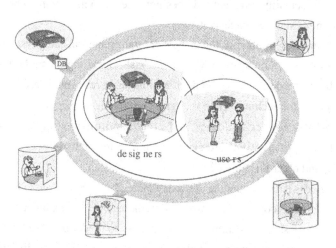

Fig. 5 Cooperative Work

A participant is in front of the screen. On the screen, images of three participants are displayed stereoscopically. In front of them, a virtual car is displayed. They can handle a car by hand gestures, spoken languages and others. They can move the car, turn it around and change the shape of the car and the color.

Fig.5 shows an image of a cooperative environment.

A group of designers are brought together into a virtual design room and talk about designing a car.

They would like to know about users' opinion. On the other hand, a group of users area brought together into a virtual meeting room. They are talking about a new car they would like to have.

Then both groups are brought together into a common place, then they can talk each other.

The designers can get opinions from users. According to them, they design a car. They show it to the users.

If it is ok, then a car is manufactured. In this way, the customization of manufacturing of a car is performed.

6 Implementation

6.1 Real-time object image recognition

(1) Image recognition

In order to display images of a highly realistic environment from the observer's perspective, cameras are placed around the targeted natural and physical objects, and a method for switching the cameras position according to observer's perspective is being considered. In actuality, this method does not offer sufficient realism because of the lack of continuity during image switching.

To overcome these problems, the images of targeted natural and physical objects are first placed into a computer using Computer Vision, and then a method that creates from the observer's perspective and displays images in real time is used.

This requires the technology to recognize targeted images, together with image creation and display technologies that create and display appropriate images, viewpoint and perspective detection technologies that detect the observer's perspective, and other technologies.

Fig. 5 provides details about the above technologies as they relate to human figures. Human figures are recognized and the wire frame model and texture are obtained and stored in the work station. Human figures are easy to model because they have many features in common, but natural and other physical objects are quite difficult to model because they come in all shapes and sizes. If a model can be created, the model is treated like a mannequin, and is then an easily recognizable target. However, new technologies will have to be developed in order to recognize natural and other physical objects that are difficult to model.

(2) Recognizing movement information

Information for head, hand and other movements by human figures will be detected in real time. Research is currently underway on non-contact movement detection methods as well as on detection methods in which sensors are placed on numerous parts of the human body.

Movement information for natural and other physical objects, on the other hand, is not easy to detect. If we target a single tree, for example, some method will have to be found to detect the movement of each individual branch.

(3) Creating images

Information related to targets acquired in (1) as well as information related to movements acquired in (2) are used to create targeted images in real time using computer graphics.

(4) Displaying images

Images created in (3) are displayed on a large screen.

Since shutter glasses are used to create a three-dimensional perspective, images corresponding to the left and right lenses are switched and displayed at high speed. A three-dimensional image is obtained by viewing the image with shutter glasses.

6.2 3-sites Virtual Space Teleconferencing Experimental System

A virtual space teleconferencing system has been constructed connecting 3 sites over digital networks.

In this system, participants can have a talk and a cooperative work together, such as building a portable shine by hand gestures and voice controls as if they were gathered in the same place.

An experiment was done among Kyoto International Conference Hall, and two ATR Labs.

At each site, 3-D human model data base and workspace model data base are stored. 3-D human model data base consists of 3-D wire frame models and colour textures. Once the facial expressions are detected, then the information are sent to each site, wire frame model is deformed, texture mapped and displayed on the screen according to the viewer's perspective.

Facial expressions such as eye movement and mouth movement are detected by the

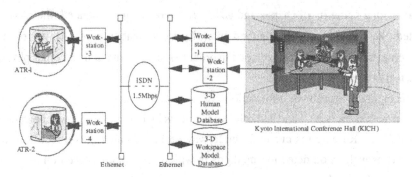

Fig.6 3-sites Virtual Teleconferencing Experimental System

marks put on the face and body motions by the sensor attached to the body. This system is shown in Fig. 6.

7 Future Prospects

We have proposed the telesensation concept in which people, animals, imaginary and other life forms work and play together in a highly realistic environment that includes real as well as unreal objects. These objects may be sent over communication lines from remote locations.

By providing an environment that goes beyond reality as just described, people will be able to perform work and activities not possible in a real environment. This system will contribute tremendously to the welfare of humankind.

An example might be providing all-night care for invalid elderly family members by artificial life forms that will notify the family, and take appropriate action at the appropriate time when there is a problem. Such a system will also provide heretofore unfathomable benefits, like helping people to function in particularly difficult environments, such as underwater, underground and in outer space.

REFERENCES

1. Terashima N.,"Basic Research on an Intelligent Communication System", IEEE Denshi Tokyo, pp.18-22, No.31, 1992.
2. Terashima N.,"Cooperative Work in Teleconferencing System with Realistic Sensation", Hypermedia '92 Bepu Bay Conference report, pp.115-119, 1992.
3. Takemura H.,"Cooperative Work Environment Using Virtual Workspace", Proc. of Virtual Computing, pp.171-181, 1992.

4. A.BENTON S., et. al.,"Interactive Computation of Display Holoranis", Proc. of Virtual Computing, pp.129-149,1992.

5. Tetsutani N.,"Stereoscopic Display Method Employing Eye-position Tracking and HDTV LCD-projector", Proc. of Int. Workshop on HDTV '92,1992.

6. Terashima N., "A New Concept for Future Telecommunication", Proc. of TAO First Int. Symp. on Three Dimensional Image Communication Technologies, S-4-5, 1993.

7. Terashima N.,"VR and Communication with a Sensation of Reality", Proc. of IEEE Denshi Tokyo, pp.21-26, No.32, 1993.

8. Terashima N.,"Virtual Space Teleconferencing System -A Distributed Virtual Environment-", Proc. of Broad band Islands '94, North-Holland, pp.35-45, 1994.

9. Terashima N.,"Telesensation -Distributed Virtual Reality- Overview and Prospects," IFIP Transactions Technology and Foundations, North-Holland, pp.49-59, 1994.

Relational Database Design
Using an ER Approach and Prolog[*]

Manuel Kolp[1] and Esteban Zimányi[2]

[1] University of Louvain, IAG-QANT, 1 Place des Doyens, 1348 Louvain-La-Neuve,
Belgium, e-mail: kolp@qant.ucl.ac.be
[2] University of Brussels, INFODOC CP 175-2, 50 Av. F.D. Roosevelt, 1050 Brussels,
Belgium, e-mail: ezimanyi@ulb.ac.be

Abstract. In the context of CASE tool development for relational database design, this paper develops a methodology that maps an enhanced Entity-Relationship (ER) schema into a relational schema and normalizes the latter into inclusion normal form (IN-NF). Unlike classical normalization that characterizes individual relations only, IN-NF concerns inter-relational redundancies. The paper formalizes sources such redundancies in ER schemas. Our methodology enhances several other proposals, in particular [10]. The paper briefly presents our implementation of the methodology using Prolog.

1 Introduction

Modern database design can be defined as the process of capturing the requirements of applications in a particular domain, mapping them onto a database management system, and tunning the implementation.

Modern methodologies generally agree on the decomposition of database design into four steps [2]: requirements specification, conceptual design, logical design, and physical design. *Requirements specification* consists of eliciting requirements from the users. *Conceptual design* develop these requirements using a conceptual model (e.g. the ER model). The output of this step is called a *conceptual schema*. *Logical design* translates the conceptual schema into the data model (e.g., the relational model) supported by the target database management system. *Physical design* transforms the logical schema into a physical schema suitable for a specific configuration.

This paper deals with logical database design. Traditionally, this activity has been based on normalization of individual relations [3]. However, classical normalization cannot characterize the relational database as a whole. Thus, redundancies and update anomalies can still exist in a set of normalized relations. Two lesser known normal forms were defined [13, 12, 10] to integrate the interaction of constraints in the database and detect these redundancies.

Nowadays, relational database design typically goes through, first, conceptual (ER) schema design and second, on translation into a relational schema.

---

[*] This work is part of the EROOS (Evaluation and Research on Object-Oriented Strategies) project, principally based at the Universities of Louvain and Brussels.

Conceptual models richer than the relational model provide a more precise and higher-level description of the data requirements and constitutes the starting point for logical design. Several methods have been proposed [18, 10] for ER to relational translations but the semantic distance between the two models can lead to anomalies in the logical schema.

In order to propose a methodology for logical relational database design, we have taken into account such anomalies, especially redundancies detected by the new normal forms, and formalized their sources in the ER schema. We improved the ER-to-relational mapping and the database normalization rules given in [10] to take into consideration enhanced ER mechanisms.

Since database design is complex, a significant research development has been the adoption of knowledge-based techniques for automating design [17]. In the context of Computer-Aided Software Engineering (CASE) tool design, we implemented in Prolog algorithms that construct a normalized relational schema from an ER one enhancing the proposals of [6, 9].

The rest of the paper is structured as follows. Section 2 defines our version of the ER model and deals with basic relational concepts. It also introduces a running example used throughout the paper. Section 3 is devoted to normalization theory and introduces the new normal forms. Section 4 formalizes some sources of redundancy in ER schemas detected by the inclusion normal form in the relational schema. Our enhancement of the design methodology based on Goh's thesis [10] is explained in Sect. 5. Section 6 introduces the implementation of these mapping rules and the normalization algorithms in Prolog. We refer to [11] for a complete description of our implementation. Finally, Sect. 7 gives conclusions and points to further work.

2   ER and Relational Concepts

The ER model describes real world concepts with *entities* — objects of the application domain with independent existence — and *relationships* among entities. In the ER schema shown in Fig. 1 (a), Department is an entity while works is a relationship. Each entity participating in a relationship is assigned one or more *roles*. Role names are omitted when there are no ambiguity. If an entity plays more than one role in a relationship, that relationship is said *recursive* and role names are mandatory. Figure 1 (a) shows a recursive relationship : supervises is defined on Professor, who plays the roles of a supervisor and a supervisee.

Cardinality constraints model restrictions to relationships. In Fig. 1 (a), every instance of Department participates in at least 1 and at most n (i.e., any number of) instances of works.

Attributes are properties of entities or relationships. For instance, Employee has an attribute EmpName. Attributes can be single- or multi-valued. Thus, as for relationships, cardinalities are attached to attributes. The most frequent cardinalities are (1,1), which are assumed as default values and omitted from figures. In Fig. 1 (a), Department has one and only one Dep#. On the contrary, Location is multivalued.

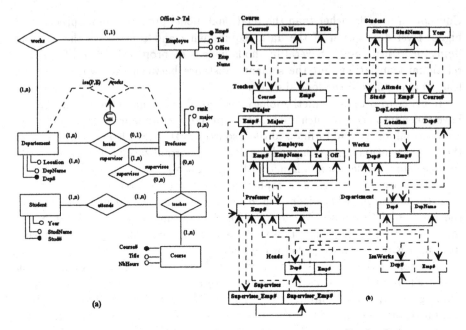

Fig. 1. An ER schema and the corresponding non normalized relations.

An attribute or combination of attributes of an entity is an *identifier* of the entity if its values exactly identify one instance of the entity. In Fig. 1 (a), `Emp#` identifies `Employee`.

Several abstraction mechanisms were added to the basic ER model. We consider weak entities, aggregated relationships, derived relationships, generalization, and subset relationships.

A *weak entity* is an entity that has no identifier of its own. Its instances are identified with respect to instances of one or more *owner entities*. A weak entity is connected to its owner entities via *identifying relationship(s)* and always has a (1,1) cardinality in these relationship(s). For example, Fig. 4 [2] shows three weak entities: `DailyTrip`, `Segment`, and `DailySeg`.

A weak entity usually has a *partial identifier*, which is the set of attributes that can uniquely identify instances related to the same owner entity(ies). Thus, the identifier of the weak entity is the combination of an identifier of an owner entity and its partial identifier. In Fig. 4, instances of `DailyTrip` are identified by the combination of `Trip#` and its partial identifier `Date`.

If a weak entity has no partial identifier, then it must define a set of identifying relationships which, when combined, uniquely identify weak entity instances. In Fig. 4, instances of `DailySeg` are identified by the combination of their owner entities `Segment` and `DailyTrip`, i.e. instances of `DailySeg` are identified by the combination of `Trip#`, `Seg#`, and `Date`.

Generalization is an abstraction mechanism involving two or more entities called superclass(es) and subclass(es). A superclass may have several subclasses and vice versa. A superclass may also be a subclass of another superclass. A subclass inherits all attributes and relationships of its superclasses. Figure 1 (a) exhibits a generalization between **Professor** and **Employee**.

Aggregated relationship models a relationship as a *participant* in another relationship. For instance, the aggregated relationship **teaches** associates **Student** via **attends** with pairs of **Professor** and **Course**. In the rest of the paper, a participant of a relationship denotes an entity or an aggregated relationship.

A relationship is called *derived* [12, 10] if it can be inferred from a combination (similar to a join) of other relationships and generalizations. Both paths (via the derived relationship and via the join of relationships or generalizations) represent the same association. The derived relationship is labeled with a logical expression using the *and* operator ($\land$) and enumerating the participating relationships or generalizations. In Fig. 1 (a), isa(P,E) $\land$ **works** labels a derived relationship combining **works** and the generalization between **Professor** and **Employee**.

A *subset relationship* models a constraint between two relationships. For instance, the subset relationship between **heads** and isa(P, E) $\land$ **works** models the constraint that every professor heading a department must work in that department. Unlike [12, 10], we differentiate generalization and subset relationships, since they correspond to different abstraction mechanisms.

Additional constraints not modeled by ER mechanisms can be explicitly attached to schemas: **Office** $\rightarrow$ **Tel** represents an FD (see below) meaning that, for every instance of **Employee**, the value of **Office** determines the value of **Tel**.

Figure 1 (b) shows a relational database schema obtained by applying an ER-to-relational mapping to the ER schema of Fig. 1 (a). A relation is associated to every entity and non derived relationship while a relational view is associated to the derived relationship. As will be shown later, this schema has redundancies and needs to be normalized. The relational view will allow to detect them.

Data dependencies are constraints on databases and relations. This paper only deals with functional and inclusion dependencies (FDs and INDs).

FDs are defined on individual relations and are represented, as in Fig. 1 (b), by continuous arrows: For instance, the FD **Stud#** $\rightarrow$ **StudName** on relation **Student** means that the value of **Stud#** determines the value of **StudName**. We denote sometimes an FD $X \rightarrow Y$ that holds on relation R by $R : X \rightarrow Y$.

INDs are interrelational constraints on pairs of relations represented, as in Fig. 1 (b), by dashed arrows: the IND **Attends**[Stud#] $\subseteq$ **Student**[Stud#] means that the set of values of **Stud#** in **Attends** is a subset of the values of **Stud#** in **Student**. INDs involving keys are referred to as referential integrity constraints.

Typically, the designer specifies a set of data dependencies. However, given a set of data dependencies F, there are other FDs and INDs that also hold on a database satisfying the dependencies in F. The set of all such data dependencies is called the *closure* of F. It may be inferred by using inference rules.

Sound and complete sets of inference rules for FDs alone [1] and for INDs alone [5] are well-known. Although there is no sound and complete set of inference rules for FDs and INDs taken together, the following rule is sound [5]:

Pullback Rule If $R[XY] \subseteq S[WZ]$ and $W \rightarrow Z$ then $X \rightarrow Y$ with $|X| = |W|$.

The specification of real world constraints in a database schema constitutes an important part of conceptual design. Important integrity constraints are directly modeled by ER mechanisms and this is especially true for FDs and INDs. When the ER schema is mapped into a relational schema, these dependencies are transferred to the corresponding relations.

An ER schema implicitly represents a set of FDs. For instance, an FD $Id(E) \rightarrow Y$ can be deduced from an entity E in an ER schema if $Id(E)$ and Y are attributes of E and $Id(E)$ is an identifier of E. In Fig. 1 (a), Stud# $\rightarrow$ StudName holds on entity Student; such constraint also holds in the corresponding relation in Fig. 1 (b). In the same way, FDs explicitly represented in an ER schema also hold in the corresponding relations.

For relationships, an FD $Id(E_1) \rightarrow Id(E_2)$ can be inferred from a relationship R if $Id(E_i)$ is an identifier of entity E_i and the maximal participation of E_1 in R is 1. In Fig. 1 (a), an FD Works : Emp# $\rightarrow$ Dep# can thus be deduced.

Similarly, ER schemas implicitly model a set of INDs. For instance, an IND $R[Id(E)] \subseteq E[Id(E)]$ can be inferred from a relationship R and a participant E of R where $Id(E)$ is the set of attributes of the identifier of E. In the example, the IND Attends[Stud#] $\subseteq$ Student[Stud#] implicitly holds.

Section 5.1 gives the mapping rules that deduce all implicit FDs and INDs in an ER schema and attach them to the corresponding relational schema.

3 Database Normalization

Normalization [3] was introduced in relational database design to avoid redundancies and update anomalies due to data dependencies. This process is based on the application of normal forms to relations and databases. Each of these forms is specific to a type of data dependency. As already said, we only deal with normal forms concerning functional and inclusion dependencies.

Third normal form (3NF) guarantees individual relations without redundancies with respect to FDs. However, even if each relation is in 3NF, redundancies and update anomalies can still exist in a database considered as a whole due to INDs and to interaction of FDs spanning several relations [13, 12].

To circumvent these problems, Ling et al. [13] introduced the *Improved third normal form* (Improved 3NF). Unlike classical normal forms, Improved 3NF considers several relations rather than individual relations and determines redundancies with respect to FDs. Normalization into Improved 3NF consists in detecting and deleting superfluous attributes. It was proven that if a database is in Improved 3NF, then each individual relation is in 3NF [13].

Inclusion normal form (IN-NF) [12, 10] was later introduced to guarantee databases without redundancies with respect to FDs *and* INDs. It was also proven that if a database is in IN-NF, it is also in Improved 3NF [12, 10].

As classical normalization theory concerns only individual relations, the choice of attribute names in different relations is not constrained. IN-NF and database normalization theory characterize a set of relations as a whole. Hence, we adopt a consequence of the Universal Relation Assumption: if an attribute appears in two or more places in a database schema, then it refers to the same notion, it represent the same semantics.

We motivates with an example the inclusion normal form and give then its formal definition.

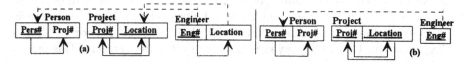

Fig. 2. A relational database not in IN-NF and normalized in IN-NF.

Figure 2 (a) shows a relational database which is not in IN-NF. A person works on a project, a project is associated to a location and an engineer (who is also a person) is associated to a location.

Suppose further that Engineer[Eng#,Location]⊆ Person ⋈ Project[Pers#, Location] hold meaning that an engineer is located at the same place as the project on which s/he is working. Then, attribute Location in Engineer is said to be restorable since Eng# → Location can be deduced from the above IND and the FDs Pers# → Proj# and Proj# → Location. It is also nonessential since it is not needed to deduce other information (it is not part of any key of Engineer). Thus, it is superfluous and it can be deleted as shown in Fig. 2 (b). Note also that all dependencies involving Location in Engineer are removed.

Inclusion Normal Form Consider a database D, a set Σ of FDs and INDs on D, a relation R and an attribute A of R. The dependencies in Σ not involving A in R, denoted $\Sigma_{R(A)}$, are the FDs $X \to Y \in \Sigma$ where $A \notin X$ and $A \notin Y$, as well as the INDs $R[X] \subseteq S[Y] \in \Sigma$ where S is not a relational view derived by join and projection from relations of D such that attribute A of R is necessary to perform a join in the construction of S.

A is *restorable* in R if its values can be deduced from $\Sigma_{R(A)}$. More precisely, A is restorable if there exists a key K of R not containing A and such that we can infer the FD $K \to A$ from $\Sigma_{R(A)}$.

A is *nonessential* in R if A is not necessary to deduce any other attribute of R. Formally, A is nonessential in R if whenever a key K of R contains A, there exists another key K' in R not containing A such that we can infer $K \to K'$ from $\Sigma_{R(A)}$.

An attribute A in a relation R of D is *superfluous* if it is both restorable and nonessential.

A database D is in IN-NF if there are no superfluous attributes in any relation schema of D.

The main difference between Improved 3NF and IN-NF is that, for inferring an FD $X \to Y$, Improved 3NF considers only FDs while IN-NF considers both FDs and INDs [3].

---

[3] For this reason, what we call restorable (nonessential, superfluous) attributes is called weakly restorable (weakly nonessential, weakly superfluous) attributes in [12, 10].

4   Relational Redundancies and ER Cycles

In this section we consider superfluous attributes and relations and relate these redundancies with the corresponding ER schemas.

IN-NF is the only normal form taking into account relational redundancies relative to inclusion constraints. As shown in Sect. 2, such constraints can be modeled in ER schemas using subset and derived relationships. For instance, Fig. 1 (a) includes the inclusion constraint that every **Professor** heading a **Department** must be an **Employee** who **works** in that **Department**. This can be written as follows in a language based on logic:

$$\forall p \in \text{Professor}, \forall d \in \text{Department} \mid \text{heads}(p, d) \Rightarrow$$
$$\exists e \in \text{Employee}\,(\text{isa}(p, e) \wedge \text{works}(e, d))$$

For this logical constraint, a subset relationship models the implication — the inclusion constraint — while a derived relationship represents the conjunction of predicates — the association of other relationships.

Since inclusion constraints are often associated with ER cycles, they become possible sources of superfluous attributes in the corresponding relational schemas. In cycles, some information can be deduced in more than one way, which is the intuition of restorability: in Fig. 1 (a), for a **Professor** who heads a **Departement**, we can deduce the **Department** in which he works either via **heads** or via **isa(P,E)** and **works**. Moreover, some information is not necessary to deduce other information, which is the intuition of nonessentiality: in Fig. 1 (a), a **Department** can be headed by more than one **Professor**. Consequently, there is no constraint holding on **heads** by which a particular **Department** determines one and only one **Professor**.

While in the ER schema **heads** captures some important semantics of the real world (a **Professor** can head a **Department**), in the relational schema of Fig. 1 (b), **Dep#** in **Heads** is restorable since we can deduce **Heads : Emp# → Dep#** from **Heads[Dep#,Emp#] ⊆ IsaWorks[Dep#,Emp#]** and from **IsaWorks : Emp# →** **Dep#**. It is also nonessential since it is not part of the left-hand side of any FD that holds on **Heads**. Therefore, **Dep#** should be removed from relation **Heads**.

On the contrary, if the cardinality between **Department** and **heads** is $(1, 1)$ instead of $(1, n)$, then **Dep#** in **Heads** becomes essential: the FD **Heads : Dep# →** **Emp#** should hold and cannot be inferred from all dependencies not involving **Dep#** in **Heads**.

ER schemas are sources of superflous attributes if they include cycles with an inclusion constraint of the kind shown in Fig. 3. Two entities **A** and **B** with identifiers **A#** and **B#** are related via, on the one hand, relationships M_i and, on the other hand, relationships N_i. All M_i and N_i, except M^* and N^*, are either 1:1 or N:1 relationships in the **B** to **A** direction and M^* and N^* are mandatory N:1 relationships in the same direction. V_1 and V_2 are derived relationships associating respectively relationships N_i and M_i. The derived and subset relationships model an inclusion constraint.

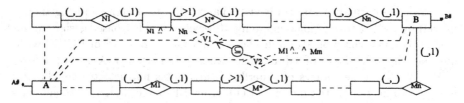

Fig. 3. A source of superfluous attributes

As already said in Sect. 2, an FD $Id(E_1) \rightarrow Id(E_2)$ can be inferred from a relationship R if $Id(E_i)$ is an identifier of entity E_i and the maximal participation of E_1 in R is 1. Consequently, we can generate, by transitivity on the corresponding relational schema, an FD B# $\rightarrow$ A# holding on a view mapping V_1 because of the (_,1) cardinality of each N_i. In the same way, the FD B# $\rightarrow$ A# also holds on a view mapping this time V_2. Thus A# is restorable.

Since we cannot generate the inverse FD A# $\rightarrow$ B#, because of the (_,>1) cardinalities of M^* and N^*, A# is nonessential and then superfluous in V_2.

We consider now redundant relations. In Fig. 1 (b), relation **Heads** (now with only Emp#) is not redundant because it contains the subset of professors heading a department while **Professor** contains all professors. This is represented by the IND Heads[Emp#] $\subseteq$ Professor[Emp#].

Suppose now that cardinality between **Professor** and **heads** is $(1,1)$ instead of $(0,1)$, meaning that all professors head a department. Now, since the participation of **Professor** is mandatory in relationship **heads**, the inverse IND Professor[Emp#] $\subseteq$ Heads[Emp#] holds. Consequently, relation **Heads** is redundant since all information contained in **Heads** is also contained in **Professor**: relation **Heads** should then be deleted.

Similarly, suppose that **Professor** only has the multivalued attribute **major**. Then, relation **Professor** in Fig. 1 (b) should also be deleted because it is redundant with respect to **ProfMajor**: ProfMajor[Emp#] $\subseteq$ Professor[Emp#] and its inverse Professor[Emp#] $\subseteq$ ProfMajor[Emp#] hold with the meaning that all information contained in **Professor** is also represented in **ProfMajor**.

Finally, notice that all attributes of **Teaches** are included in relation **Attends**. Since both INDs Teaches[Course#, Emp#] $\subseteq$ Attends[Course#, Emp#] and Attends[Course#, Emp#] $\subseteq$ Teaches[Course#, Emp#] hold, then relation **Teaches** is redundant and should be removed. On the contrary, if relation **Teaches** had an attribute, such as **semester**, then it would not be redundant.

5 Design Methodology

Traditionally, database design has been accomplished using normalization. However, since the adoption of conceptual models in the mid-70's, normalization theory ceased to be the main logical design step. In fact, working first with ER or another rich conceptual model directly produces 3NF relations in most cases.

Nowadays normalization is only viewed as a verification step removing anomalies left by the ER-to-relational mapping. However, usual ER-based design methodologies remain focused on attaining classical normal forms (3NF or BCNF) without removing other kinds of redundancies studied in this paper. As shown in Sect. 4, ER cycles can be sources of superfluous attributes not detected by classical normalization. Hence, the interest of enhanced ER-based design methodologies that remove anomalies due to cycles and inclusion constraints.

Ling and Goh proposed an integrated design methodology including normalization into IN-NF [Goh92]. These algorithms comprise three main steps. The following sections develop and enhance these steps :

(1) ER-to-Relational Mapping : an ER schema is mapped into a set of non normalized relations. Further, data dependencies are generated to represent implicit constraints of the ER schema.
(2) Relation normalization and key generation: each relation is decomposed into a set of 3NF relations and at least one key is found for each 3NF relation by using Bernstein algorithm [4] for decomposition into 3NF. Since several papers present in detail this algorithm and its implementation (see, e.g., [6]), we do not develop this phase in the paper.
(3) Database normalization: superfluous attributes and relations are deleted lending a database in IN-NF.

5.1   ER to Relational Mapping

Entities Each entity E is mapped into a non normalized relation of the same name comprising all single valued attributes of E. For instance, entity **Student** in Fig. 1 (a) is mapped into relation **Student** in Fig. 1 (b). As said in Sect. 2, FDs implicitly present in an entity E hold on the relation representing E. In the example, **Student : Stud#** $\rightarrow$ **StudName** is generated.

Similarly, FDs explicitly added to an entity E also hold on the relation representing E. In our example, **Employee : Office** $\rightarrow$ **Tel** also holds.

Weak Entities Given a weak entity W and its identifying entities $I_1, \ldots, I_n$, add to the relation R representing W the identifiers $Id_1, \ldots, Id_n$ where Id_i is an identifier of I_i. Consider the example shown in Fig. 4 (a). **Trip#** is added to both relations **Segment** and **DailyTrip**.

Each weak entity either has a partial identifier (as in **Segment** and **DailyTrip**) or can be identified by a combination of its identifying entities (as in **DailySeg**). Therefore, an FD $R : X \rightarrow Y$ holds if Y belongs to R and either (1) X is the combination of Id_i and the partial identifier of W, or (2) X is the combination of the identifiers $Id_{i_1}, \ldots, Id_{i_j}$ of its identifying entities.

In the example, the identifier of **DailyTrip** (resp., **Segment**) is **Trip#Date** (resp., **Trip#Seg#**). Further the FDs **DailyTrip : Trip#Date** $\rightarrow$ **Time** and **Segment : Trip#Seg#** $\rightarrow$ **Price** also hold.

On the other hand, **DailySeg** has no partial identifier but can be identified by a combination of its identifying entities **DailyTrip** and **Segment**. Therefore, the identifiers of both entities must be added to relation **DailySeg**. Fur-

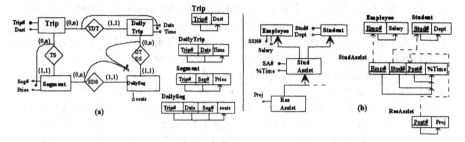

Fig. 4. Mapping weak entities and multiple inheritance.

ther, `Trip#Seg#Date` is the identifier of `DailySeg` and the FD `DailySeg` : `Trip#Seg#Date` → `seats` holds.

Generalizations Given a subclass E and its direct superclasses $S_1, \ldots, S_n$, add to the relation R representing E one of the identifiers $Id(S_i)$ of S_i for each i. In our running example, relation `Professor` inherits `Emp#` from `Employee`.

Consider now Fig. 4 (b) to illustrate multiple inheritance. Since `StudAssist` is a direct subclass of both `Employee` and `Student`, `Emp#` and `Stud#` must be added to relation `StudAssist`. Thus, all (own or inherited) attributes of `StudAssist` can be obtained using natural join on `Emp#` and `Stud#`. `StudAssist` has three identifiers but only one of them is needed to be inherited to relation `ResAssist` to be able to access all its own or inherited attributes. It is not necessary to add the other two. In the example, `SA#` is chosen.

Further, an FD $R : Id(S_i) \rightarrow Y$ holds if Y belongs to R. For instance, the FD `Professor` : `Emp#` → `Rank` holds in Fig. 1 (b).

An IND $E[Id(S)] \subseteq S[Id(S)]$ is generated from a subclass E and one of its direct superclasses S where $Id(S)$ is the common identifier of E and S. In our example, the IND `Professor[Emp#]` $\subseteq$ `Employee[Emp#]` holds.

Relationships Each non derived relationship R is mapped into a non normalized relation of the same name comprising all single valued attributes of R and the identifiers of R, i.e., the identifiers of all entities participating in R directly or indirectly through aggregated relationships. Consider a relationship R and a set of participants (entities or aggregated relationships) $E_1, \ldots, E_n$ of R. The identifier of R, $Id(R)$, is recursively defined by $Id(R) = Id(E_1) \cup \ldots \cup Id(E_n)$ where $Id(E_i)$ is as follows:

- if Ei is an entity, then $Id(Ei)$ is the set of attributes of the identifier of Ei.
- otherwise, Ei is an aggregated relationship over $A_1, \ldots, A_n$ then $Id(E_i) = Id(A_1) \cup \ldots \cup Id(A_n)$.

In Fig. 1 (b), the identifier of relation `Teaches` (where every participant is an entity) is {`Course#`, `Emp#`}. On the other hand, the identifier of relation `Attends` is {`Stud#`, `Course#`, `Emp#`}, i.e., the identifier of `Student` and the identifier of the aggregated relationship `teaches`.

Given a relationship R and two participants E_1 and E_2 of R, an FD $Id(E_1) \rightarrow Id(E_2)$ can then be inferred if the maximal participation of E_1 in R is 1. In our example, two FDs Emp# → Dep# on relations Works and Heads are deduced.

Moreover, for every participant E_i which is an aggregated relationship over $A_1, \ldots, A_n$, then propagate to R those FDs $E_i : Id(A_i) \rightarrow Id(A_j)$ provided that $Id(A_i)$ and $Id(A_j)$ are also attributes of R. For instance, suppose the cardinality between Course and teaches is $(1,1)$ instead of $(1,n)$; then an FD Course# → Emp# should be generated on Teaches and propagated to Attends.

For recursive relationships role names are introduced to avoid ambiguity. Our practical solution consists in concatenating role and identifier names in the relation mapping R. In the running example, relation Supervises is obtained with attributes Supervisee_Emp# and Supervisor_Emp#. The FD Supervisee_Emp# → Supervisor_Emp# holds on the relation.

Multivalued Attributes For each multivalued attribute A of an entity or a relationship E, create a new relation M that includes the identifier $Id(E)$ of E and A. In the running example, a relation DepLocations represents the multivalued attribute Location of entity Department.

An IND $M[Id(E)] \subseteq R[Id(E)]$ holds on relations M and R mapping respectively a multivalued attribute A and an entity E; the inverse IND also holds if A is mandatory. In our example, the IND DepLocation[Dep#] $\subseteq$ Department[Dep#] and its inverse Department[Dep#] $\subseteq$ DepLocation[Dep#] are obtained.

INDs for Relationships Given a (aggregated) relationship R and a participant E of R, an IND $R[Id(E)] \subseteq E[Id(E)]$ can be inferred where $Id(E)$ is recursively defined as shown previously. In our example, Works[Dep#] $\subseteq$ Department[Dep#] (Department is an entity) and Attends[Emp#, Course#] $\subseteq$ Teaches[Emp#, Course#] (teaches is an aggregated relationship) are deduced.

The inverse IND $E[Id(E)] \subseteq R[Id(E)]$ can also be deduced if the minimal participation of E in R is greater than 0. Hence, we can also deduce from Fig. 1 (a) the IND Department[Dep#] $\subseteq$ Works[Dep#].

For recursive relationships role names in INDs are also taken into account. Fig. 1 (b) involves three INDs on relation Supervises.

Derived Relationships Each derived relationship V is mapped into a relation R comprising only the attributes composing the identifiers of the participants of V since V possesses no proper attributes. For example, derived relationship isa(P, E) $\wedge$ works is mapped into relation IsaWorks of Fig. 1 (b).

An FD $X \rightarrow Y$ holds on R if it belongs to the closure of all FDs valid on any relation representing a component of V provided that X and Y are also attributes of R. In our example, we can add to IsaWorks the FD Emp# → Dep# which is valid in relation Works.

Subset Relationships For two relationships R and S such that R is a subset of S, an IND $R[X] \subseteq S[X]$ is generated from R and S where X is the set of common attributes from R and S. In our example, we can add to the relational schema the INDs Heads[Dep#, Emp#] $\subseteq$ IsaWorks[Dep#, Emp#].

Further, all FDs valid on S are also attached to R. Therefore, the FD Emp# → Dep# is also attached to Heads.

Minimal Covers Construct a minimal cover for the FDs attached to each relation.

5.2 Relation Normalization

The algorithm for decomposition into 3NF [4] consists of the following steps:

- Make sure that the set of FDs is minimal,
- Partition the set of FDs into groups such that all FDs in each group have equivalent left-hand sides,
- Construct a relation for each group of FDs, and
- Generate keys from left-hand sides of FDs.

In Fig. 1 (b), relation Employee is normalized in two 3NF relations: (1) Employee_a(Emp#,EmpName,Tel) with FDs Emp# → Tel and Emp# → EmpName; and (2) Employee_b(Office,Tel) with the FD Office → Tel.

Further in this step, the following relation keys are added: Course# on Course, Stud# on Student, {Course#,Emp#} on Teaches, {Stud#,Emp#,Course#} on Attends, Emp# on Employee_a, Office on Employee_b, Emp# on Professor, Dep# on Department, Emp# on Heads, Emp# on IsaWorks, {Dep#,Location} on DepLocation and Emp# on ProfMajor.

5.3 Database Normalization

Specialize INDs If relation R has been decomposed into 3NF relations R_i, then replace each IND of the form $R[X] \subseteq S[Y]$ (resp. $Q[Y] \subseteq R[X]$) by a set of INDs of the form $R_i[X'] \subseteq S[Y']$ (resp. $Q[Y'] \subseteq R_i[X']$) where X' is the intersection of X and the set of attributes of R_i, and Y' the subset of Y corresponding to X'. In our example, Employee_a replaces Employee in the following INDs

Works[Emp#] $\subseteq$ employee[emp#]   employee[emp#] $\subseteq$ Works[emp#]
professor[emp#] $\subseteq$ employee[emp#].

Eliminate redundant attributes For each original or decomposed relation R, eliminate the superfluous attributes as well as the FDs and INDs involving these attributes by using the algorithm presented in Sect. 5.4. In our example, attribute Dep# in relation Heads, Heads : Emp# → Dep#, and Heads[Dep#] $\subseteq$ IsaWorks[Dep#] are removed.

Add INDs For each non normalized relation R decomposed into a set of 3NF relations $R_1,\ldots,R_n$, add to the database all INDs of the form $R_i[X] \subseteq R_j[X]$ where X is the set of attributes common to R_i and R_j. In our example, the inclusion dependencies Employee_a[Tel] $\subseteq$ Employee_b[Tel] and Employee_b[Tel] $\subseteq$ Employee_a[Tel] are added.

Eliminate redundant relations An all-key 3NF relation R is redundant with respect to another relation S if the INDs $R[U] \subseteq S[X]$ and $S[X] \subseteq R[U]$ hold where U comprises all the attributes of R. Then, every relation R redundant with respect to a relation S must be eliminated, as well as the two INDs relating R and S. Further, attach all FDs of R to S and replace R by S in all INDs having R in its left- or right-hand side. In our example, Teaches is redundant with respect to Attends. Thus, the two INDs relating Teaches and Attends are removed; Attends replaces Teaches in the three INDs.

5.4   Detecting Superfluous Attributes in a Relation

For each attribute A of a relation R perform the following four steps:

Initialization Construct the set K of keys K_i of R. If K only consists of a key containing all attributes of R (i.e. R is all-key), then no attribute A of R is not superfluous. Otherwise construct K', the set of keys of R not including A, temporarily remove all FDs involving A in R and all INDs involving a view V in its right-hand side where attribute A of R is necessary to perform a join in the construction of V.

In Fig. 1 (b), for relation Heads and the attribute Dep#, $K = K' = \{[\text{Emp\#}]\}$ and the FD Emp# $\rightarrow$ Dep# is temporarily removed.

Restorability Test If K' is not empty then choose any key K_i from K. If $K_i \rightarrow A$ cannot be deduced from the dependencies valid on the database (those that are not temporarily deleted), then A is not superfluous, otherwise A is restorable. In our example, the FD Emp# $\rightarrow$ Dep# can be deduced from the dependencies valid on the database. Thus Dep# is restorable.

Nonessentiality Test If $K - K'$ is empty, then A (found to be restorable in the previous step) is superfluous. Otherwise if there exists a key K_i of R containing A such that $K_i \rightarrow U$, where U is the set comprising all attributes of R, then A is superfluous. Otherwise, let C be the closure of K_i, and reinsert the dependencies temporarily removed. If $(C \cap U) - \{B\} \rightarrow U$ cannot be deduced then A is not superfluous. Otherwise A is superfluous and insert into K' any key of R contained in $(C \cap U) - \{B\}$. In our example, since $K - K'$ is empty Dep# is nonessential and thus superfluous.

Reinsert or Specialize Dependencies If the attribute is not superfluous then reinsert the dependencies temporarily removed in the first step. Otherwise add the INDs that can be deduced by transitivity using attribute A of R. Given the INDs $Q[X] \subseteq R[Y_1]$ and $R[Y_2] \subseteq S[Z]$ if $Y = Y_1 \cap Y_2$ and if $A \in Y$, then add the IND $Q[X'] \subseteq S[Z']$ where X' and Z' are the attributes corresponding to Y.

Then, replace every IND of the form $R[X] \subseteq S[Y]$ or $S[Y] \subseteq R[X]$ where $A \in X$ by $R[X'] \subseteq S[Y']$ and $S[Y'] \subseteq R[X']$ where $X' = X - \{A\}$ and Y' is the set of attributes corresponding to X' provided that X' is not empty.

6   Database Design Methodology in Prolog

In this section we briefly describe our implementation of the logical design methodology. For a complete description we refer to [11].

The use of Prolog [7] for building our logical schema generator is essential to our approach. The declarative nature of Prolog gives it several advantages (clarity, modularity, conciseness, and legibility) over conventional programming languages and make it more suitable for CASE prototype design. It also makes possible an integration between static knowledge of the world, or facts, and deductive statements, or rules.

Figure 5 shows a printout of our system. As shown in the figure, an ER conceptual schema can be draught directly on the screen using a Graphical User

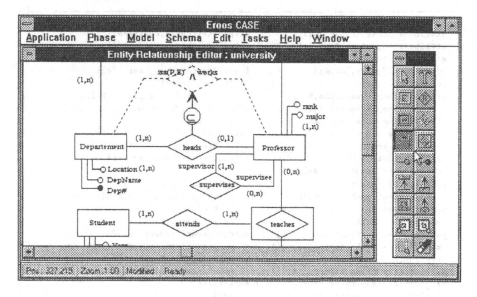

Fig. 5. ER editor.

Interface. However, this graphical representation is encoded by a set of Prolog predicates which are introduced in the base of facts. These predicates form the input to the logical schema generator which automatically generate a relational schema normalized in IN-NF. Figure 6 gives the encoding of the example ER schema of Fig. 1 (a). We refer to [11] for a complete description of these predicates and all its arguments.

Unlike ER facts constituting the original input of the logical schema generator, relational facts are generated when the ER schema is mapped into a relational schema, during the first step. This relational schema is then modified during the second step for 3NF normalization, and during the third step for IN-NF normalization.

Figure 7 shows the result of applying the ER to relational mapping to our example. This figure corresponds to the relational schema of Fig. 1 (b).

The second step of our methodology normalizes each 3NF relation and generates keys. With respect to our example, the predicates given in Fig. 8 are added. Notice that relation `employee` is decomposed into `employee_a` and `employee_b`.

Database schema normalization algorithms produce the final base of facts representing a database schema in IN-NF. Referring to our example, `depNo` is detected superfluous in `heads`, and thus `rel_attrs(heads, [empNo,depNo])` is replaced by `rel_attrs(heads,[empNo])`. Further, the following predicates are removed.

```
fd(heads,[empNo],[depNo]). ind(heads,[depNo],depart,[depNo]).
        ind(depart,[depNo],heads,[depNo]).
```

Also, since relation `teaches` is redundant with respect to relation `attends`, the following predicates are removed:

```
entity(depart).                          entity(course).
attribute(depart,depNo,1,1).             attribute(course,courseNo,1,1).
attribute(depart,depName,1,1).           attribute(course,title,1,1).
attribute(depart,location,1,n).          attribute(course,nbHours,1,1).
identifier(depart,[depNo]).              identifier(course,[courseNo]).
entity(employee).                        entity(professor).
attribute(employee,empNo,1,1).           attribute(professor,status,1,1).
attribute(employee,tel,1,1).             attribute(professor,major,1,n).
attribute(employee,empName,1,1).         entity(student).
attribute(employee,office,1,1).          attribute(student,studNo,1,1).
identifier(employee,[empNo]).            attribute(student,studName,1,1).
er_fd(employee,[office],[tel]).          attribute(student,year,1,1).
identifier(student,[studNo]).
relationship(teaches).                       relationship(attends).
participates(teaches,professor,0,n,''). participates(attends,student,1,n,'').
participates(teaches,course,1,n,'').        participates(attends,teaches,1,n,'').
relationship(works).                         relationship(isaWorks).
participates(works,employee,1,1,''). participates(isaWorks,professor,1,1,'').
participates(works,depart,1,n,'').          participates(isaWorks,depart,1,n,'').
relationship(heads).                         participates(heads,professor,0,1,'').
participates(heads,depart,1,n,'').          relationship(supervise).
participates(supervise,professor,0,1,supervisor).
participates(supervise,professor,0,n,supervisee).
generalization(employee,partial,exclusive,'','').isa(professor,employee,'','').
derived_relationship(isaWorks,[professor,works]). subset_of(heads,isaWorks).
```

Fig. 6. Prolog facts representing the schema of Fig. 1 (a).

```
rel_attrs(depart,[depNo,depName,location]).       rel_attrs(works,[empNo,depNo]).
rel_attrs(employee,[empNo,tel,empName,office]).  rel_attrs(heads,[empNo,depNo]).
rel_attrs(student,[studNo,studName,year]).       rel_attrs(isaWorks,[empNo,depNo]).
rel_attrs(attends,[empNo,courseNo,studNo]).  rel_attrs(professor,[empNo,major]).
rel_attrs(course,[courseNo,title,nbHours]). rel_attrs(teaches,[empNo,courseNo]).
rel_attrs(supervisor,[supervisor_empNo,supervisee_empNo]).
fd(employee,[empNo],[empName]).          fd(depart,[depNo],[depName]).
fd(employee,[empNo],[tel]).              fd(depart,[depNo],[location]).
fd(employee,[empNo],[office]).           fd(student,[studNo],[studName]).
fd(employee,[office],[tel]).             fd(student,[studNo],[year]).
fd(course,[courseNo],[title]).           fd(course,[courseNo],[nbHours]).
fd(professor,[empNo],[major]).           fd(works,[empNo],[depNo]).
fd(isaWorks,[empNo],[depNo]).            fd(heads,[empNo],[depNo]).
view(isaWorks,[professor,works]).
ind(heads,[depNo],depart,[depNo]).       ind(attends,[studNo],student,[studNo]).
ind(depart,[depNo],heads,[depNo]).       ind(student,[studNo],attends,[studNo]).
ind(teaches,[courseNo],course,[courseNo]).   ind(works,[depNo],depart,[depNo]).
ind(course,[courseNo],teaches,[courseNo]).   ind(depart,[depNo],works,[depNo]).
ind(works,[empNo],employee,[empNo]).         ind(isaWorks,[depNo],depart,[depNo]).
ind(employee,[empNo],works,[empNo]).     ind(isaWorks,[empNo],professor,[empNo]).
ind(teaches,[empNo],professor,[empNo]). ind(professor,[empNo],employee,[empNo]).
ind(heads,[empNo],professor,[empNo]).
ind(heads,[empNo,depNo],isaWorks,[empNo,depNo]).
ind(attends,[courseNo,empNo],teaches,[courseNo,empNo]).
ind(teaches,[courseNo,empNo],attends,[courseNo,empNo]).
```

Fig. 7. ER to relational mapping: results of the first step.

```
tnfdecomp(employee,employee_a). rel_attrs(employee_a,[empNo,tel,empName]).
tnfdecomp(employee,employee_b).        rel_attrs(employee_b,[tel,office]).
fd(employee_a,[empNo],[empName]).         fd(employee_a,[empNo],[tel]).
fd(employee_a,[empNo],[office]).          fd(employee_b,[office],[tel]).
key(depart,[depNo]).   key(employee_a,[empNo]).  key(employee_b,[office]).
key(student,[studNo]). key(course,[courseNo]).   key(professor,[empNo]).
key(works,[empNo]).    key(heads,[empNo]).        key(isaWorks,[empNo]).
key(depart,[depNo]).   key(employee_a,[empNo]).  key(employee_b,[office]).
key(student,[studNo]). key(course,[courseNo]).   key(professor,[empNo]).
key(works,[empNo]).    key(heads,[empNo]).        key(isaWorks,[empNo]).
key(teaches,[courseNo,empNo]).       key(attends,[studNo,empNo,courseNo]).
```

Fig. 8. 3NF relational normalization: results of the second step.

```
rel_attrs(teaches,[empNo,courseNo]). key(teaches,[empNo,courseNo]).
       ind(attends,[empNo,courseNo],teaches,[empNo,courseNo]).
       ind(teaches,[empNo,courseNo],attends,[empNo,courseNo]).
```

and the following predicates

```
ind(teaches,[empNo],professor,[empNo]). ind(teaches,[courseNo],course,[courseNo]).
                ind(course,[courseNo],teaches,[courseNo]).
```

are replaced by

```
ind(attends,[empNo],professor,[empNo]). ind(attends,[courseNo],course,[courseNo]).
                ind(course,[courseNo],attends,[courseNo]).
```

7 Conclusion and Further Work

The main goal of our work has been to test and enhance the methodology for relational database design of [10], especially with respect to its viability in the context of computer-aided software engineering (CASE) tools development.

We demonstrated the usefulness of the inclusion normal form (IN-NF). As said in Sect. 4, since ER cycles and inclusion constraints are often present in conceptual schemas, IN-NF normalization is needed to safely translate ER schemas into relational schemas.

We then presented our methodology for relational database design. It improved the algorithm of [10] in the following respects:

- take into account multivalued attributes, weak entities and recursive relationships
- distinguish generalization and subset relationships, in particular to be able to represent parallel generalizations
- take into account several identifiers for entities, in particular due to the inheritance of identifers for (multiple) generalization
- generate implicit FDs during the ER to relational mapping
- use minimal covers instead of original sets of FDs
- project the INDs for 3NF decompositions

– generate INDs that can be deduced by transitivity, before removing superfluous attributes and specialization of INDs after removing each superfluous attribute

Several databases have been tested with our Prolog code and have provided convincing results. These tests have allowed us to improve the mapping rules and the IN-NF normalization algorithms.

Another important result of our work is the development of an environment supporting relational database design. It helps detecting errors as soon as possible in the development life-cycle, which constitutes a necessity in software engineering. Our system validites the ER specifications introduced by the user by performing integrity checking based on the syntax and semantics of the ER abstractions. Whenever errors are detected, the user is informed with appropriate explanations. Then our system allows to automatically generate the corresponding relational database schema, normalized in IN-NF.

This work has been carried out in a larger project in which a prototype CASE tool for Object-Oriented Information Systems Development is being constructed. The system, developed in a PC platform using an LPA Prolog compiler, generates C++ code and relational database schemes in SQL for Oracle. It integrates concepts and models from many object-oriented methods (e.g. [16, 8]).

An application is described by several complementary models capturing static, dynamic, and functional aspects. The CASE tool was conceived with a modular architecture: different abstractions can be selectively incorporated in each model, thus allowing to customize the conceptual languages used to describe the system throughout the development life-cycle. In this context, we formalized a new abstraction for the description of object models, called *materialization* [15]. In [20] we also described another module of our CASE, devoted to dynamic specification using the formalism of statecharts.

Several issues need to be further investigated. Concerning the ER formalism, abstractions like aggregation and materialization could also be implemented.

Our mapping rules produce relations that keep track of the distinction between entities and relationships. Other more optimized rules can generate fewer relations but lose this semantic classification. Our sytem could be used to compare pros and cons of each method. Optimization of our mapping rules might be for instance realised by implementing a relation merging algorithm [14].

Normalization algorithms implemented in our system only deal with FDs and INDs. The methodology could be enhanced by taking into account less common data dependencies like multivalued and join dependencies, take fourth and fifth normal forms into consideration, and thus shed light on their user-oriented semantics. We could also analyze the consequences of normalization into Boyce-Codd normal form (BCNF). This might be achieved with the algorithm of Tsou of Fisher [19]. Every relation of a database that is in IN-NF is only guaranteed to be in 3NF. However, as is well-known, it is sometimes impossible to reach BCNF for a 3NF relation without losing dependency preservation.

References

1. W. Armstrong. Dependency structures of database relationships. In *Proceedings of the IFIP Congress*, pages 580–583, Geneva, Switzerland, 1974.

2. C. Batini, S. Ceri, and S. Navathe. *Conceptual Database Design: An Entity-Relationship Approach*. Benjamin/Cummings, 1992.

3. C. Beeri, P. Bernstein, and N. Goodman. A sophisticate's introduction to database normalisation theory. In *Proc. 4th Int. Conf. on Very Large Databases*, pages 113–124, 1978.

4. P. Bernstein. Synthesising third normal form relations from functional dependencies. *ACM Trans. on Database Systems*, 1(4):277–298, 1976.

5. M. Casanova, R. Fagin, and C. Papadimitriou. Inclusion dependencies and their interaction with functional dependencies. *Journal of Computer and System Sciences*, 28(1):29–54, 1984.

6. S. Ceri and G. Gottlob. Normalization of relations and Prolog. *Comm. of the Assoc. for Computing Machinery*, 29(6):524–544, 1986.

7. W. Clocksin and C. Mellish. *Programming in Prolog*. Springer-Verlag, 1984.

8. D. Coleman, P. Arnold, S. Bodoff, C. Dollin, H. Gilchrist, F. Hayes, and P. Jeremaes. *Object-Oriented Development: The Fusion Method*. Prentice Hall, 1994.

9. A. Doğaç, B. Yuruten, and S. Spaccapietra. A generalized expert system for database design. *IEEE Trans. on Software Engineering*, 15(4):479–491, Apr. 1989.

10. C. Goh. Towards a viable methodology for logical relational database design. Master's thesis, National University of Singapour, 1992.

11. M. Kolp and E. Zimányi. Enhanced ER to relational database design and its implementation in Prolog. Technical Report RR 95-01, INFODOC, Université Libre de Bruxelles, Belgium, Apr. 1995. Submitted to publication.

12. T. Ling and C. Goh. Logical database design with inclusion dependencies. In *Proc. of the 8th IEEE Int. Conf. on Data Engineering, Tempe, Arizona*, Feb. 1992.

13. T. Ling, F. Tompa, and T. Kameda. An improved third normal form for relational databases. *ACM Trans. on Database Systems*, 6(2):329–346, 1981.

14. M. Markowitz. Merging relations in relational databases. In *Proc. of the 8th IEEE Int. Conf. on Data Engineering, Tempe, Arizona*, pages 428–437, 1992.

15. A. Pirotte, E. Zimányi, D. Massart, and T. Yakusheva. Materialization: a powerful and ubiquitous abstraction pattern. In J. Bocca, M. Jarke, and C. Zaniolo, editors, *Proc. of the 20th Int. Conf. on Very Large Databases*, pages 630–641, Santiago, Chile, 1994. ACM Press.

16. J. Rumbaugh, M. Blaha, W. Premerlani, F. Eddy, and W. Lorensen. *Object-Oriented Modeling and Design*. Prentice Hall, 1991.

17. V. Storey. A selective survey of the use of artificial intelligence for database design systems. *Data & Knowledge Engineering*, 11:61–102, 1993.

18. T. Teorey. *Database Modeling and Design. The Entity-Relationship Approach*. Morgan Kaufmann, 1990.

19. D. Tsou and P. Fischer. Decomposition of a relation scheme into Boyce-Codd normal form. *ACM-SIGACT*, 14(3):23–29, 1982.

20. E. Zimányi. Statecharts and object-oriented development: a CASE perspective. In *Proc. of the 3rd Int. Conf. on Practical Application of Prolog*, pages 697–718, Paris, France, Apr. 1995.

Supporting Distributed Individual Work in Cooperative Specification Development

Motoshi Saeki   Saeeiab Sureerat   Kotaro Yoshida

Dept. of Computer Science
Tokyo Institute of Technology
Ookayama 2-12-1, Meguro-ku, Tokyo 152, Japan
E-mail : saeki@cs.titech.ac.jp

Abstract. This paper discusses a supporting tool for distributed individual work to develop specifications in cooperative situation. Our supporting tool consists of two parts — one is *method base* for individual use, and another is a structured electronic mail system for communication use. *Method base* has various catalogued specification & design methods, so a worker can select a suitable method for his problem domain. It can integrate the products developed by team members with the different methods into one. Our structured electronic mail system supports 1) sending and receiving the products with messages or with comments, 2) automatically notifying the modification of products and the progress status of development activities, 3) composing comprehensive mails by using templates, and 4) storing mails in structured mail folders and retrieving them.

1   Introduction

Designing software specifications effectively is important for developing high quality software since specification & design phases are the early steps in software development processes. Software development activities, including designing specifications, are essentially cooperative and performed by a collaboration of a team. In such cooperative situation, different roles of workers such as customers, users, analysts, designers, managers, and so on, actually participate in the activities and communicate with each other.

Conventional CASE tools for supporting specification development[11] are for a single designer. Thus it seems to be difficult to apply effectively the existing CASE tools to the activities which are performed by a team. The members cooperate, collaborate, coordinate, and communicate with each other to develop a product.

The tools, so called groupware, for supporting various kinds of group work have been developed[7]. However they are for general purpose. As Curtis pointed out[6], the specialization to specification development application allows us to develop a more effective supporting tool.

Communication with the members of a development team, e.g. conversation, is one of the most important features for these cooperative work, and communication style or characteristics depends on which phase the members are in.

To develop an effective tool, first of all, we must observe what communications are made in each phase of actual specification processes. By these observations from the viewpoint of communication, we found that typical cooperative specification processes can be divided into three phases — 1) making a working plan by a team, 2) performing an assigned task by each member distributively and in parallel, and 3) reviewing and integrating the products which have developed by members. This paper focuses on the support for the second phase, i.e. distributed individual tasks.

The organization of the paper is as follows. First, we discuss the requirements to our supporting tool for distributed individual work. These have been obtained from the empirical case studies. The overview of a tool for supporting effectively distributed individual work is discussed in this section. We will introduce two parts of our tool — one is called *method base*, and another is a *structured E-mail system*. *Method base* is a kind of database system where various methods for software specification & design are stored. An user can select a design method suitable for his problem domain, and be navigated according to the selected method. It allows the users to integrate their products developed by using the different methods into a final specification. The structured E-mail system is used for composing and filing E-mails in addition to sending and receiving them. Sections 3 and 4 show the functions of the method base system and the E-mail system respectively by using example. Related work will be discussed in section 5.

2 Overview of Our Supporting Tool

In distributed individual work, each worker performs his design task according to his design method. In this situation, he may use the design methods different from ones others use, because he select his methods which are suitable for him and for his assigned part. The tool can hold various kinds of design methods, and integrate the products developed with the different methods into one. In addition, it provides user-friendly interface such as graphical notation and manipulation of the products.

In the phase of distributed individual work, communication is made in loosely coupled style. In this style, the workers do not communicate with each other so much and they often communicate locally with the relevant members. This fact involves that the communication facility based on electronic mail can be sufficient.

Supporting distributed individual design work requires tools where the CASE tool facility to manage various kinds of design methods and communication facility are tightly combined. Our supporting tool consists of two parts — *Method Base* for an individual design task and a *structured E-mail system* for communications among the members.

Method base is a kind of CASE tools and can hold the following information to navigate an individual worker according to the methods that he selects for his problem domain.

1. documents produced during development (call them products), and
2. histories of the performed activities[14], and
3. multiple design methods and semantical relationships among them.

Assume that a worker has selected JSD (Jackson System Development) method. His method base suggests what he should do next, i.e. it provides procedural information such as "identify entities and actions in the real world". Furthermore method base can integrate the products developed with the multiple methods into one.

We have developed a kind of a structured electronic mail system for interaction with other members. In addition to usual message passing by electronic mail, it supports 1) sending and receiving the products with messages or with comments, 2) automatically notifying the modification of products and the progress status of development activities, 3) composing comprehensive mails easily by using templates, and 4) storing received mails in structured mail folders and retrieving them.

A worker often accesses to product parts which the other members are developing, and to obtain or confirm interface or boundary among the other members. In our environment, however, he cannot update the products of the other members and can see them from his session.

3 Method Base System

3.1 Meta Model

To hold information about design methods as data in the method base system, we should develop formal representations of various kinds of design methods, i.e. meta modelling technique. To apply a meta model in cooperative specification processes, it needs to provide the mechanism for integrating the products which have been developed by different methods. We have focused on the common concepts included in the methods, and developed the common meta model of the methods. Though the methods contain many constitutional concepts which are expressed by the different technical terms, there are several common concepts among them. For example, the "object" concept in OOA and OOD (Object Oriented Design) can semantically involve a "source & sink (external entity)" or a "data store" concept in Data Flow Diagram (DFD) in SA as discussed in [2]. Thus we can consider that these concepts have the same concept type. We named this type *object type*. We have explored these concepts and had the other 5 concept types — *process, data, state, event,* and *association* types, and their relationships as a common meta model of the methods. Figure 1 illustrates the overview of our meta model having common meta model and method-specific meta model parts. If we define a method, we import a necessary part of the common meta model and then describe the method-specific part incrementally. Method-specific parts such as notation of diagrams and method procedures are defined separating from common meta model part. The integration of various CASE tools can be done through the common meta model. More details of our

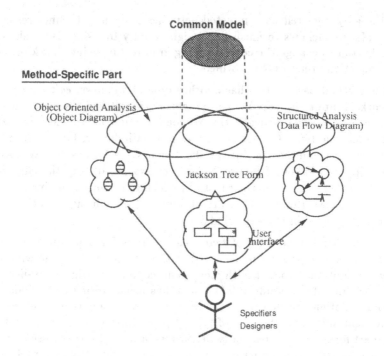

Fig. 1. Overview of Common Meta Model and Method-Specific Parts

meta model is out of scope of this paper, and it was discussed in [16] and its implementation with PCTE in [17].

3.2 Functions of Method Base System — Example

Before the integration mechanism by using common meta model, we should clarify the functions of our method system. It has two types of windows where the workers do their individual tasks — one is called *overview window* and another is *edit window*. Overview windows displays development activities, their assigned workers, the methods being used, the current status, and so on in a table form as shown in the bottom left window of Figure 2. In this example, we have two workers, Jane and Mary, who are involved in designing a lift control system. Before starting their individual tasks, the lift control system has been decomposed into three parts — Lift Box (shortened to Lift), Floor button, and Scheduler. Mary has been assigned to the design task for Lift and will use OOA (strictly speaking, Object Diagram of Coad & Yourdon's OOA[5]), SA, and STD (State Transition Diagram). The decomposition, assignment and used methods were already determined. Note that the design of scheduler will be duplicatedly performed by both of Jane and Mary, i.e. their activities are replicative and either of them would be redundant. In the overview window, the black rectangle and the gray ones tell us completed activities and on-going ones respectively. When Mary concludes to finish her design activity by using OOA, she clicks a menu button to change the corresponding gray rectangle into black one so that the completion of her task is notified.

The method base system has another type of windows, *edit window*, where the workers input and edit their products. Each edit window is for an activity and it can be opened by using the overview window as a menu. For example, Mary points the Lift-OOA area in the overview window with her mouse to open an edit window, and then the lift specification in OOA form is displayed as shown in the top left window of Figure 2. The workers can display the edit windows where the other workers do their tasks by clicking the corresponding areas of the overview window. However they are prohibited from editing the other workers' products displayed in the edit windows.

An edit window consists of two areas : one is for a product and another for activities. The product is displayed in the left side area of the window, and Mary can edit the product with icon menu commands which are displayed in the upper line of the window. This menu has usual editing commands such as creation, deletion, movement of graphical objects, resizing them, and text input. In the right side area, the flow of typical procedures of OOA is displayed in a flow chart form. Mary is currently performing a procedure expressed by a black oval surrounded with a larger oval in the figure, i.e. "identifying structures".

Mary opens her edit windows for SA (Data Flow Diagram) and STD successively. Figure 2 shows three edit windows for a lift specification. Thus she has been developing a lift specification from three viewpoints : OOA, SA, and STD.

3.3   Product Integration — Example

Let's turn to the explanation of the integration mechanism of the products developed by the different methods. The product elements which have the same identifier and the same type can be common in the products. For example, the objects identified in OOA process can be considered as source & sinks or data stores from the viewpoint of SA. You have identified the object "lift button" and the service "turn light (on or off)" in the "lift button" object in OOA as shown in the window of Figure 2. You also identified the source&sink "lift button" and the process (bubble) "turn light" in SA. These can be considered as the "common" elements for integrating the OOA specification and the SA. Figure 3 shows the example of the "common" elements, lift button and turn light, and the integration mechanism through these elements. "Turn light" belongs to the *process type* in the meta model because "service" concept in OOA and "process (bubble)" concept in SA can be considered as *process type*.

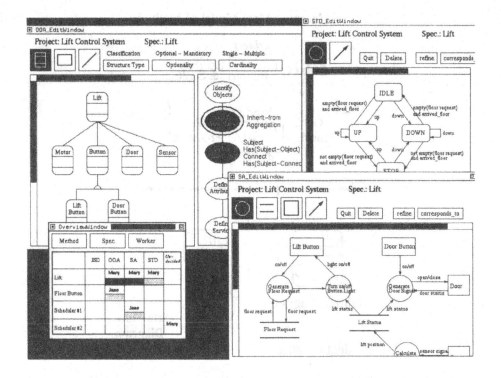

Fig. 2. Example of An Overview Window and Edit Windows

4 Structured E-mail System

4.1 Requirements to E-mail Systems

Electronic mail is one of asynchronous communication styles and it has several merits and disadvantages which synchronous communication styles such as verbal conversation do not have. Thus, before developing the system, we observed communication histories of electronic mails in actual specification processes and analyzed them. In the analyzed processes, our designers found the problems of electronic mail communication which are shown in the following :

1. Text in the mails, who was composed by the others, might be not comprehensive to the readers.
2. Conclusion to which the members came might be ambiguous in the electronic discussion, or it might be unclear where the conclusion was in the set of the mails.
3. All of the members did not discuss a topic very much but the discussions might be one-to-one.
4. It is often delayed to read the messages. Delayed readers might be left behind or out of the discussions.

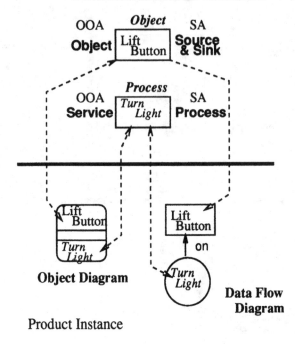

Fig. 3. Integration of OOA and SA products through "Common" elements

To make electronic mail communication efficient in actual specification processes, our system provides the support for the senders to compose comprehensive mails rapidly so that the readers can understand the senders' intents clearly, and the support for the readers to store the mails according to their contents and to retrieve them easily. More concretely, it has the following functions :

1. Supporting message sending and receiving (incl. developed products, and products with messages or comments) among the members.
2. Automatically notifying the state of progress of development activities and the modification of the specifications to the relevant members.
3. Facilitating to compose by using templates electronic mails easy for the other members to understand.
4. Storing in the structural manner the received electronic mails and retrieving the mails from mail folders.
5. Facilitating viewing the other members' products from his session in any time of his development.

We focus on the templates for composing electronic mails and the structure of our mail folders for storing the received mails in the following subsections.

4.2 Electronic Mails and Templates

Figure 4 shows the slot structure of a typical electronic mail. The received mails can be linked to the product elements which they refer to. It has the slots "To", "Cc", "From", and "Subject" which our usual electronic mails have. This part looks like an envelope. The text of an electronic mail (called mail text) can be composed by using templates. The sender fills up the slots of his selected template. He can attach several mail texts composed by the templates to an electronic mail as shown in the figure.

We have seven templates "proposal", "question", "answer", "request", "problem", "notification", and "amendment" as shown in Figure 5. The template types allow the readers to understand the sender's intent quickly. These templates have been obtained as abstractions of senders' intent included in the electronic mails that we analyzed. That is to say, the categorization of the template types is based on what the sender wants the receivers to do, i.e. what kinds of response the sender hopefully looks forward to hearing from the receivers. For example, the type "question" is used for asking yes-no questions or wh-questions (what, when, where, who, which, why and how), so the mails of this type should be replied with "yes", "no", or other sentences. The template "question" contains these three answer patterns so that the receivers can make response quickly and easily. It is similar to return post cards. The receivers that intends to reply to the received mails select the answer patterns in the mails, or fill with a text the slot of the answer template included in the mails. A mail of the type "proposal" needs as a response the receivers' positions to it — "support" (agreement) or "not support" (disagreement). Thus this type contains an answer template standing for the receiver's position.

Note that the template "notification" can be used for confirming conclusions of electronic discussions to the workers. Clicking the button "Decided" in "notification" means that the mail text is a conclusion. The receivers can retrieve what have been already decided in their mail folders.

4.3 Mail Folders

Figure 6 shows how to store the received mails in structural manner. The mail texts are extracted from the received mail and they are linked to each other in a mail folder. The mail folder is automatically created from the topic of the received mails, i.e. the value of their "Subject" slot. Consider that two workers A and B would communicate with each other about a topic. The worker A sends to B the text whose type is "problem". The worker B sends back the answer text where the sentences are quoted from a part of A's text. A's text and B's text are connected with "reply_to" and "quote_from" links in the mail folder. The worker can retrieve the mails by using a mail browser as mentioned later.

Figure 7 shows the logical schema for managing electronic mails in Entity Relationship Model. The type "product element" is a super type of the types appearing in the method meta models such as "object", "state", and so on. Thus mails can have relationships to the product elements being developed. All

240

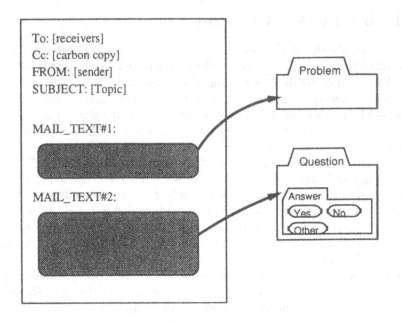

Fig. 4. Electronic Mail

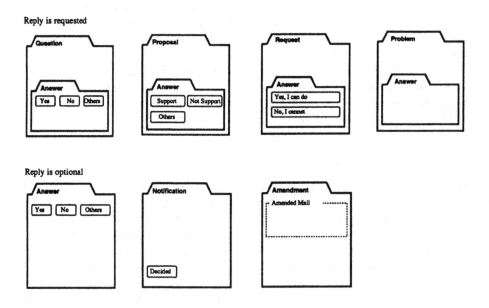

Fig. 5. Template for Composing Mail Texts

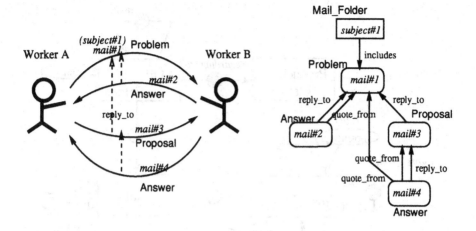

Fig. 6. Storing the Received Mails

electronic mails have their senders, receivers, cc (carbon copy) and date as attributes. Mails consist of a set of the texts which the senders compose from the templates. The type "mail_text" in the figure is a super type of the seven template types "proposal", "question", "answer", "request", "notification", "problem", and "amendment". An amendment mail is used for correcting the mails previously sent or for adding something to the previous mails. An "amendment" type text has a specific relationship "amend_to" to other mail texts. Thus our system has the simple function of version control for received mails.

4.4 Sending and Receiving Electronic Mails — An Example

The workers may often review and check the products which the other workers have developed or are developing. By using our system, they can view the other workers' products in their workstations and attach the comments to the products directly. However, they cannot update the products which the other workers are developing. These comments are sent to the workers who are developing the products by electronic mail.

To illustrate an electronic mail linked to a product element as a comment, let's turn our example of the two workers, Mary and Jane. Jane reviews the specification of a lift which Mary has developed by using OOA. The reviewer Jane checks the object diagram and finds that the "emergency button" object have been missing in it. She composes a comment by using the template of "problem" type. After she fills the text into the template, she links it to Mary's button object by pointing the button object in Mary's Lift-OOA edit window with her mouse. The "To" slot is automatically filled by pointing the object. And then she sends her comment with the linked product by electronic mail. That is to say, our system can handles not only linear text but also hypertext and graphical information.

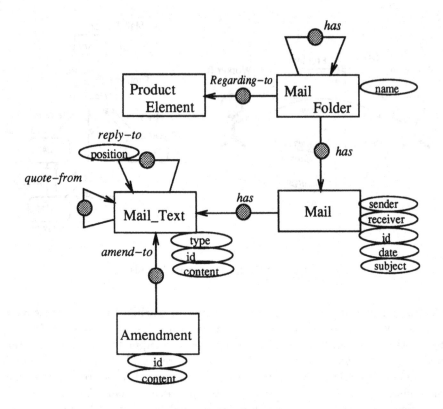

Fig. 7. Logical Schema for Electronic Mails

Mail folders used for filing are automatically created according to the values of the slot "Subject". The receivers can retrieve the received mails by the mail browser which displays hierarchically mail folders, mails, and their relationships. Figure 9 shows that Mary, the receiver of Jane's comment, reads the comment to the "button" object by using the mail browser. Since the mail sent by Jane has the value "emergency button" in the "Subject" slot, it has been stored in the sub folder "emergency button". All the mails discussing about the emergency button will be stored in this sub folder. She clicks the icon expressing an mail on the browser, and the corresponding mail text is displayed in a new window "ReadMail". The browser displays the semantical relationships among the mails such as "reply-to", so the workers could understand easily what their discussions reach. If Mary wants to reply to the comment, she fills the answer template included in it as shown in the ReadMail window of the figure.

Without the browser, Mary can also read the mails regarding a product element by pointing the corresponding object in her edit window, because every electronic mail for products is associated with the relevant products elements in the edit windows as shown in Figure 8. After pointing the "button" object, the

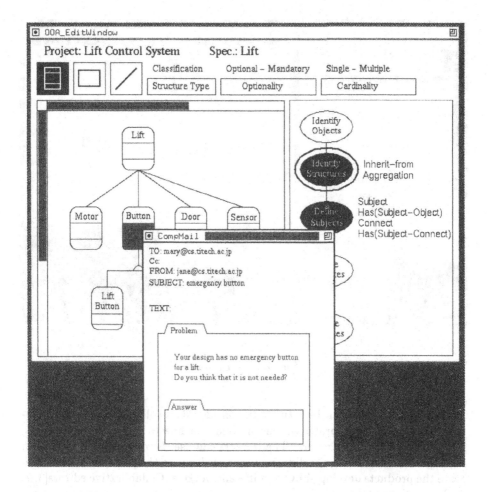

Fig. 8. Sending an Electronic Mail

list of its sub folders, i.e. the "Subject" list is displayed and then she selects a sub folder, e.g. emergency button. The list of the mails which are stored in the sub folder is displayed.

5 Related Work

Providing the support tools for cooperative work is the main problem in CSCW (Computer Supported Cooperative Work) and Software Process communities. They have studied this topic independently and produced research results. CSCW community focuses on the tools only for general purpose, e.g. electronic meetings, decision making support, collaborative editing and so on, and does not consider special applications such as software development yet.

244

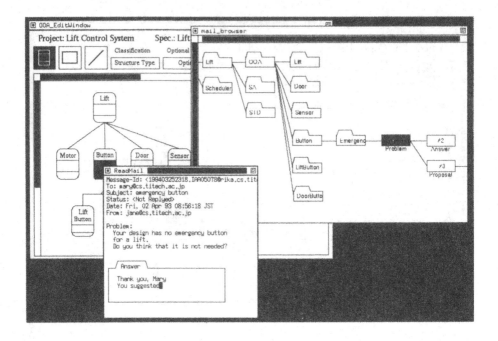

Fig. 9. Mail Browser

Our method base can be intuitively considered as collaborative editing systems. However it incorporates various methods, i.e. how to compose the products, into collaborative editing work. The workers can use the methods different from the other workers. Using the different methods leads to the problem how to integrate the products developed by the different workers. Collaborative editors[3, 7] deal with linear texts which are partitioned into editable grains such as chapters, sections, sentences, words or characters, and integrate the grains edited by workers in linear order. Thus the integration mechanism is very simple. How to integrate the products developed according to the different methods depend on the used methods. Our method base provides the integration mechanism according the used methods by common meta model.

Structured electronic mail systems such as Object Lens[10] have been developed for general applications. To provide effective support in specific applications such as specification development, their templates, procedures and action rules should be re-designed and customized according to the applications. Our system can be considered as a customization and an extension of Object Lens. To develop it, we explored the characteristics of communication in the actual specification processes. Coordinator[21] based on Speech Act Theory restricts the mail types which the receivers of a mail can reply. This restriction might be useful to process the semantical meaning of the mails by computers. However, it is too strong in software development and results in inconvenience. In actual software development processes, the accidental situations that do not meet the restriction often appear.

Several meta modelling techniques based on Entity Relationship Model[20, 18, 4], and based on attribute grammer[19] have been proposed to represent design methods. However, they did not consider the relationship among the methods formally to integrate the products developed by multiple methods. Our meta model can support method integration, i.e. integrating method fragments into a new method. We also mentioned an example of using multiple methods OOA, DFD, and STD in Figure 2. It might be considered as a simple example of method integration of OOA, SA, and STD to a new method. Viewpoint approach[12] is very interesting and similar to our system. It supports development by multiple methods but has not supported communication among workers yet.

Recent research in Software Process community have produced process-centered software develop environments (SDEs), e.g. Merlin[13] and Marvel[1]. Supporting cooperative work in them is mainly concurrency control of multiple access to a product to maintain consistency and to avoid interference. In the phase which our tool supports, the only worker that is developing a product can update it, and the others can only read it. Thus we have no automatic concurrency control mechanism of accesses to a product now. Support for communication and providing work environment for individual workers to use easily are very important in this phase. Our focus is on the personal process support[9] with loosely coupled communication.

6 Discussion and Directions to Future Work

This paper discussed the characteristics of distributed individual work in cooperative specification processes and proposed the tool for supporting the work based on the CASE tool with multiple methods and the structured electronic mail system. Users of our method base system requested that its guidance or navigation during a method use should be more powerful. The edit window of a method displays the only simple flow chart of the typical procedures which appears in text books. It was useful for our users to confirm the procedures. However the users wanted to illustrate the know-how of applying the procedures. Displaying small examples can be considered as useful guidance.

The consistency check among the products developed by different methods is a crucial problem to this kind of tool as discussed in [12]. Our common meta model has been developed not to represent all methods precisely and completely but to use product integration. How to define the method-specific parts depends on tool users, in particular so called method engineers. Thus the supporting tool for designing the methods, a CAME (Computer Aided Method Engineering) tool[8], would be needed. Graphical notation of the products is one of method-specific parts and specific graphical editors for each method are needed. Many of these graphical editors have the common parts because almost graphical notations consists of edges and nodes. Automatic or semi-automatic generation of the graphical editors would be one of the functions of the CAME tool.

The integration that our method base system can deal with is product integration and it cannot integrate procedural aspect of different methods yet. Suppose that we integrate Object Diagram (OD) and STD into a new method. We have the typical procedure for composing an OD and the procedure for a STD. What is the procedure for developing a product following the new method? It must integratedly consist of the two procedures of OD and STD. How to integrate them into the procedure of the new method depends on semantical connection between OD and STD. Assume that an object in a OD has internal state and its state transitions are specified a STD. In this case, the new method naturally starts with the procedure "identify objects" which comes from the first procedure of OD. We will explore the patterns of semantical connections on different methods and their relation to the way of integrating procedures.

The practical assessment of this tool is one of future work. The scope of our prototype tool is asynchronous communication tasks, e.g. an individual design activity. Supporting synchronous communication in face-to-face session such as elicitation and review of user requirements[15] is one of the future research.

Acknowledgements

The authors wish to thank K. Iguchi, M. Shinohara, W. Kuo and A. Tanaka for helpful support to the development of the prototype system.

References

1. N.S. Barghouti. Supporting Cooperation in the MARVEL Process-Centered SDE. In *SIGSOFT'92 : Proc. of the Fifth ACM SIGSOFT Symposium on Software Development Environments*, pages 21–31, 1992.
2. G. Booch. *Object Oriented Design With Applications*. Benjamin Cummings, 1991.
3. U. Borghoff and G. Teege. Application of Collaborative Editing to Software-Engineering Projects. In *ACM SIGSOFT, Software Engineering Notes*, number 3, pages A56–A64, 1993.
4. S. Brinkkemper. *Formalisation of Information Systems Modelling*. Thesis Publisher, 1990.
5. P. Coad and E. Yourdon. *Object-Oriented Analysis*. Prentice Hall, 1990.
6. B. Curtis. Implication from Empirical Studies of the Software Design Process. In *Proc. of Int. Conf. by IPSJ to Commemorate the 30th Anniversary*.
7. C.A. Ellis, S.J. Gibbs, and G.L. Rein. Groupware : Some Issues and Experiences. *Commun. ACM*, 34(1):38–58, 1991.
8. F. Harmsen and S. Brinkkemper. Computer Aided Method Engineering. In *Proc. of the 4th Workshop on the Next Generation of CASE Tools*, pages 125–140, 1993.
9. W.S. Humphrey. The Personal Software Process – Rationale and Status. In *Proc. of the 8th International Software Process Workshop*, pages 102–103, 1993.
10. K. Lai, T. W. Malone, and K. Yu. Object Lens : A Spreadsheet for Cooperative Work. *ACM Trans. on Office Information Systems*, 6(4):332–353, 1988.

11. T.G. Lewis. *CASE : Computer-Aided Software Engineering.* Van Nostrand Reinhold, 1991.

12. B. Nuseibeh, J. Kramer, and F. Finkelstein. Expressing the Relationships between Multiple Views in Requirements Specification. In *Proc. of the 15th ICSE*, pages 187–196, 1993.

13. B. Peuschel and W. Schafer. Concepts and Implementation of a Rule-based Process Engine. In *Proc. of 14 th ICSE*, pages 262–279, 1992.

14. C. Potts. A Generic Model for Representing Design Methods. In *Proc. of 11 th ICSE*, pages 217–226, 1989.

15. C. Potts, K. Takahashi, and A. Anton. Inquiry-Based Requirements Analysis. *IEEE Software*, 11(2):21–32, 1994.

16. M. Saeki, K. Iguchi, K. Wen-yin, and M. Shinohara. A Meta-Model for Representing Software Specification & Design Methods. In *Information System Development Process*, pages 149–166. North-Holland, 1993.

17. M. Saeki and K. Wenyin. PCTE based Tool for Supporting Collaborative Specification Development. In *Proc. of PCTE'94 Conference*, pages 121–134, 1994.

18. K. Smolander, K. Lyytinen, V.P. Tahvanainen, and P. Marttiin. MetaEdit — A Flexible Graphical Environment for Methodology Modelling. In *Proc. of 3rd International Conference CAiSE91, LNCS 498*, pages 168–193, 1991.

19. X. Song and L.J. Osterweil. Experience with an Approach to Comparing Software Design Methodologies. *IEEE Trans. on Soft. Eng.*, 20(5):364–384, 1994.

20. P. Sorenson, J. Tremblay, and A. McAllister. The Metaview System for Many Specification Environments. *IEEE Software*, 2(5):30–38, 1988.

21. T. Winograd. Where the action is. *BYTE*, 13(13):256–260, 1988.

Handling Changes in Dynamic Specifications in Object Oriented Systems

Parimala N.
Birla Institute of Technology and Science
Pilani 333031
INDIA

ABSTRACT

One of the benefits of object oriented analysis is that the specifications are resilient to change in the requirements. However, we show in this paper, that a change in the dynamic part of the requirements may lead to inconsistent changes in the analysis product. This is essentially due to the nature of the state transition diagrams. We present an augmentation to state transition diagrams. The main benefit of this augmentation is the ability to handle change in the dynamic requirements of the systems.

1.0 Introduction

Object oriented analysis views the system in terms of objects and information organized around these objects. The analysis proceeds along two dimensions. These are the static and the dynamic. The static part of the system analysis is concerned with identifying the static nature of the system which includes the identification of the classes, their structural properties and the class hierarchy. The dynamic analysis deals with the behavioural aspects of objects. This includes state transition of objects, events and event trace.

All systems are susceptible to change and object oriented systems are no exception. However, object oriented systems are to provide stability over change in the requirements. The impact of change is supposed to be easily identified, bounded and assessed[Coa91]. Even though different methods of object oriented analysis have been studied and compared [Goo92], the study does not include the efficiency with which these methods handle changes in the requirements. Our main work has been to identify the changes that may take place in the requirements and study how easily these changes can be incorporated in the object oriented analysis of the system. The change itself, can be in the static aspect or in the dynamic behaviour of the system. In this paper we are concerned with a change in the dynamic behaviour of the system.

Modelling the dynamic behaviour in object oriented systems has several aspects to it. These different aspects are modelled using different models. The major aspects that are dealt with are the state transition nets, the event trace, and sometimes process modelling using data flow diagrams. The state transition nets show the different states of objects of a given class; the events that cause the transition from one state to another and the actions that are to be performed. State transition nets are varyingly called state transition models [Emb92], state diagrams [Rum92], state transition diagrams [Sul93, Boo94]. The exact sequence of events is also specified in some methodologies using timing diagrams [Fir93, Boo94] or scenarios [Rum92] or object charts[Col92].

Whenever there are multiple diagrams, then any change in the system may imply that many diagrams will have to undergo modifications. In such a situation, there is the possibility that partial modification may take place leaving the set of diagrams in an inconsistent state.

We postulate that when analysing the dynamic behaviour of an object the different aspects of this behaviour must not be segragated and analysed separately but all aspects must be considered together. The dynamic requirements analysis must consider the intra as well as inter object behaviour together. That is, the analysis must deal with an integrated behaviour of objects covering state transitions that an object undergoes and the events that are sent and received by objects. If this were to be done, then any change to be effected would be in one place. This would eliminate partial modification which leads to inconsistent diagrams.

The purpose of the paper is to show that if all the different behaviourial aspects are combined and an integrated diagram is drawn then the modifications to be made to the diagrams to reflect a change in the system would be localised. This localisation implies that changes are to be made in only one place and therefore, do away with partial modification. In this paper we have considered integrating the state transition diagrams and events that are sent and received by objects, to demonstrate the feasibility of the approach.

We adopt standard definitions for the static part of the object model viz. object class, class diagrams and class hierarchy. In the dynamic part we combine the state transition diagram and the event flow to define an integrated diagram.

The layout of this paper is as follows. In section 2, we show, with the help of an example how partial modification may happen leaving the analysis diagrams inconsistent. The integrated diagram for the dynamic behaviour is defined in section 3. A detailed example is worked out in the next section. Here, it is also shown how changes to analysis cause local changes in the integrated diagram. Section 5 is the concluding section.

2.0 The Partial Modification Problem

In this section we perform analysis of an example system and arrive at the different diagrams of the analysis document. Later, we assume that requirements undergo a change and show how partial modifications can take place.

Consider a library system where books are borrowed, renewed and returned by authorized borrowers of the library. The object classes are Book, Borrower and Counter. Every borrower has a limit on the number of books he may borrow. Let us assume that the requirements analysis has been performed and the analysis document which consists of a suite of class diagrams and diagrams for the dynamic behaviour have been arrived at. Of interest to us in this paper, are the diagrams reflecting the dynamic behaviour of the classes. The state transition diagrams for the Borrower, Book and Counter are given in Fig. 2.1 using the notation of [Rum92].

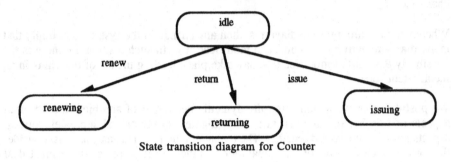

State transition diagram for Counter

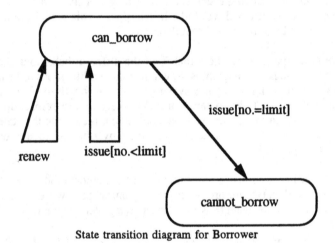

State transition diagram for Borrower

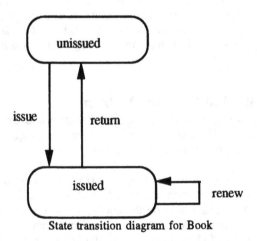

State transition diagram for Book

Fig. 2.1. State Transition Diagrms for Counter, Borrower, and Book

The event trace for borrowing a book could be as depicted in Fig. 2.2.

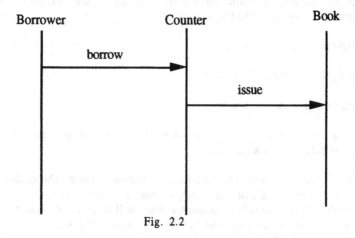

Fig. 2.2

Similar event traces can be drawn for returning and renewing borrowed books.

Let us, now, assume that there is a change in the policy of lending books in the library. The new rule is that no book may be renewed. Notice that the change is in the dynamic behaviour of the system. If this change is to be incorporated then modifications to different diagrams has to be made to reflect the situation. In particular, the state transition diagram of Borrower has to be modified; in the class Counter the transition to the state renewing has to be absent; similarly in the class Book self transition to the state 'issued' upon renew should be removed. The event trace for the scenario of renewal will have to be removed.

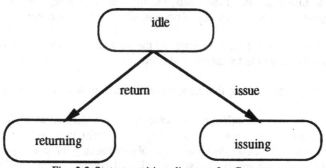

Fig. 2.3 State transition diagram for Counter

If it happens that only the state transition diagram of Counter is modified to the one in Fig. 2.3 then we see that the suite of state transition diagrams is inconsistent. There is nothing in the system to prevent this situation. Clearly if the changes were to be made in just one place then there would be no inconsistency.

3.0 The Model

We accept the standard conventions for arriving at the definition of the static behaviour of the system. The broad steps are

1) Identify the classes

2) Identify the attributes of each class

3) Identify the inheritance hierarchy

As the 'state' of an object is crucial for defining the behaviourial model we explicitly include the definition of a state here.

State The state of an object is the collective values of the attributes that are defined in the class to which the object belongs. During analysis, the state of an object is identified by a name. This later in design stage, will be mapped to the exact values of the attributes. An object, from the time it is created to the time it is destroyed must be in some specified state.

3.1 Dynamic analysis

Dynamic analysis is concerned with identifying the behaviour of objects. This encompasses

1) the state transition that an object goes through as a result of events or conditions and

2) the event trace.

In our approach we combine both the above aspects of the dynamic behaviour and draw a single integrated diagram. Thus, the integrated diagram relates states of the object changing states and the states of objects which cause the event.

We distinguish between events internal to the system from those external to it. In this paper, we are concerned with internal events.

Event An event is something that happens at a point in time. It causes a transition.

Transition State transition[Rum92, Boo94, Shl88] is defined as a change of state which is caused by an event. Even though the event causing the transition is specified, it is usual to leave unspecified the object causing the event and the state in which it can cause the event. As suggested by [Rum92], in the later stages the various diagrams have to be combined to arrive at such information.

We propose to augment the state transition diagram to include not only the state transition of objects of a given class C1 but also the states of objects of class C2....Cn which cause the state transitions of objects of C1. To achieve this, first we define the augmented state transition diagram (ASTD), then the scenario state transition diagram (SSTD) and lastly the integrated diagram (ID).

Augmented State Transition Diagram

Let an object O1 belonging to class C1 in state OS1 make a transition to state OS2 when the event ev1 takes place. Let ev1 be an event caused by an object O2 belonging to class C2 in state OS. Then, a dotted line with the label ev1 is drawn from OS of O2 to the transition of O1 from OS1 to OS2 as shown in Fig. 3.1. This augmented state transition diagram specifies

1) the state transition of an object of a class C1,

2) the class C2, an object of which causes the event, and

3) the state in which it can cause the event.

It may be possible that the event ev1 can be caused by objects belonging to more than one class. In such a situation, the ASTD consists of

1) the state transition of an object of a class C1,

2) all the classes, an object of each of which cause the event, and

3) the state in which each of these objects can cause the event.

Consider the library example defined in section 2.0. The ASTD for the state transition of Counter is shown in Fig. 3.2. Borrower in state can_borrow causes the event 'borrow' when the counter is in idle state. The counter then moves to on_issual state.

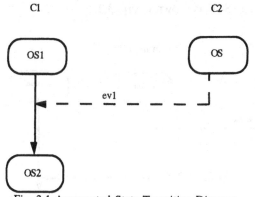

Fig. 3.1 Augmented State Transition Diagram

Scenario State Transition Diagram

An event scenario [Rum92] is a sequence of events occuring during one particular execution of the system. It gives the event trace.

The scenario state transition diagram is drawn as follows:

Start with the first event of the event trace and draw the ASTD for the object which goes through a state transtition when this event taken place. Pick up the next event and extend the existing state transition diagram with states, transitions and events to include the ASTD for the state transition caused by the new event. Continue till all the events in the event trace are exhauted.

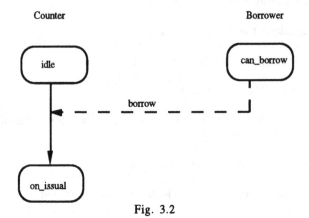

Fig. 3.2

For the library system, consider the event trace for borrowing a book as shown in Fig. 2.2. The first event is borrow. This causes a state transition in Counter. The ASTD for Counter will be as shown in Fig. 3.2. The next event is issue. This causes a state transition in Book from the state unissued to issued. This implies, that a new class Book has to be added and the ASTD for Book has to be included in our earlier diagram. The SSTD is shown in Fig. 3.3.

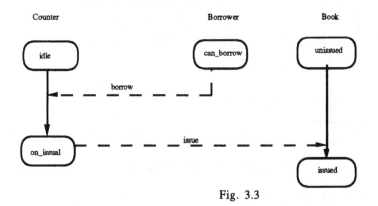

Fig. 3.3

Integrated Diagram

An integrated diagram consists of all the scenario state transition diagrams of a given system. That is, it includes the scenario state transition diagrams for all the event traces of the system. An example integrated diagram is given in section 4.0.

Graphical notation

A state is denoted by a rounded rectangle with the name of the state of the object in it. The class name to which the object belongs is written above the ASTD of the object. Each object's state transition is in a vertical line. State transitions are denoted as solid lines with an arrow head. The head points to the new state of the object. The events are denoted as dashed lines with an arrow head. The line originates from the state of the object which causes the event and it points to the transition of the object which undergoes the particular transition as a result of the event.

Solution to Partial Modification Problem

The integrated diagram gives the complete specification of the dynamic behaviour of the system which includes the state transitions of objects of various classes and the states of objects of other clases causing the event. When the system undergoes a change in its dynamic behaviour then the entire gamut of changes to be made in the analysis diagrams is in one place. Thus, any change can be reflected without giving rise to any partial modification.

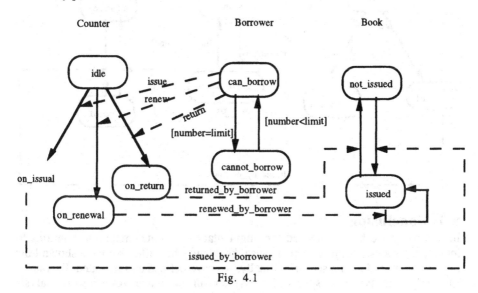

Fig. 4.1

4.0 The Library - An Integrated Diagram

In this section we consider the library example of section 2 again and draw an integrated diagram for the dynamic behaviour. The static analysis remains the same. Therefore, we have the object classes - Borrower, Book and Counter. The integrated diagram is shown in Fig. 4.1. The Counter in the idle state can recieve from Borrower in the state 'can_borrow' the event - renew. The counter then moves to the next state - on renewal'.

The event renewed_by_borrower is sent to the book which is in the 'issued' state. The transition takes the book back to the same 'issued' state. This completes the event trace for renew.

Similarly, we can trace the sequence of events and state changes for issue and return.

Borrower moves from the state 'can_borrow' to the state 'cannot borrow' when the condition 'number = limit' is reached and goes back to 'can_borrow' when 'number < limit'.

Assume now, as before, that there is the change in renewal policy. Here, we trace the sequence of events and state changes for renew. All events in this sequence are dropped. If any event causing the state transition is the only one causing it, then the new state of the recieving object is also dropped. Applying this to renew we get the modified integrated diagram as shown in Fig. 4.2.

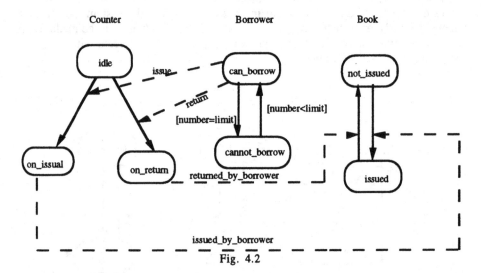

Fig. 4.2

5.0 Conclusion

In this paper we have proposed the augmentation of state transition diagrams to include the states of objects which cause the event. Thereafter, we have shown how this can be extended to a scenario. Finally, we have defined an integrated diagram to capture all the dynamic aspects of the system in one place. As a result, analysis diagrams have information localised and thereby do away with inconsistency when requirements undergo a change. We have also shown how change in reuqirements can be easily accomodated.

Integrated diagrams themselves may become large and unwieldy if drawn for an entire system. Currently, we are in the process of defining abstraction of integrated diagrams so that integration can be visualized at different levels. As is well known, this permits developing integrated diagrams at various levels of abstraction.

References

[Col92] Coleman D. et al, Introducing Object charts or How to use Statecharts in Object oriented Design, IEEE Trans. on SE, Vol. 18, No. 1, 1992

[Boo94] Booch G., Object Oriented Analysis and Design with applications, The Benjamin/Cummings Publishing Company

[Coa91] Coad P. and Yourdon E., Object Oriented Analysis, Yourdon Press, 1991

[Emb92] Embley D.W. et al, OO Systems Analysis - A model driven approach, Yourdon Press, 1992

[Fir86] Firesmith D.G., Object Oriented Requirements Analysis and Logical Design - A Software Engineering Approach

[Goo92] Goor G. van den et al, A comparison of six Object Oriented Analysis and Design Methods, Research project report, Method Engineering Institute, Netherlands

[Rum91] Rumbaugh J. et al, Object_oriented Modeling and Design, Prentice Hall, 1991

[Sha88] Shlaer S., Mellor S.J., Object Oriented Systems Analysis: Modelling the world in Data, Yourdon Press, Englewood Cliffs, N.J. 1988

[Sul93] Sully P., Modelling the world with objects, Prentice Hall, 1993

The Rejuvenation of Materialized Views

(Extended Abstract)

Inderpal Singh Mumick

AT&T Bell Laboratories
600 Mountain Avenue
Murray Hill, NJ 07974, USA.
mumick@research.att.com

Abstract. This is a short summary of a talk presented at the sixth International *Conference on Information Systems and Management of Data* (CISMOD 95) held in Bombay in November 1995.

The summary describes some of the applications that are causing a renewed interest in materialized views, the problems in supporting the applications, and sketches how the current work in this area is addressing some of the problems.

1 Introduction

What is a view? A view is a derived relation defined in terms of base (stored) relations. A view thus defines a function from a set of base tables to a derived table; this function is typically recomputed every time the view is referenced.

What is a materialized view? A view can be materialized by storing the tuples of the view in the database. Index structures can be built on the materialized view. Consequently, database accesses to the materialized view can be much faster than recomputing the view. A materialized view is thus like a cache – a copy of the data that can be accessed quickly.

Why use materialized views? Like a cache, a materialized view provides fast access to data; the speed difference may be critical in applications where the query rate is high and the views are complex so that it is not feasible to recompute the view for every query.

What is view maintenance? Just as a cache gets *dirty* when the data from which it is copied is updated, a materialized view gets dirty whenever the underlying base relations are modified. The process of updating a materialized view in response to changes to the underlying data is called view maintenance.

What is incremental view maintenance? In most cases it is wasteful to maintain a view by recomputing it from scratch. Often it is cheaper to use the heuristic of inertia (only a part of the view changes in response to changes in the base relations) and thus compute only the changes in the view to update its materialization. Algorithms that compute changes to a view in response to changes to the base relations are called *incremental view maintenance* algorithms. A classification and survey of several view maintenance algorithms appears in [GM95].

Materialized views were investigated in the 1980's as a tool to speed up queries on views. Several view maintenance algorithms were proposed [NY83, Pai84, SI84, BLT86, LHM+86, BCL89, SP89, QW91], analytical models for the cost of view maintenance were developed [Han87, SR88, SF90], and the impact of materialized views on the performance of queries was studied [Han87, BM90]. The use of view maintenance algorithms in maintaining integrity constraints was also recognized [BC79]. However, implementations of materialized views in research prototypes and commercial systems were missing (with the sole exception of ADMS [Rou91]). The benefit of materialized views to applications was never proven, and no killer application came up to force the implementation of materialized views.

2 Applications for Materialized Views

Over the last couple of years, there has been a renewed interest in materialized views in both the research and industrial communities. Several factors are contributing to this:

- New and novel applications for materialized views and view maintenance techniques are emerging. These include data warehousing, mobile systems, data visualization, data replication, and distributed CD-ROM services. More applications remain to be discovered !
- A new look at some traditional applications such as banking, retailing, and billing shows that these can be managed more efficiently, and with fewer errors, using materialized views.
- There is an interest from the database vendors to implement materialized views, and a desire by large database customers for functionality that can be obtained through materialized views.
- Disk prices have come down dramatically, so the disk space needed to store materialized views is cheaper.

We discuss a couple of these applications, and some of the problems the applications present.

Data Warehousing: A database system that collects and stores data from several databases is often described as a data warehouse. The warehouse usually integrates data from multiple sources, and provides a different way of looking at the data than the databases being integrated, so that copying each database into the warehouse is wasteful.

Materialized views provide a framework within which to collect information into the warehouse from several databases without copying each database in the warehouse. Queries on the warehouse can then be answered using the materialized views without accessing the remote databases. Provisioning (modification of the data) still occurs on the remote databases. The modifications are then transmitted to the warehouse. Incremental view maintenance techniques can be used to maintain the materialized views in response to these modifications.

Banking and Retailing: Banking, retailing, and billing systems deal with a continuous stream of transactional data. One characteristic of such data is that it can get very large, and it can be beyond the capacity of any database system to even store, far less access, all of this data for answering queries. Consequently, the data is stored for only a specific period of time (this can vary from a few weeks to a few months), and even then it typically forms the largest data set in the system.

Decision support applications typically involve aggregate queries over the data. For example, one may be interested in total cash withdrawals from each branch of a bank on each of the last 30 days. Transactional applications also may need to access aggregate data. For example, a withdrawal transaction would like to know the balance in the customer's account, and the balance may be obtained by adding up the credits and debits from *all* the past banking transactions in the account.

Applications that need a fast response time and high throughput on queries with aggregation build summary tables by aggregating the current transactional data sets along several dimensions, and store the summary tables as if they were base data. By asking queries against the summary tables whenever possible, the response time and throughput for queries is improved. Similarly, summaries that encapsulate the interesting features of old transactional data (such as the account balance) are defined, and stored as base data. One of the problems with this approach is that the applications also have the burden of maintaining the summary tables after every update to the transactional data.

Materialized Views can be used to move the functionality of summary tables into the database system, and out of the application code. The maintenance of summary tables is then automated. Consequently, the maintenance process can be more efficient, as well as error-free (an error in the maintenance code was responsible for Chemical bank ATM withdrawals causing incorrect updates to customer balances on February 18, 1994, leading to several bounced checks and frustrated customers [NYT94]). More summary tables can be defined, and each summary table can be defined quickly. Further, since the system knows that the summary tables are really views, even queries that reference the base transactional data can be optimized to reference the summary tables instead.

3 Problems

We give a sampling of new view materialization/maintenance problems due to these application domains. Citations to the recent work addressing some of the problems are included.

Data Warehousing:

- What should the view language support to enable an integration of data from different sources? Outer-joins [GJM94] and general matching [ZHKF95] have been proposed.
- While the materialized views are available for view maintenance in a data warehouse, access to the remote databases may be restricted or expensive. It is thus beneficial if the view maintenance can proceed without accessing the remote databases. A view that can be so maintained is called self-maintainable in [GJM94] and autonomously computable in [BCL89]. When is a view self-maintainable? What is the impact of foreign keys, functional dependencies, and aggregation on self-maintainability?
- The remote databases need to communicate their updates to the warehouse. Depending on the capabilities of the remote databases, they may send the actual changes, the SQL insert/delete/update statements, or database snapshots. Can data replication facilities in database products be used to send a log of changes? How does one run view maintenance in each scenario?
- When a view is not self-maintainable, what are the other views that can be materialized at the warehouse so that the collection of views can be *self-maintained*?
- When a view is not self-maintainable, and a remote database needs to be accessed to maintain it, how do we guarantee consistency of the materialized view? [ZGHW95]
- A materialized view in a warehouse may be modified directly at the warehouse, either to make annotations or to clean up erroneous data. Thus, the materialization is no longer derived exactly according to the definition of the view. How does one maintain such views without losing the warehouse-specific modifications?

Banking and Retailing:

- How do you deal with aggregation during view maintenance? [GMS93, GL95]
- In a batch environment, where updates to the base relation and hence view maintenance are done in batch, a desirable goal is to minimize the amount of time for which the view is unavailable for queries? It is thus important that the view be write locked by the view maintenance process for the minimum possible time. Can we implement view maintenance algorithms such that they compute the update to the materialized view separately from the materialization, without locking the data, and then quickly apply the update to the materialization, holding the lock for the short duration needed to apply the update.
- In a transactional environment, the view must be maintained after each transaction, so the view maintenance algorithm must be very efficient. What is the class of views that can be maintained in a high throughput transaction system? [JMS95]

- Given a set of queries over the transactional data, what aggregate views must be materialized, so that the queries can be answered even after some of the transactional data has been discarded?
- A user of the database system may not know about all the views being materialized by the system. How can a query over base data be automatically optimized to use the materialized views instead [RSU95, LMSS95, CKPS95, DJLS95, GHQ95]. For instance, consider a query in a retailing application that wants to compute the number of items sold for each item. A query optimizer can optimize this query to access a materialized view that stores the number of items sold for each item and store, and avoid access to a much larger sales-transactions table.

Implementations: The applications are forcing a closer look at implementation issues surrounding materialized views and view maintenance. When are materialized views maintained – before the transaction that updates the base relation commits (immediate maintenance), or after the transaction commits (deferred maintenance [LHM+86, SP89, SF90])? Is view maintenance part of the transaction or not? What transaction consistency guarantees are possible when the view maintenance is deferred? Several implementation efforts are under way. There are two efforts at AT&T Bell Laboratories, one investigating how to implement materialized views inside an object-oriented database, while the second tries to implement materialized views on top of an existing relational database. A project at Columbia University is implementing view adaptation [GMR95] on top of Sybase. The WHIPS project at Stanford is attempting to integrate data from multiple sources into a data warehouse using materialized views [HGW+95].

Acknowledgements

I thank Dallan Quass and Timothy Griffin for comments on a draft of this paper, and Ashish Gupta for exploring several of these topics with me.

References

[BC79]   Peter O. Buneman and Eric K. Clemons. Efficiently monitoring relational databases. *ACM Transactions on Database Systems*, 4(3):368–382, September 1979.

[BCL89]   J. A. Blakeley, N. Coburn, and P. Larson. Updating derived relations: Detecting irrelevant and autonomously computable updates. *ACM Transactions on Database Systems*, 14(3):369–400, September 1989.

[BLT86]   J. A. Blakeley, P. Larson, and F. W. Tompa. Efficiently Updating Materialized Views. In [Sig86], pages 61–71.

[BM90]   Jose A. Blakeley and Nancy L. Martin. Join index, materialized view, and hybrid hash join: A performance analysis. In [DE90], pages 256–263.

[CKPS95]   Surajit Chaudhuri, Ravi Krishnamurthy, Spyros Potamianos, and Kyuseok Shim. Optimizing queries with materialized views. In *Proceedings of the Eleventh IEEE International Conference on Data Engineering*, Taipei, Taiwan, March 6-10 1995.

[DE90]   *Proceedings of the Sixth IEEE International Conference on Data Engineering*, Los Angeles, CA, February 5-9 1990.

[DEB95]   Jennifer Widom, editor. *IEEE Data Engineering Bulletin, Special Issue on Materialized Views and Data Warehousing*, 18(2), June 1995.

[DJLS95]   Shaul Dar, H.V. Jagadish, Alon Y. Levy, and Divesh Srivastava. Answering SQL queries with aggregation using views. Technical Memorandum, AT&T Bell Laboratories, 1995.

[GHQ95]   Ashish Gupta, Venkatesh Harinarayan, and Dallan Quass. Generalized projections: A powerful approach to aggregation. In Umeshwar Dayal, Peter M.D. Gray, and Shojiro Nishio, editors, *Proceedings of the 21st International Conference on Very Large Databases*, Zurich, Switzerland, September 11-15 1995.

[GJM94]   Ashish Gupta, H. V. Jagadish, and Inderpal Singh Mumick. Data integration using self-maintainable views. Technical Memorandum, AT&T Bell Laboratories, November 1994.

[GL95]   Timothy Griffin and Leonid Libkin. Incremental maintenance of views with duplicates. In [Sig95].

[GM95]   Ashish Gupta and Inderpal Singh Mumick. Maintenance of Materialized Views: Problems, Techniques, and Applications. In [DEB95], pages 3–19.

[GMR95]   Ashish Gupta, Inderpal Singh Mumick, and Kenneth A. Ross. Adapting materialized views after redefinitions. In [Sig95].

[GMS93]   Ashish Gupta, Inderpal Singh Mumick, and V. S. Subrahmanian. Maintaining views incrementally. In *Proceedings of ACM SIGMOD 1993 International Conference on Management of Data*, Washington, DC, May 26-28 1993.

[Han87]   Eric N. Hanson. A performance analysis of view materialization strategies. In Umeshwar Dayal and Irv Traiger, editors, *Proceedings of ACM SIGMOD 1987 International Conference on Management of Data*, pages 440–453, San Francisco, CA, May 27-29 1987.

[HGW+95]   Joachim Hammer, Hector Garcia-Molina, Jennifer Widom, Wilburt Labio, and Yue Zhuge. The Stanford Data Warehousing Project. In [DEB95], pages 41–48.

[JMS95]   H. V. Jagadish, Inderpal Singh Mumick, and Avi Silberschatz. View maintenance issues in the chronicle data model. In [POD95], pages 113–124.

[LHM+86]   Bruce Lindsay, Laura Haas, C. Mohan, Hamid Pirahesh, and Paul Wilms. A snapshot differential refresh algorithm. In [Sig86], pages 53–60.

[LMSS95]   Alon Y. Levy, Alberto O. Mendelzon, Yehoshua Sagiv, and Divesh Srivastava. Answering queries using views. In [POD95], pages 95–104.

[NY83]   J. M. Nicolas and K. Yazdanian. An Outline of BDGEN: A Deductive DBMS. In *Information Processing*, pages 705–717, 1983.

[NYT94]   New York Times. Bug in chemical bank's ATM software, February 18 1994. Front page article.

[Pai84]   R. Paige. Applications of finite differencing to database integrity control and query/transaction optimization. In H. Gallaire, J. Minker, and J. Nicolas, editors, *Advances in Database Theory*, pages 170–209, New York, 1984. Plenum Press.

[POD95]   *Proceedings of the Fourteenth Symposium on Principles of Database Systems (PODS)*, San Jose, CA, May 22-24 1995.

[QW91]   Xiaolei Qian and Gio Wiederhold. Incremental recomputation of active
            relational expressions. *IEEE Transactions on Knowledge and Data En-
            gineering*, pages 337–341, 1991.

[Rou91]   Nick Roussopoulos. The incremental access method of View Cache: Con-
            cept, algorithms, and cost analysis. *ACM Transactions on Database Sys-
            tems*, 16(3):535–563, September 1991.

[RSU95]   Anand Rajaraman, Yehoshua Sagiv, and Jeffrey D. Ullman. Answering
            queries using templates with binding patterns. In [POD95], pages 105–
            112.

[SF90]   Arie Segev and Weiping Fang. Currency-based updates to distributed
            materialized views. In [DE90], pages 512–520.

[SI84]   Oded Shmueli and A. Itai. Maintenance of Views. In *Proceedings of
            ACM SIGMOD 1984 International Conference on Management of Data*,
            pages 240–255, 1984.

[Sig86]   Carlo Zaniolo, editor. *Proceedings of ACM SIGMOD 1986 International
            Conference on Management of Data*, Washington, D.C., May 28-30 1986.

[Sig95]   Michael Carey and Donovan Schneider, editors. *Proceedings of ACM
            SIGMOD 1995 International Conference on Management of Data*, San
            Jose, CA, May 23-25 1995.

[SP89]   Arie Segev and Jooseok Park. Updating distributed materialized views.
            IEEE Transactions on Knowledge and Data Engineering, 1(2):173–184,
            June 1989.

[SR88]   Jaideep Srivastava and Doron Rotem. Analytical modeling of material-
            ized view maintenance. In *Proceedings of the Seventh Symposium on
            Principles of Database Systems (PODS)*, pages 126–134, Austin, TX,
            March 21-23 1988.

[ZGHW95]   Yue Zhuge, Hector Garcia-Molina, Joachim Hammer, and Jennifer
            Widom. View maintenance in a warehousing environment. In [Sig95].

[ZHKF95]   G. Zhou, R. Hull, R. King, and J-C. Franchitti. Using object matching
            and materialization to integrate heterogeneous databases. In *Proc. of
            3rd International Conference on Cooperative Information Systems*, pages
            4–18, 1995.

Normalization of Linear Recursions Based on Graph Transformations

Xiaoyong Du, Naohiro Ishii

Department of Intelligence and Computer Science
Nagoya Institute of Technology, Nagoya 466, Japan
E-mail:{duyong, ishii }@egg.ics.nitech.ac.jp

Abstract. In this paper, we propose a new approach to generate normal form formulas for linear recursions based on graph transformations. We first extend the graph model proposed in [17] for representing linear recursive definitions completely, coupling with graph equivalence definitions. The new graph model is called IE-graph. Then three basic equivalence-preserving graph transformation techniques are newly defined on IE-graphs: (1) realigning; (2) reducing; (3) expanding. Based on these graph transformation techniques, we show that a general IE-graph can always be transformed equivalently into a set of disjoint unit cycles, called Normal IE-graph. The formula generated by our method is more efficient than that generated by Han and Zeng's method [7], because the formula generated by our method contains usually less variables in the recursive predicate.

Key words: deductive databases, linear recursions, compilation and optimization, normalization

1 Introduction

Compilation is a powerful preprocessing technique in the evaluation of recursions in deductive database systems. It has been studied extensively in the past decade. Henschen-Naqvi method [5], Magic Sets and Counting [2, 1] etc. are some typical examples of compilation strategies. Many recent studies on compilation focus on normalization of linear recursion [12, 13, 6, 7, 17, 10]. A normal form of a linear recursion is one which recursive rule has the form as follows:

$$P(x_1, x_2, ..., x_n) :- A_1(x_1, y_1), A_2(x_2, y_2), ..., A_n(x_n, y_n),$$
$$P(y_1, y_2, ..., y_n)$$

It is also called a n-chain recursion.

The compilation of linear recursion into highly regular normal form has several advantages. Firstly, it facilitates the quantitative analysis of recursive queries and generation of efficient query evaluation plans. When a complex recursion is compiled into normal form, it is straightforward to select an appropriate algorithm from a set of candidate query processing algorithms applying on bounded recursions, single chain recursions(partial transitive closure), and multi-chain recursions respectively. Secondly, it facilitates capture of more bindings for efficient query evaluation. Some bindings which are difficult to be captured by other techniques, e.g. Magic Sets, can be captured naturally by the normalization compilation technique, so as to reduce the set of relative data of the query. Thirdly, it will have notable implications on nonprocedural implementation of logical programs. Recursive rules are implemented in most systems using some procedural interpretations, which may distort the declarative semantics and cause problems of nontermination and inefficient processing. For example, a PROLOG programmer must clearly understand the underlying PROLOG impleementation mechanism, and carefully arrange the order of rules as well as predicates in each rule. Therefore normalization is a powerful tool for the analysis and evaluation of complex linear recursions in deductive databases.

Graphs are widely used in compilation of deductive rules. For example, α-graph [8] and its variants V-graph [6], I-graph [17]; A/V(Argument/Variable) graph [11] and Augmented A/V graph [13]; Rule/Goal graph [16]; and V-Matrix [7] etc.. These graphs are powerful for analyzing structure properties of recursive formulas, as well as understanding the behaviors properties of formula expansion.

However, all of these graphs have some common shortages when they are used in compilation and optimization of recursive rules. First, some optimization techniques can not be represented properly by these graph model. For example, as well known, there are two kinds of techniques utilized in normalization of linear recursion, that is, formula expansion [6, 7, 10, 17], and variable vectorization [17, 7]. Although formula expansion can be represented by graph expansion [17], nothing is for variable vectorization. Second, from the point of view of graph transformation, the process of compilation and optimization is also a process of graph transformation. That is, the graph corresponding a normal form formula should be equal to that corresponding original formula. Clearly, a definition of graph equivalence is necessary, however lack in the previous graph models.

The main contributions of this paper are to propose a graph model, coupling with graph equivalence definitions, and a set of equivalence-preserving graph transformations. Based on the graph model, a generation algorithm of normal form for recursive formulas is proposed. It is shown that this algorithm is better than the one proposed in [7], because the normal form it generated contains possibly less variables in the recursive predicate.

The paper is organized as follows: A graph model, called IE-graph, is introduced in Section 2 for representing linear recursions. An IE-graph is consisted of two parts, that is, I-graph and E-graph which are used to model the recursive rule and exit rules respectively. In Section 3, we propose three basic equivalence-preserving graph transformation operations. Based on these operations, we develop an algorithm which optimizes and transforms a general IE-graph to a Normal IE-graph in Section 4. The paper also gives a short comparison with related work in Section 5, and conclusion in Section 6.

2   The Graph Model

As same as most studies in this field, we restrict our discussion on function-free, negation-free, safe, single linear recursions. Moreover no constants and repeated variables are existed in the definitions. The later assumptions can be relaxed straightforwardly as discussions in [17], if we introduce a special predicate "EQUAL", and treat it as a special EDB relation.

Graph models are widely used for understanding behaviours and properties of recursive formulas. Our graph model for linear recursions includes two graphs, that is, I-graph and E-graph, to represent the recursive rule and the exit rule respectively. I-graph is not a new one. It was invented originally in [8] (called α-graph) and redefined in [17]. However I-graph is not enough for our purpose that rule rewriting can be treated as graph transformation. Hence, A new graph, called E-graph, as well as graph equivalence are introduced in the model.

Let r be a single linear recursion definition. r can be represented completely by I-graph and E-graph as follows:

Definition 1 (I-graph). I-graph $G_i(r) = (V_i, E_u, E_d, \Gamma_i)$ is a labeled hybrid graph defined for the recursive rule in r, where

a) V_i is a finite, nonempty set of nodes such that a node is defined and labeled by x if x is a variable in the recursive rule.
b) E_u is a finite, nonempty set of undirected edges such that an undirected edge (x, y) is defined and labeled by Q for every pair of variables x, y in Q. Specially, if Q is an unary base predicate, then a self-loop (x, x) will be defined.
c) E_d is a finite, nonempty set of directed edges such that for each variable x in the recursive predicate P in the head, a directed edge is defined between x and its corresponding variable y in the P in the body, denoted by $\langle x, y \rangle$.
d) Γ_i is the set of labels attached to edges in E_u and E_d.

Definition 2 (E-graph). E-graph $G_e(r) = (V_e, E, \Gamma_e)$ is a labeled weighted undirected graph defined for the exit rule in r, where

a) V_e is a finite, nonempty set of nodes such that a node is defined and labeled by x if x is a variable in the exit rule.
b) E is a finite, nonempty set of undirected edges such that an undirected edge (x, y) is defined and labeled by Q for every pair of variables x, y in Q. Specially, if Q is an unary base predicate, then a self-loop (x, x) will be defined.
c) Γ_e is the set of labels attached to edges in E.

If there are more than one exit rules, then the E-graph is union of such kind of graphs.

Example 1. Consider the following recursive formula.

$$P(x, y, z) : - E(x, y, z)$$
$$P(x, y, z) : - A(x, y), B(y, z_1), C(z, z_1), D(x_1, z_1), P(x_1, y_1, z_1)$$

Its I-graph and E-graph is showed in Fig. 1 (a) and (b) respectively. □

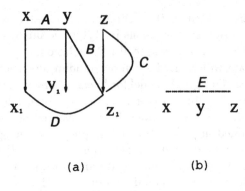

Fig. 1. IE-graph

Traversals are allowed in I-graphs. When a directed edge is encountered during a traversal , the directed edge has weight of 1 if it is along the direction of the traversal and -1 if against the direction of the traversal. The weight of an undirected edge is 0.

A traversal may form a cycle. A cycle is called *one-directional cycle* if all directed edges in the cycle is along in the same direction, otherwise it is *multiple-directional cycle*. A cycle is called *unit cycle* if it is a one-directional and its weigh is equal to one.

Nodes appearing in the head of an arrow are called *C-node*, otherwise called *A-node*. They are denoted as V_c, and V_a respectively. Obviously $V = V_c \cup V_a$. Node x is called *derived node* of y if x and y appears in the head and tail of a directed edge respectively. I-graph with n C-nodes is called *n-D I-graph*. I-graph which consists of a set of disjoint unit cycles is called *Normal I-graph*

From a pair of I-graph and E-graph (called IE-graph for shorthand), it is possible to generate a set of directed or undirected graphs based on the following basic operations.

Definition 3 (Breeding). Let G be an I-graph, and H an I-graph or E-graph which has same C-variables with G. Breeding H according to G , denoting as $\beta_G(H)$, means to replace all C-nodes in H by their derived nodes according to G, and all other nodes by new nodes.

Clearly, the operation $\beta_G(H)$ can not change the category of H. That is $\beta_G(H)$ is still an I-graph or E-graph if H is an I-graph or E-graph respectively.

Definition 4 (Gluing). Let G_1 and G_2 be I-graph or E-graph. Gluing G_1 and G_2, denoting as $G_1 * G_2$, is a graph where the same node in G_1 and G_2 is merged. If there are two directed edges $P < x, x_1 >$, and $P < x_1, x_2 >$, then replace them by a new directed edge $P < x, x_2 >$. Moreover,if G_1 or G_2 be a set of disjoint graphs, then $G_1 * G_2 = \{x * y | x \in G_1, y \in G_2\}$.

Clearly, $G_1 * G_2$ is an I-graph if G_1 or G_2 is an I-grtaph.

Example 2. In Example 1, let G be the I-graph in Fig. 1 (a)(denoted as I_1), and H be the graph in Fig. 1 (b)(denoted as E_1) respectively. Then operation $\beta_{I_1}(I_1)$, $\beta_{I_1}(E_1)$, $I_1 * I_1$, and $I_1 * E_1$ are showed in Fig. 2. $\qquad\square$

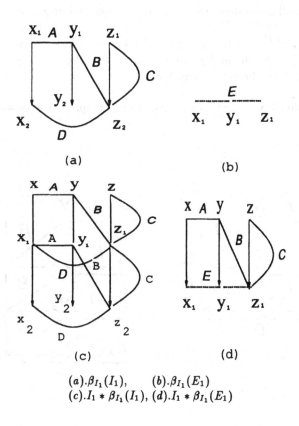

$(a).\beta_{I_1}(I_1),\qquad (b).\beta_{I_1}(E_1)$
$(c).I_1 * \beta_{I_1}(I_1),\ (d).I_1 * \beta_{I_1}(E_1)$

Fig. 2. Breeding and Gluing of IE-graph

Definition 5 (k-th I-graph). Let G_i be an I-graph. The k-th I-graph of G_i, denoted as G_r^k, is defined as follows:

$$G_i^1 = G_i.$$
$$G_i^k = G_i^{k-1} * \beta_{G_i^{k-1}}(G_i)$$

The k-th I-graphs are also I-graph. The result can be infered by the fact that G_i as well as $\beta_{G_i^{k-1}}(G_i)$ is I-graph.

Definition 6 (k-th Generated Graph). Let $G(r) = < G_i, G_e >$ be an IE-graph. The k-th generated graph, denoted as $R^k(r)$, is an undirected graph defined as follows:

$$R^k(r) = gen(G_i^k * \beta_{G_i^k}(G_e)).$$
$$R^0(r) = G_e.$$

where function gen(G) means to eliminate all directed edges from G.

The set of all generated graphs is called generated graph or R-graph [17] of G.

Node set $V(R_r^k)$ can be arranged as k layers, where the nodes in the first layer are all C-nodes in the I-graph G_r^1, and the nodes in the k-th layer are all nodes generated by the k-th expansion.

Example 3. The 2nd, 3rd generated graph of Example 1 are shown in Fig.3 (a), (b). □

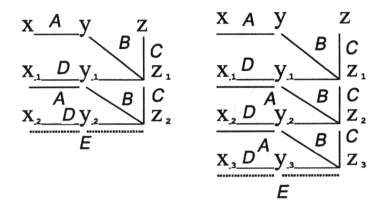

2nd R-graph   3rd R-graph

Fig. 3. R-graphs

In an IE-graph, each label corresponds to a predicate, and all labels form conjunctive relation. Hence we can define the concept of graph equivalence as follows:

Definition 7 (Equivalence). Let $G(r_1) = < G_i, G_e >$ and $H(r_2) = < H_i, H_e >$ be two IE-graphs defined for two linear recursive formula r_1 and r_2 respectively.

Let R_G and R_H be R-graph of G and H respectively. G and H are strong equivalence, denoted as $G = H$, if $R_G = R_h$. G and H are weak equivalence, denoted as $G \approx H$, if $R_G = A \cup B * R_H$, where A and B are two undirected graph.

Clearly, if there is a 1:1 mapping f from the subgraphs of R_G to the subgraphs of R_H, such that for each R-graph $x \in R_G$, there is $\bigwedge_{l \in \Gamma(x)}(l) = \bigwedge_{l \in \Gamma(f(x))}(l)$, then $R = H$.

Example 4. The following recursive formula is obtained by expanding the recursive rule in Example 1 one time.

$$P(x, y, z) :- E(x, y, z)$$
$$P(x, y, z) :- A(x, y), B(y, z_1), C(z, z_1), D(x_1, z_1), E(x_1, y_1, z_1)$$
$$P(x, y, z) :- A(x, y), B(y, z_1), C(z, z_1), D(x_1, z_1),$$
$$A(x_1, y_1), B(y_1, z_2), C(z_1, z_2), D(x_2, z_2), P(x_2, y_2, z_2)$$

Clearly, there is a 1:1 mapping between the generated subgraph of Example 1 and Example 4. For each R-graph $x \in R_G$, there is, $\bigwedge_{l \in \Gamma(x)}(l) = \bigwedge_{l \in \Gamma(f(x))}(l)$. Therefore, these two definitions are strong equivalence. □

3 Graph Transformations

In this section, we introduce some equivalence-preserving graph transformation techniques, which can be used to normalize linear recursions.

3.1 Realigning

Definition 8 (Subgraph). Let G_i be an I-graph. $K \subseteq G_i$ is called a subgraph of G_i if the following condition is satisfied: if A-node x_1 is in K, then $x \in K$ and all edges which contain node x are in K, where x_1 is the derived node of x.

Definition 9 (Realigning). Let $G =< G_i, G_e >$ be an IE-graph of a linear recursive formula r, and K a subgraph of G_i. Realigning K in G means to transform G to $H =< H_i, H_e >$ as follows:

- H_i is obtained by replacing every edge $C(x, y) \in K$ by $C(x_1, y_1)$ in G_i, where x_1, y_1 are derived variables of x and y respectively. If x or y does not connect with any other nodes, then it is also eliminated.
- H_e is obtained by eliminating all edges $C(x, y) \in K$ from $G_i * \beta_{G_i}(G_e)$.

Example 5. The IE-graph of the recursion defined in Example 1 is converted into Fig.4 by realigning A(x,y). □

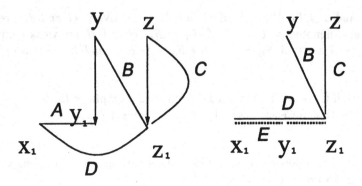

Fig. 4. Realigning A(x,y) in Example 2.3

Theorem 10. *Let $G = <G_i, G_e>$ be an IE-graph of a linear recursive formula r, and K subgraph in G_i. Let H be the IE-graph obtained by realigning K from G, then $R_G \approx R_H$, and $R_G = G_e \cup K * R_H$*

Proof. Let mapping f be:

$$f(R_G^{k+1}) = R_H^k * K$$
$$f(R_G^0) = G_e$$

Clearly, it is a one-to-one mapping, and $R_G^0 = G_e$.

Moreover, considering the difference between G_i^{k+1} and H_i^k for any non negative integer k, there are extra labels K and $\beta_{G_i^k}(G_i/K)$. However, these extra labels just appear in K and the E-graph H_e. Therefore, $\bigwedge_{l \in \Gamma(R_G^{k+1})}(l) = \bigwedge_{l \in \Gamma(R_H^k * K)}(l)$, for k= 0, 1, 2, ... $\qquad\Box$

Therefore realigning can be viewed as a kind of equivalence-preserving graph transformation operation.

3.2 Reducing

Definition 11 (Reducing). Let $G = <G_i, G_e>$ be an IE-graph where $G_i = (V_i, E_u, E_d, \Gamma_i)$ and $G_e = (V_e, E, \Gamma_e)$. Assume C-node x_1 and x_2 are connected, and their derived nodes are y_1 and y_2 respectively. Then they can be vectorized into a new node X, and G is transformed into a new IE-graph $H = <H_i, H_e>$, called reduction of G by this vectorization, as follows:

Let Y be a vector of y_1, and y_2, the derived nodes of x_1 and x_2. Replacing x_1 and x_2 by X, and replacing y_1 and y_2 by Y. Then, if there are two same directed edges $P < X, Y >$, then merging them to one. If there are two labels $P_1(u, v)$ and $P_2(u, v)$, which have same variables, then defining a new label $P(u, v) = P_1(u, v) \bowtie P_2(u, v)$ to replace them.

Example 6. The IE-graph in Fig.5 (a) can be converted into (b) by vectorizing x and y. ☐

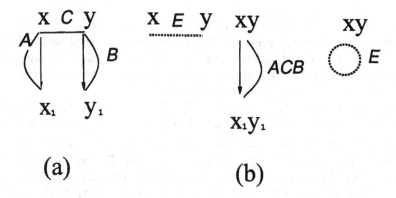

(a)   (b)

Fig. 5. Reducing

Theorem 12. *Let G and H be two IE-graphs, and C-node x_1 and x_2 be connected in G. If H is obtained by one reduction operation by vectorizing x_1 and x_2, then $G = H$.*

Proof. Assuming C-nodes of G are $\{x_1, x_2, ..., x_n\}$, A-nodes are $\{y_1, y_2, ..., y_n\}$, and the graph reduction is caused by a connected edge $C(x_1, x_2)$. Let $X = (x_1, x_2), Y = (y_1, y_2)$. Clearly, there is one to one mapping from R_G to R_H, that is, $f(R_G^i) = R_H^i$. We prove now $\bigwedge_{l \in \Gamma(R_G^i)}(l) = \bigwedge_{l \in \Gamma(R_H^i)}(l)$. by inducing on i.

Basis: It is easy to prove $G_e = H_e$. Now we prove $G_i^1 = H_i^1$.

Assume only one label $P(u, v) \in G_i$ will be changed to a new label $Q(U, V) \in H_i$. Then by the definition of reduction,

$$Q(U,V) = \begin{cases} Q(X,v) = P(x_i, v) \times Dom(x_j) \\ \quad \text{if } u = x_i \in \{x_1, x_2\}, v \notin \{y_1, y_2\} \\ Q(u,Y) = P(u, y_i) \times Doom(y_j) \\ \quad \text{if } u \notin \{x_1, x_2\}, v = y_i \in \{y_1, y_2\} \\ Q(X,Y) = P(x_i, y_i) \times Dom(x_j) \times Dom(y_j) \\ \quad \text{if } u = x_i \in \{x_1, x_2\}, v = y_i \in \{y_1, y_2\} \\ P(X) \quad \text{if } u, v \in \{x_1, x_2\} \\ P(Y) \quad \text{if } u, v \in \{y_1, y_2\} \end{cases}$$

where x_i (resp. y_i) $\neq x_j$(resp. y_j).

Obviously, in each cases, there is

$$Q(U,V) \bowtie C(X) = P(u,v) \bowtie C(x_1,x_2)$$

or

$$Q(U,V) \bowtie C(X) \bowtie H_e = P(u,v) \bowtie C(x_1,x_2) \bowtie G_e$$

Therefore $\bigwedge_{l \in \Gamma_e(G_i^1)}(l) = \bigwedge_{l \in \Gamma_e(H_i^1)}(l)$, or $G_i^1 = H_i^1$.

It should be noted that predicate $Dom(x)$ is just introduced for proving the theory. In fact, it can be eliminated from every formulas obtained by expansions, because the variable x, which is restricted by $Dom(x)$, is always restricted by some other base predicates in the same formula.

Induction: Assume $\Gamma(R_G^{m-1}) = \Gamma(R_H^{m-1})$. We now prove $\Gamma(R_G^m) = \Gamma(R_H^m)$. Let y_j^k be A-nodes generated in the k-th I-graph. Obviously, it is enough if we can prove the conjunction of all labels , which contain y_j^{m-1} as their variables, are equal in $\Gamma(R_G^m)$, and $\Gamma(R_H^m)$. Because y_1^{m-1}, and y_2^{m-1} are now connected by $C(y_1^{m-1}, y_2^{m-1})$, which is generated at the m-th expansion step, there are similar results as case i=1, except E should be replaced by $C(y_1^{m-1}, y_2^{m-1})$. Hence, $\Gamma(R_G^m) = \Gamma(R_H^m)$.

Therefore, $R_G = R_H$ □

3.3 Expanding

Definition 13 (Expanding). Let $G = <G_i, G_e>$ be an IE-graph of a linear recursive formula r. The k-expanding is a graph transformation technique that converts G to $H = <H_i, H_e>$ as follows:

- $H_i = G_i^k$;
- $H_e = G_e \cup gen(G_i^1 * \beta_{G_i^1}(G_e)) \cup ... \cup gen(G_i^{k-1} * \beta_{G_i^{k-1}}(G_e))$

Clearly, 1-expanding of G_i is itself.

Example 7. The IE-graph in Fig.6 (a) can be converted into (b) by k-expanding.

□

Theorem 14. *Let G and H be two IE-graphs. If H is obtained by k-expanding on G, then $G = H$.*

Proof. Let $R_H^k(1) = gen(H_i^k * \beta_{H_i^k}(G_e))$, and $R_H^k(2) = gen(H_i^k * ren(E_1))$, where $E_1 = gen(G_i * \beta_{G_i}(G_e))$. Then,

$$\begin{cases} R_G^{2k} = R_H^k(1) \\ R_G^{2k+1} = R_H^k(2) \end{cases}, k = 0, 1, ...$$

Therefore, G=H. □

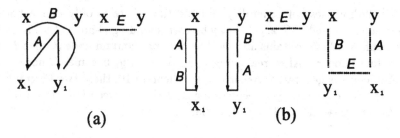

Fig. 6. 2-Expanding

4 Optimization and Normalization

Any general IE-graph can be transformed to a Normal IE-graph by following three steps: First, by realigning some edges, we can detach some variables, so as to reduce the arity of the IE-graph. It means that the corresponding formula contains less variables in its recursive predicate. Then, all C-nodes can be classified by connection relation, and the IE-graph can be transformed to a Simple IE-graph by reducing operation. Lastly, we show that a Simple IE-graph can be transformated into a Normal IE-graph by finite expanding operation.

4.1 Node Detaching

The first step of optimization and normalization is to eliminate all detachable nodes such that the arity of the graph can be reduced. The work of [15] showed that this kind of optimization is very meaningful.

Definition 15 (Detaching). A C-node $x \in G$ is detachable if it is connected only with some other C-nodes.

Example 8. In Fig. 1, node x is detachable. However node z is not detachable. □

Theorem 16. *A C-node $x \in G$ is detachable if one of the following conditions is satisfied:*

(1).All nodes connected to x are C-nodes.
(2).If there is an A-node y_1 connected to x, and there is a subgraph K which contains node y_1 but does not contain node x.

Proof. Without loss of generality, assume that I-graph is a connected graph.
(1). If x is connected with itself, that is, x is an isolated node, then x is detachable [3]. Without loss of generality, assume x is connected with only one C-variable y and $y \neq x$. Do IE-graph transformation by realigning $C(x, y)$. Then x is converted into an isolated node, therefore can be detachable.

(2). Without loss of generality, assume that there is only one A-node y_1 connected to x by edge $C(x, y_1)$. Let y be corresponding C-node of y_1. If there is a subgraph K which contains node y_1 but does not contain node x, then K does not contain node x_1. After realigning K in G, node y_1 becomes C-node, and a new A-node y_2 is generated. According to Theorem 10, these two IE-graphs are equal, and the edge $C(x, y_1)$ becomes a connection between C-nodes in the new IE-graph. By case (1), node x is detachable. □

By the theorem, we can redefine the concept of detachable to include those C-nodes which satisfying the condition of Theorem 16.

Clearly, in order to detect if a C-node is detachable, an algorithm to construct a subgraph which contains some specified node x is necessary.

Algorithm 1 (Constructing subgraph K)
Input: An I-graph G_i, and a node $x \in G_i$
Output: A subgraph K of G_i
Method:
begin
$push(x, S)$;
while S is not empty do
      $pop(u, S)$;
      if u is a C-node then do
            if $C(u, y) \in G_i$ and $C(u, y) \notin K$ then insert $C(u, y)$ into K;
            if y is an A-node, then $push(y, S)$;
            enddo;
      if u is an A-node then do
            $push(v, S)$; /* v is the corresponding C-node of u */
            insert $P < u, v >$ into K;
            enddo;
      enddo;
end

Example 9. In Fig. 1, $K(x)$ contains C-node y and edge $A(x, y)$; $K(y)$ contains C-node x, z, and A-node z_1, as well as edge $A(x, y)$, $B(y, z_1)$ and $C(z, z_1)$; $K(z)$ contains node z, z_1 and edge $C(z, z_1)$, $P < z, z_1 >$. Therefore, node x and y are detachable, but node z is not detachable. □

The following algorithm can be used to reduce an IE-graph. It first detects every C-node to see if it is a detachable one, and constructes realigned component for every detachable C-node at same time. It then detaches all detachable node by applying realigning.

Algorithm 2 (Reducing IE-graph)
Input: An IE-graph $G = (G_i, G_e)$

Output: An new IE-graph $H = (H_i, H_e)$
Method:
begin
R = empty;
for each C-node x do
    flag = .TRUE.;
    for each edge $C(x, y) \in G_i$ connected to node x do
        if y is a C-node then insert $C(x, y)$ in $L(x)$;
        if y is an A-node, then do
            constructing a subgraph for y,
            denoted as $K(y)$;
            if $x \notin K(y)$ then do;
                insert $C(x, y)$ in $L(x)$;
                insert $K(y)$ in $L(x)$;
                enddo;
            else flag = .FLUSE.;
            enddo;
        enddo;
    if flag = .TRUE. then insert L(x) into R else L(x) = empty;
    enddo;
realigning R;
end

4.2   Converting to Simple IE-graphs

Definition 17 (Simple IE-graph). An IE-graph is simple if for each pair of C-node x, y in its I-graph, there is no any connection between x and y.

To convert an IE-graph to a Simple IE-graph, those connected nodes should be vectorized, hence reducing operation defined in Section 3 can be utilized in the following algorithm.

Algorithm 3 (Reducing A General IE-graph)
Input: The IE-graph of a recursive formula
Output: The Simple IE-graph
Method:
begin
flag := 0;
while flag=0 do
    flag := 1
    for each C-variable x in the I-graph, do
        for each C-variable y connected to x do
            flag = 0;
            reducing IE-graph by vectorizing x, y ;
            enddo;

          enddo;
      enddo;
end

Theorem 18. *Algorithm 3 is correct. That is, (1). Algorithm can reduce a general IE-graph to a Simple IE-graph in finite steps. (2). The two recursive formulas corresponding to the IE-graph and its Simple IE-graph have the same result.*

Proof. (1a). If there is a reduction occurred in the FOR loop, then the number of nodes in the reduced IE-graph is less than the original IE-graph. Hence, the algorithm will be terminated at most after n loops, where n is the arity of the IE-graph.

(1b). The termination condition of the algorithm is no C-node connection is found in the I-graph. Therefore, the IE-graph output by the algorithm contains no any node connection between C-nodes,i.e. a Simple IE-graph.

(2). it can be proved from Theorem 12.   □

4.3 Expanding IE-graphs

A Simple IE-graph may not be a Normal IE-graph, however it contains at most a one-directional cycle. Moreover it is a pure one-directional cycle if the Simple IE-graph is obtained by applying node detaching operations and then reducing operations. Some existed results shown that a one-directional cycle can be transformed into n unit cycles by finite expanding operations. That is, a Simple IE-graph can always be transformed into a Normal IE-graph by expanding operations.

Theorem 19. *A simple IE-graph is a pure one-directional cycle if it is obtained by applying node detaching operations and then reducing operations*

Proof. (1). We firstly prove that there exists at most a one-directional cycle in every connected components of a Simple IE-graph. Firstly, cycles in a Simple IE-graph are always one-directional cycles. In fact, if there is a multi-directional cycle, then there is at least one undirected edge which connects two C-nodes. This is contradictory to the definition of the Simple IE-graph.

Then, assume there exist two cycles C_1 and C_2 in a connected component of a Simple IE-graph. we prove it will deduce a contradiction. By the definition of the Simple IE-graph, there are only two possible cases:

Case 1: There is an undirected edge, let $C(x,y)$, to connect cycles C_1 and C_2. Then edge C should certainly connect to an A-node and a C-node in C_1 and C_2 respectively(see Fig.7 (a)). Because C_1 is a one-directional cycle, there is a C-variable, let u, which connects to x by an undirected edge. Therefore, y and u are connected. It is contradictory to the definition of the Simple IE-graph;

Case 2: Cycle C_1 and C_2 have a common directed edge $\langle x,y \rangle$(Fig.7 (b)). In this case, there exist undirected edges $C(u,y)$ and $D(y,v)$ in C_1 and C_2 respectively. Obviously, u,v are C-nodes because C_1 and C_2 are one-directional cycles. That means u and v are connected, which is contradictory to the definition of the Simple IE-graph.

Therefore, there exists at most a one-directional cycle in every connected components of a Simple IE-graph.

(2). Then we prove it is a pure one-directional cycle if it is obtained by applying node detaching optimization repeatedly and then reducing. In fact, any noncyclic elements has been detached by detaching operation in previous step.   □

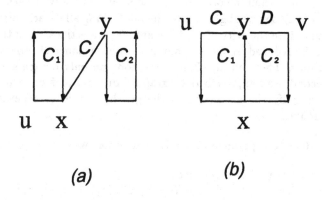

Fig. 7. Expanding

Theorem 20. *A m-D one-directional cycle can be transformed to m unit cycles by expanding exactly m times. [17]*   □

Theorem 21. *Assume G to be a n-D Simple I-graph, which consists of p disjoint w_i-D one-directional cycle (i=1,...,p). Then G will decomposed into n unit cycles after $lcm(w_i)$ times expansions. [17]*   □

5   Related Work

Graphs in [8, 6, 17] are utilized as a representation tool for understanding the structural properties of recursive formulas easily. In our paper, we treat the graph as objects on which some operations and graph equivalence are defined. Therefore, the process of normalization of recursive formulas is viewed as process of graph transformation. To our knowledge, the methodology is new in study of normalization of recursive formulas.

Yong etc. [17] proposed a method to generate a normal form for linear recursions based on classification on the I-graph. Some meaningful results have been proved in that paper. Unfortunately, it seems there is no meaningful result for complex cases, for example, general multidirectional cycles, dependent cycles, and mixed components.

Our paper gives a general algorithm which can transform any IE-graph to normal IE-graph. This result is also proved by some other authors [7, 10].

Han and Zeng [7] proposed an algorithm to generate a compiled normal form for any linear recursive formula. Their algorithm utilized a matrix to describe variable connection relation in a formula, and simulated expansion of the formula by a set of matrix expansion rules. By classifying distinguished variables by connection relation, and aggregating them into a vectorized variable, the algorithm generated a normal form for any recursive formula.

Compared with their algorithm, our approach has advantages as follows: First, by realigning some subgraph and then detaching all detachable nodes, our algorithm can reduce the arity of the IE-graph. It means the arity of the recursive predicate can be reduced, so as to enhance the performance of processing of the recursive formula. The Han and Zeng's method can only detach some isolated variables. Second, our approach does graph reduction and node detaching at first, it will simplify the expansion of the graph, as well as generation of the compiled formula.

Example 10. Recursive predicate P is defined as follows:

$$\begin{cases} P(x,y,z) :- E(x,y,z) \\ P(x,y,z) :- A(x,y), B(y,z), C(z,z_1), P(x_1,y_1,z_1) \end{cases} \quad (1)$$

By Han and Zeng's method, this rucursion will be transformed into

$$\begin{cases} P(u) :- E(u) \\ P(u) :- D(u,u_1), P(u_1) \end{cases} \quad (2)$$

where, $u =< x,y,z >$ and $D(u,u_1) = A(x,y), B(y,z), C(z,z_1)$. It is a single chain recursion.

However, it can be transformed into the following form as:

$$\begin{cases} P(x,y,z) :- E(x,y,z) \cup A(x,y), B(y,z), Q(z) \\ \\ Q(z) \quad :- C(z,z_1), E(x_1,y_1,z_1) \\ Q(z) \quad :- C(z,z_1), F(z_1), P(z_1) \end{cases} \quad (3)$$

where $F(z) = A(x,y), B(y,z)$.

Clearly, Formula (3) is more efficient than (2) because P is a three-ary recursive predicate and Q is an unary one. □

Lu etc. [10] also proved the similar result based on a graph model. They calssified a graph into trees, cycles, connected components etc. and studied the structure property of expansion of each class of graph. Chains can be constructed from analyzing the structure property. In their study, however, only expansion is considered. Optimization like realigning and reducing in our paper, which can simplify the formula, is not included.

Our algorithm contains a step to detach some variables from the formula, so as to reduce the arity of the recursive predicate. There is also some other work which has similar purpose, such as recursively redundancy removability [11, 9], factability [14], and detaching isolated variables [3] etc.. The isolated variables(nodes) in [3] are special cases of detachable nodes in this paper. Our previous paper [4] have also showed that the concept of detachable nodes is different with the concept of recursively redundancy removability in [11, 9] and factability in [14].

6   Conclusion

In this paper, we extended the existing graph model for presenting linear recursions, and defined newly strong and weak graph equivalence for graph transformation. Based on this graph model, three basic graph transformation operators are defined and proved that all of them are equivalence-preserving. We also proved that an IE-graph, which corresponds to a general recursive formula, can be optimized by detaching some variables, and then be transformed into a Simple IE-graph by variable vectorization. Furthermore, a Simple IE-graph can be transformed into a Normal IE-graph. That is, a general normal form compiled formula can be derived. Therefore, graph transformation can be used for noramlization and optimization of linear recursions.

As we have stated that the purpose of finding a general compiled formula is improving the efficiency of query evaluation. Therefore, it is necessary to study how to apply the results to query processing.

References

1. F.Bancilhon and R.Ramekrishnan: An Amateur's Introduction to Recursive Query Processing Strategies" Proc. ACM SIGMOD, 1986
2. Bancilhon F., Maier D., Sagiv Y., Ullman J.D. : "Magic Set and Other Strange Ways to Implement Logic Programs". *Proceedings. ACM SIGACT-SIGMOD-SISART Symposium on Principles of Database Systems(PODS), (1986).*
3. X.Du and N.Ishii: "Optimizing Linear Recursive Formulas by Detaching Isolated Variables". *IEICE TOIS Vol.E78-D No.5, May, 1995*
4. X.Du and N.Ishii: "Reducing the Arity of Recursive Predicates by Realigning Some Predicates", *Proc. ICLP'95 Workshop on Deductive Database and Logic Programming, June 17, 1995*
5. L.Henschen and S.Naqvi: On Compiling Queries in Recursive First-Order Data base, *JACM vol.51 (1984)*
6. Jiawei Han: Compiling General Linear Recursions by Variable Connection Graph Analysis, *Comput. Intell.* 5,12–31, (1989)
7. Jiawei Han, Kangsheng Zeng: Automatic Generation of Compiled Forms for Linear Recursions, *Information Systems,* Vol.17, No.4, pp.299–322, (1992)
8. Y.Ioannidis: A Time Bound on the Materialization of some Recursively Defined Views, *Proceedings of Very Large Data Bases(VLDB), (1985)*

9. Lakshmanan,L.V.S., Hernandez,H.J.: "Structural Query Optimization – A Uniform Framework for Semantic Query Optimization in Deductive Databases", *Proc. ACM Symposium on Principles of Database Systems(PODS)*, (1991), pp.102-114.

10. W.Lu, D.L.Lee, J.Han: A Study on the Structure of Linear Recursion, *IEEE Trans. Knowledge and Data engineering*, Vol.6, No.5, pp.723–737 (1994)

11. J.Naughton: Data Independent Recursion in Deductive Database, *Proceedings ACM SIGACT-SIGMOD-SIGART Symp. Principles of Database Systems(PODS)*, (1986).

12. J.Naughton: One-Side Recursions, *Proc. 6th ACM SIGACT-SIGMOD-SIGART Symp. Principles of Database Systems(PODS)*, pp.340–348 (1987)

13. J.Naughton: Minimizing Function-Free Recursive Inference Rules. *J. ACM* Vol.36, No.1, pp.69–91 (1989)

14. J.Naughton, R.Ramakrishnan, Y.Sagiv, J.D.Ullman: Argument Reduction by Factoring, *Proc. the 15th Int. Conf. on Very Large Data Bases(VLDB)*, pp.173-182 (1989)

15. S. Seshadri, J.F.Naughton: "On the Expected Size of Recursive Datalog Queries", *Proc. ACM SIGACT-SIGMOD-SIGART Symposium on Principles of Database Systems (PODS)*, pp.268–279 (1991).

16. J.D.Ullman: Principles of Database and Knowledge-Base Systems, Vol.II, *Computer Science Press*, (1989)

17. C.Yong, H.J. Kim, L.J.Henschen, and J. Han: Classification and Compilation of Linear Recursive Queries in Deductive Databases. *IEEE Transactions on Knowledge and Data Engineering*, Vol.4, No.1, pp.52–67 (1992)

Data Retrieval and Aggregates in SQL^*/NR

Yiu-Kai Ng and Nael Qaraeen

Computer Science Department, Brigham Young University, Provo, UT 84602, U.S.A.

Abstract. Standard SQL is incapable of handling recursive database queries and nested relations. A proposed solution to allow recursion in SQL was given in SQL^* [KC93], while a solution to allow nested relations in SQL was given in SQL/NF [RKB87]. However, these two problems with SQL were handled separately, and an extended SQL that handles both recursive queries and nested relations is still lacking. To overcome this shortcoming, we propose an extended SQL, called SQL^*/NR, that not only can handle both recursive queries and nested relations, but also allows aggregate operators. A query Q in SQL^*/NR is processed by first transforming Q into rule expressions in LDL/NR, a logic database language for nested relations, and the transformed rule expressions are evaluated for retrieving the desired result of Q. Transforming Q into rule expressions in LDL/NR is desirable since LDL/NR handles recursion on nested relations with a built-in mechanism for recursive query processing. In this paper, we define SQL^*/NR and include an approach for transforming SQL^*/NR queries into rule expressions in LDL/NR. SQL^*/NR, as defined, enhances the expressive power of standard SQL and SQL/NF and has the expressive advantage over SQL^*.

1   Introduction

SQL (Structured Query Language), a database query language that has been standardized, has gained lots of popularity and has been supported by most relational database systems since it was first introduced in the 1970's. SQL is widely accepted because it is a "user-friendly" query language that has a simple, declarative syntax and semantics. However, some database applications have revealed at least two limitations of SQL. First, standard SQL is incapable of handling recursive database queries (queries in which a relation is defined in conjunction with its own definition [KC93]). Second, standard SQL cannot handle nested relations that allow non-atomic (i.e., decomposable) attribute values. These deficiencies in standard SQL limit its expressive power and capability of handling complex data, respectively.

A proposed solution to the first problem of SQL was given in [KC93] who defined SQL^* that extends standard SQL to allow recursive queries. A solution to the second problem of SQL was proposed in [RKB87] where SQL/NF, which extends standard SQL to handle nested relations, was defined. However, these two problems with standard SQL are not handled by a single SQL-type language. An extended SQL, called SQL^*/NR, that can handle both recursive queries and nested relations, is to be defined in this paper. The development of SQL^*/NR

was motivated in part by the fact that the nested relational model can be used for storing and retrieving complex data, while an extended SQL that allows recursive queries on nested relations enriches the expressive power of standard SQL and SQL/NF, and has the expressive advantage of the nested relational structure over SQL^* that operates on flat relations. Furthermore, SQL^*/NR allows aggregate operators, such as min, max, sum, etc., that makes the language similar to standard SQL and more powerful. SQL^*/NR, as defined, enhances database query languages in [RKB87, Uni91] and recursive query languages for flat relations [KC93] by providing better versatility and a richer functionality for expressing complex data. Moreover, SQL^*/NR incurs most, if not all, of the standard SQL's advantageous features.

We present the details of SQL^*/NR as follows. In Section 2 the basic set of constructs in SQL^*/NR are described. Extended SQL constructs in SQL^*/NR, including subquery components that specify the basis and recursive definitions of a recursive query, are used for creating an SQL^*/NR query. To evaluate an SQL^*/NR query Q, we first transform Q into rule expressions in LDL/NR, a logic database language for nested relations. Rule expressions in LDL/NR are chosen as an internal representation of Q because they support languages of *declarative* nature and handle *recursion* and *nested relations*. Formal definitions of these rule expressions and justification for processing SQL^*/NR queries by using a logic database language are given in Section 2.3. In Section 3 SQL^*/NR queries with aggregate operators are introduced. In Section 4 we give a concluding remarks. In addition, we include in Appendix A the formal syntax of SQL^*/NR in BNF notations. (An algorithm for the transformation of SQL^*/NR queries with(out) aggregate operators into rule expressions in LDL/NR is given in [Qar95].)

2 SQL^*/NR Queries and LDL/NR Rule Expressions

In this section we first describe SQL^*/NR subqueries which comprise the basic constructs of an SQL^*/NR query. Hereafter, we present the structure of an SQL^*/NR query, and show how SQL^*/NR subqueries are embedded in an SQL^*/NR query. Finally, rule expressions in LDL/NR, which are used for processing a given SQL^*/NR query, are introduced. LDL/NR is of interest because it captures the constraints of nested relations precisely. Furthermore, since for each complex-object type there is a nested-relation type with the same "information capacity", LDL/NR can handle complex-data type queries.

2.1 SQL^*/NR Subqueries

An SQL^*/NR subquery (or subquery for short) allows the user to retrieve data from a set of relations S. The subquery may comply with specific conditions or constraints that are applied to certain attributes of the relations in S. This is similar to the functionality provided by *SELECT, FROM,* and *WHERE* clauses of standard SQL.

The *Select-From-Where* structure of an SQL^*/NR subquery, which is a major construct of SQL^*/NR and is referred as a Select-From-Where-Expression or SFW-Expression for short [RKB87], is made up of the following clauses:

SELECT attribute-list
FROM relation-list
[WHERE boolean-expression]

This expression can handle queries for nested relations. Since an SQL^*/NR subquery is written as an SFW-Expression, a subquery and an SFW-Expression will be interchangeably used throughout the remaining of this paper. Furthermore, we assume that all (atomic and non-atomic) attribute names appearing at all levels of nesting in a nested relation are unique, and apostrophe (') is an invalid symbol in any subquery since it is reserved for the transformation process of an SQL^*/NR query.

The Select Clause   The **Select** clause, which is the first component of an SFW-Expression, allows the user to specify data items to be included in the final result of a subquery by choosing attributes in different relations referenced in the SFW-Expression. The **Select** clause is of the following format:

SELECT *atomic-attr*$_1$, ..., *atomic-attr*$_n$, *non-atomic-attr*$_1$, ...,
non-atomic-attr$_m$

The **Select** clause consists of the keyword SELECT, followed by a list of atomic and non-atomic attributes. A non-atomic attribute can be a nested relation that is constructed by an SFW-Expression called an *incremental subquery* [RKB87]. (See Example 4.) The **Select** clause is used for specifying all the desired (atomic and non-atomic) attributes to be included in the result of a subquery. It corresponds to the *extended projection* operation in the extended relational algebra [RKS88].

Assume that *Attr* is an (atomic or non-atomic) attribute in nested relation r. If *Attr* appears at the top level scheme of r, we can either include *Attr* in the **Select** clause, or concatenate r (or its reference name which is also called *tuple variable* in [KS91]) with *Attr* separated by a dot in the **Select** clause. Otherwise, *Attr* must be embedded one level deep in r, and we specify *Attr* in the **Select** clause with its preceding (embedded) relations (or their reference names) separated by dots. (To reference attributes embedded more than one level deep in a nested relation, we use nested incremental subqueries as discussed below.) In addition, each chosen (atomic or non-atomic) attribute C can be referenced by a distinct, new column name to distinguish C from other chosen attributes. A new column name is specified by including the keyword **AS** and the new name following a chosen attribute name.

Example 1. Let Student be the nested relation in Figure 1. To retrieve all the

286

departments in Student, we enter

> SELECT Dept ... or SELECT Student.Dept ... .

To retrieve all the children of a student, we enter

> SELECT Children ... or SELECT Student.Children ... .

To retrieve all students' names and their children names, and reference a student's name and his/her children names by *Sname* and *Child-Name*, respectively, we enter

> SELECT S.Name AS Sname, (SELECT Cname ...) AS Child-Name ...

where S is the reference name of Student. □

Student			
Name	SID	Children	Dept
		Cname	Age	
John	5678	David	5	CS
		Chris	10	
Bill	2134	Denise	8	Chemistry
		Jim	16	

Has_taken		
SID	Courses	
	Course#	Crhrs
5678	CS 220	3
	PE 100	1
2134	Math 411	3

Fig. 1. Nested Relations *Student* and *Has_taken*

If only certain attributes of a relation r (which is embedded in a nested relation s) along with some attributes in relations $s_1, \ldots, s_n$ are to be included in an embedded relation in the result of a subquery Q, we could specify an *incremental subquery SQ* in Q. (It is assumed that s, $s_1, \ldots$, and s_n are specified in the **FROM** clause of SQ or Q.) An example of an incremental subquery is *(SELECT Cname ...)* in Example 1. Upon evaluating SQ, the result of SQ includes the desired attributes in $r, s_1, \ldots, s_n$. SQ is specified as an argument in the **Select** clause of Q, and the resultant relation of SQ is given a reference name (i.e., column-name) which is specified after SQ and the keyword **AS** in the **Select** clause.

The From Clause The **From** clause includes a set of relations used for computing the result of an SFW-Expression. The format of the **From** clause is:

$$\textbf{FROM } rel_1, \ldots, rel_n$$

The **From** clause consists of the keyword FROM followed by a list of constructs $rel_i, 1 \leq i \leq n$, where rel_i denotes either a (an embedded) relation name (which could be followed by an optional *reference name*), a $NEST$ operation, or an $UNNEST$ operation. (Each of the $NEST$ and $UNNEST$ operations yields a relation.) The order in which the relation names and the $NEST$ and $UNNEST$ operations may be arranged in the **From** clause is of no importance. The **From** clause corresponds to the *extended cartesian product* operation in the extended relational algebra [RKS88].

Example 2. To use the relation Student in Figure 1 in an SQL^*/NR subquery Q, we include FROM *Student* in Q. To use Student, and Course# and Crhrs of the relation Courses embedded within the nested relation Has_taken in Figure 1, we include

FROM Student S, (UNNEST Has_taken ON Courses)

in a subquery, where S becomes the reference name of Student. □

(Embedded) Relations that are listed in the **From** clause are referenced during the evaluation of the SFW-Expression. An embedded relation r (such as *Courses* in *Has_taken*) must be specified in the **FROM** clause of an incremental subquery (such as *(SELECT Cname from Courses)*) in a subquery Q, and the nested relation (such as *Has_taken*) within which r is embedded must be included in the **FROM** clause of Q. Reference names of relations (such as S for Student in Example 2) allow the user to specify multiple copies of the same relation and same attribute name appeared in different relations that are included in the **From** clause. *NEST* (resp. *UNNEST*), an aggregating (resp. disaggregating) operation, allow the user to restructure a relation making it more aggregated (resp. flatter), before such a relation is used by the **Select** or **Where** clause in the SFW-Expression. The *NEST* and *UNNEST* operations have the following formats:

(NEST <nested-relation-name> ON attribute-list AS <column-name>) and
(UNNEST <nested-relation-name> ON <attribute-list>)

where nested-relation-name and attribute-list in NEST and UNNEST reference a nested relation r and the attributes in r on which NEST and UNNEST are applied, respectively. The keyword AS in NEST allows the user to give a reference name (i.e., column-name) to the resultant relation of the NEST operation.

The Where Clause   We use the **Where** clause in an SFW-Expression to specify conditions on attributes that need to be satisfied. The **Where** clause is of the following format:

WHERE Boolean-Expression

The **Where** clause, which is optional in an SQL^*/NR subquery, consists of the keyword WHERE followed by a boolean expression that might include k different components grouped together using a combination of m different logical ANDs and n different logical ORs ($0 \leq m, n \leq k - 1$, and $m + n \leq k - 1$). The clause references atomic and non-atomic attributes of the (embedded) relations in the **From** clause. These attributes are compared to other attributes or constants using (set) comparison operators. The results of such comparisons are grouped together using logical ANDs or logical ORs (if there are any), and its evaluation yields the final result of the boolean expression of the **Where** clause. The **Where** clause corresponds to the *extended selection* (except the *set membership* operators, **in** and **notin**) in the extended relational algebra that is recursively applied to deal with selections on different levels of nesting in a nested relation, if such selections are required.

Example 3. To construct a subquery Q based on Student in Figure 1 in which one of the two conditions must be satisfied: either the age of a student's child is in between 1 and 5, or in between 12 and 15, inclusively, we include

WHERE ((C.Age $\geq$ 1 AND C.Age $\leq$ 5) OR (C.Age $\geq$ 12 AND C.Age $\leq$ 15))

in Q, where C is the reference name of the relation Children embedded within Student. $\square$

The selections also handle the comparisons of atomic and non-atomic attributes of two nested relations involved in the *extended natural join* in the extended relational algebra. In the *extended natural join* of two nested relations s and s', two tuples $t \in s$ and $t' \in s'$ are *joinable* if the extended intersection on the projections over common (atomic and non-atomic) attributes of t and t' is non-empty. This constraint is similar to the constraint of the traditional natural join operation, i.e., two tuples contribute to the join if they agree on common attributes.

2.2   *SQL*/NR* Queries

This subsection describes the structure of an SQL^*/NR query (or query for short) which is constructed by using subqueries described in section 2.1.

An SQL^*/NR query Q, which can handle recursion on nested relations, consists of an INSERT INTO statement [KC93] that includes the keyword *INSERT INTO* followed by the relation R (that is computed recursively), one subquery called the *Basis-subquery*, and n ($0 \leq n$) different *ALSO* statements, each of which contains a subquery called *Recursive-subquery*. The format of Q is as follows:

$$\begin{aligned}
&\text{INSERT INTO } R \\
&\qquad \textit{Basis-subquery} \\
&\text{ALSO} \\
&\qquad \textit{Recursive-subquery}_1 \\
&\qquad \cdots \\
&\text{ALSO} \\
&\qquad \textit{Recursive-subquery}_n
\end{aligned}$$

The relation R referenced in the INSERT INTO statement of an SQL^*/NR query Q is the relation to be computed recursively using itself and other relations $r_1, r_2, \ldots, r_n$ that are specified in the *Basis-subquery* or *Recursive-subqueries* of Q.

A recursive query in SQL^*/NR consists of at least two subqueries, a *Basis-subquery* and a *Recursive-subquery*, which are SQL^*/NR subqueries. The main difference between the *Basis-subquery* and the *Recursive-subqueries* in a query Q is that any relation specified in Q can be referenced by any of these subqueries, except the computed relation R which can only be referenced in the *Recursive-subqueries* but not in the *Basis-subquery*. Furthermore, all subqueries must yield *compatible* relation schemes[1].

Example 4. Let Connected be the nested relation in Figure 2 which contains the flight information about the group of cities that are connected directly with a particular city by a flight and their respective distances. We specify the following SQL^*/NR query that generates the relation *Reachable* in Figure 2. *Reachable* contains the information about a group of cities that can be reached, either through a direct or an indirect flight, from a particular city.

    INSERT INTO Reachable
        SELECT Source-City, (SELECT Dest-City
                                FROM Route) AS Cities
        FROM Connected
    ALSO
        SELECT R.Source-City, (SELECT Dest-City
                                FROM Route
                                WHERE Dest-City ≠ R.Source-City) AS Cities
        FROM Connected C, Reachable R
        WHERE C.Source-City in R.Cities

Note that the SFW-Expression, (SELECT Dest-City FROM Route ...) AS Cities, is an incremental subquery. □

Connected		
Source-City	Route	
	Dest-City	Dist
Chicago	New York	750
New York	Chicago	750
	Los Angeles	1970
Boston	Chicago	550

Reachable	
Source-City	Cities
	Dest-City
Chicago	{ New York, Los Angeles }
New York	{ Chicago, Los Angeles }
Boston	{ Chicago, New York, Los Angeles }

Fig. 2. Nested Relations *Connected* and *Reachable*

Further note that tuples in the resultant relation R generated by the evaluation of an SQL^*/NR query have unique atomic components, i.e., tuples with the same atomic components are merged, and duplicate tuples are removed with set equality holding on non-atomic attributes. These constraints are applied to each level of nesting in R.

---

[1] Two relation schemes $R_1(A_1, \ldots, A_n)$ and $R_2(B_1, \ldots, B_n)$ are *compatible* if $A_i \in R_1$ and $B_i \in R_2$ ($1 \leq i \leq n$) have the same domain and $A_i = B_i$.

2.3 Rule Expressions in *LDL/NR*

Rule expressions in *LDL/NR* are chosen as the internal representation of an *SQL*/NR* query *Q* since there exists an efficient implementation of *recursion* in higher-order logic database systems (*LDL/NR* is a higher-order logic database language) that guarantees termination and preserves completeness of *Q* [LN95]. A logic (deductive) database system, which has a built-in reasoning capability, may be used both as an inference system and as a representation language [GN90, GM92]. Moreover, the syntax and semantics of higher-order logic, which form the theoretical foundation of a deductive database language, are simple, well-understood, and formally well-defined. We take the advantages of the built-in recursive query processing mechanism provided by deductive database systems to process our *SQL*/NR* queries through the transformation of an *SQL*/NR* query *Q* into rule expressions in *LDL/NR* which are evaluated to yield the answer of *Q*.

LDL/NR restricts *HILOG* [CC90] to nested relations, and is simpler in notation than *HILOG-R* [CK91] which requires type and named attribute to be attached to each argument of a type declaration and named attribute to each data value in a tuple. Moreover, *LDL/NR* is more complete than *LDL* [STZ92] which allows only nested tuples rather than nested sets. Furthermore, the semantics of sets in the logic database language defined by [Kup90] is different from the semantics of set terms as defined in *LDL/NR*. From the theoretical point of view, set terms as defined in [Kup90] cannot handle the nested relational structure since these set terms do not capture the constraint of alternating tuple terms and set terms in a nested relation precisely and cannot represent the constraints of a nested relation explicitly. On the other hand, *LDL/NR* is a logic database language for nested relations, and the semantics of sets in *LDL/NR* follows the definition of nested relations [Hul89] which extends the semantics of first-order logic languages to handle recursion. Form the implementation point of view, [Kup90] has not addressed how to implement set(-term) matching and set(-term) unification and their respective semantics as does *LDL/NR* for processing queries with set terms.

Syntax of Rule Expressions in *LDL/NR*   Rule expressions *RE*s in *LDL/NR* are based on the notions of type, term, and formula which in turn are defined on an alphabet in *LDL/NR*. Constants and variables in *RE*s are of *atomic type*. There are two other types, *set type* and *tuple type*, in *RE*s.

Definition 1. A *type* in *RE*s is inductively defined as follows: (i) an *atomic type* is a type, (ii) a *set type* is a type, and for a set-type $s\{r\}$, r is of either atomic type or tuple type, and (iii) a *tuple type* is a type, and for a tuple-type $p(s_1, \ldots, s_n)$, s_i $(1 \leq i \leq n)$ is of either atomic type or set type. □

Tuple type and set type can be used alternatively to form complex data types, as in [CC90, CK91]. The type declaration *dept(Dname, projects{Pname}, employees{ employee(Ename, EID)})* defines *dept* which is of tuple type with components *Dname*, which is of atomic type, and *projects* and *employees*, which are of set type.

Definition 2. Objects of atomic type are called *atomic terms*. A *term* is inductively defined as follows: (i) a constant is a term, (ii) a variable is a term, (iii) for a set-type $s\{r\}$, an instance $s\{t_1, \ldots, t_m\}$ is a term called *set term*, where t_i $(1 \leq i \leq m)$ is of type r, and (iv) for a tuple-type $p(s_1, \ldots, s_n)$, an instance $p(t_1, \ldots, t_n)$ is a term called *tuple term*, where t_i is of type s_i, $1 \leq i \leq n$.□

Example 5. An instance of the tuple-type *dept* is

$$dept(cs, \, projects\{db, se, lp\}, \, employees\{employee(smith, 123),$$
$$employee(jones, \, 567)\}). \; □$$

Definition 3. A *(well-formed) formula*, which is constructed by terms, is inductively defined as follows: (i) a tuple term or set term is a formula, (ii) if T and S are terms which form the arguments of a (set) comparison operation θ, then $T\theta S$ is a formula, written as $\theta(T, S)$, (iii) if F is a formula and X is a variable, then $\exists X \; F$ and $\forall X \; F$ are formulas, and (iv) if F and G are formulas, so are $\neg F$, $F \vee G$, $F \wedge G$, $F \rightarrow G$, and $F \leftrightarrow G$. □

A tuple term, set term, or (set) comparison operator with arguments is called an *atom*. A *ground formula (term)* is a formula (term) without variables. A *closed formula* is a formula with no free occurrence of any variable.

Definition 4. A *rule (expression)* is of the form *head* : – *body*, where *head* is an atom and *body* is a conjunction of atoms. A *unit rule* is a rule with an empty body. A *fact* is a ground unit rule. □

Example 6. The following rule retrieves all employees who work on one other project besides the *db* project:

works_on_pj(Ename) : – employee(Ename, EID),
works_on(EID, wprojs{*Pname, db*}). □

Semantics of Rule Expressions in *LDL/NR*   The declarative semantics of rule expressions in *LDL/NR* is given by the usual semantics of formulas in *LDL/NR*. Meaning for each symbol in a formula should be assigned in order to discuss the truth or falsity of the formula. The various quantifiers and connectives have a fixed meaning, but the meaning assigned to each term can vary [Llo87]. We first define *LDL/NR* universe and *LDL/NR* base.

Definition 5. Let L be an *LDL/NR* program. The *LDL/NR universe* U of L, denoted U_L, is the set of all ground atomic terms (constants) of L, and the *LDL/NR base* B of L, denoted B_L, is the set of all ground unit rules of L. □

To define formally the meaning of a fact as a logical consequence of a set of facts and *LDL/NR* rules, we introduce the concepts of *LDL/NR* interpretation and *LDL/NR* model.

Definition 6. Let L be an LDL/NR program. An interpretation for L is an LDL/NR *interpretation* if the following conditions are satisfied: (a) the domain of the interpretation is the LDL/NR *universe* U_L, (b) constants in L are assigned to "themselves" in U_L, (c) for a set term $s\{t_1,\ldots,t_m\}$ in L, the assignment of s is a mapping from U_L^m to $\{true, false\}$, and (d) for a tuple term $p(s_1,\ldots,s_n)$ in L, the assignment of p is a mapping from U_L^n to $\{true, false\}$. $\square$

Definition 7. Given an LDL/NR interpretation I of an LDL/NR program L and a closed formula F of L, I is an LDL/NR *model*, which is a subset of B_L, for F if F is true with respect to I (or I is a model for F). If S is a set of closed formulas of L, then I is an LDL/NR *model* for S if I is a model for every formula of S. $\square$

A set term $s\{X\}$, where X is a variable, denotes a set term of arbitrary cardinality. As in [CC90], it is assumed that the satisfaction of $s\{e_1,\ldots,e_n\}$ by an LDL/NR interpretation I implies the satisfaction of $s\{X\}$ by I, where $X \subseteq \{e_1,\ldots,e_n\}$. Furthermore, the satisfactions of $s\{a_1,\ldots,a_n\}$ and $s\{b_1,\ldots,b_m\}$ by I imply the satisfaction of $s\{a_1,\ldots,a_n,b_1,\ldots,b_m\}$ by I.

2.4 Transforming SQL^*/NR Queries to LDL/NR Rule Expressions

We give a few examples below; each of these examples includes an SQL^*/NR query Q and the rule expressions in LDL/NR transformed from Q. (An algorithm which transforms an SQL^*/NR query into LDL/NR rule expressions is given in [Qar95].)

Example 7. The transformed rule expressions for the SQL^*/NR query in Example 4 are:

> reachable($Source\text{-}City_1$, cities$\{Dest\text{-}City_1\}$) :-
> connected($Source\text{-}City_1$, route$\{$route'($Dest\text{-}City_1$, $Dist_1$)$\}$).

> reachable($Source\text{-}City_2$, cities$\{Dest\text{-}City_1\}$) :-
> connected($Source\text{-}City_1$, route$\{$route'($Dest\text{-}City_1$, $Dist_1$)$\}$),
> reachable($Source\text{-}City_2$, cities$\{Dest\text{-}City_2\}$),
> in($Source\text{-}City_1$, cities$\{Dest\text{-}City_2\}$),
> $Dest\text{-}City_1 \neq Source\text{-}City_2$.

Subscripted attributes, such as $Source\text{-}City_1$ and $Source\text{-}City_2$, can be used for distinguishing attributes with the same name from different relations. $\square$

Example 8. Let Parent be the nested relation in Figure 3, and let Q be the following SQL^*/NR query that retrieves all the people who are related to others. The resultant relation Related of Q is shown in Figure 3. It is assumed that Person A is related to person B if either (i) A and B are sibling, (ii) B's parent is related to A, or (iii) A's parent is related to B.

Note that any reference name N assigned to an atomic attribute A in a **Select** clause does not actually change A to N in the rule expressions transformed from Q. It is because N is used in Q only to satisfy the constraint of compatible relation scheme or denote a new reference name of A in Q. This assumption holds for each of the following examples.

INSERT INTO Related
    SELECT P_1.Children.Person, (SELECT Person AS Relative
                       FROM P_2.Children
                       WHERE Relative $\neq$ P_1.Children.Person)
                       AS Relatives
    FROM Parent P_1, Parent P_2
    WHERE P_1.Pname $= P_2$.Pname
ALSO
    SELECT R.Person, (SELECT Person AS Relative
                       FROM Children) AS Relatives
    FROM Parent P, Related R
    WHERE P.Pname in R.Relatives
ALSO
    SELECT Children.Person, (SELECT Relative
                       FROM Relatives) AS Relatives
    FROM Parent P, Related R
    WHERE Pname $= R$.Person

The transformed rule expressions of Q are:

$related(Person_1, relatives\{Person_2\})$:-$parent(Pname_1, children\{Person_1\})$,
                              $parent(Pname_2, children\{Person_2\})$,
                              $Pname_1 = Pname_2$,
                              $Person_2 \neq Person_1$.

$related(Person_2, relatives\{Person_1\})$:-$parent(Pname_1, children\{Person_1\})$,
                              $related(Person_2, relatives\{Relative_2\})$,
                              $in(Pname_1, relatives\{Relative_2\})$.

$related(Person_1, relatives\{Relative_2\})$:-$parent(Pname_1, children\{Person_1\})$,
                              $related(Person_2, relatives\{Relative_2\})$.
                              $Pname_1 = Person_2$. $\Box$

Example 9. Let Parent be the nested relation in Figure 3, and let Married be the flat relation in Figure 4. Let Q be the following SQL^*/NR query that retrieves all groups of people who are of the same generation, and the resultant relation Same-Generation of Q is shown in Figure 4.

It is assumed that a person P and a group of people G are of the same generation if P's parent and a parent of a person in G are of the same generation

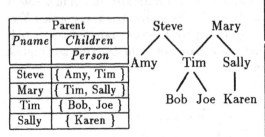

The Related table:

Related		
Person	*Relatives*	
	Relative	
Amy	{ Tim, Bob, Joe }	
Tim	{ Amy, Sally, Karen }	
Sally	{ Tim, Bob, Joe }	
Bob	{ Joe, Amy, Sally, Karen }	
Joe	{ Bob, Amy, Sally, Karen }	
Karen	{ Tim, Bob, Joe }	

The Parent table:

Parent	
Pname	*Children*
	Person
Steve	{ Amy, Tim }
Mary	{ Tim, Sally }
Tim	{ Bob, Joe }
Sally	{ Karen }

Fig. 3. Nested Relations *Parent* (and Its Graph Form) and *Related*

or are respectively of the same generation as a married couple. It is further assumed that P is of the same generation as himself, and every person who is either a child or a parent is in the flat relation People (which is a relation with a single attribute *person*).

In Q, a set operation $UNION$, which is an infix operator in SQL^*/NR, is used in the following format:

SFW-Expression UNION SFW-Expression.

(The syntax, semantics, and transformation steps of UNION in SQL^*/NR are easily determined.)

```
INSERT INTO Same-Generation
    SELECT (SELECT Person AS Sg-Person /* tuples with singular value */
            FROM P-Grp) AS Sg-Grp
    FROM (NEST People on Person AS P-Grp)
ALSO
    SELECT (SELECT Person AS Sg-Person
                FROM P₁.Children
                UNION
                SELECT Person AS Sg-Person
                FROM P₂.Children) AS Sg-Grp
    FROM Parent P₁, Parent P₂, Same-Generation SG
    WHERE P₁.Pname in SG.Sg-Grp AND P₂.Pname in SG.Sg-Grp
ALSO
    SELECT (SELECT Person AS Sg-Person
                FROM P₁.Children
                UNION
                SELECT Person AS Sg-Person
                FROM P₂.Children) AS Sg-Grp
    FROM Parent P₁, Parent P₂, Same-Generation SG₁,
            Same-Generation SG₂, Married
    WHERE P₁.Pname in SG₁.Sg-Grp AND P₂.Pname in SG₂.Sg-Grp AND
            Spouse-A in SG₁.Sg-Grp AND Spouse-B in SG₂.Sg-Grp
```

Married	
Spouse-A	*Spouse-B*
Steve	Mary
Mary	Steve

Same-Generation
Sg-Grp
Sg-Person
{ Steve }
{ Mary }
{ Amy, Tim, Sally }
{ Bob, Joe, Karen }

Fig. 4. A Flat Relation *Married* and a Nested Relation *Same-Generation*

The transformed rule expressions of Q, in which sg denotes the resultant nested relation Same-Generation, are:

sg(sg-grp$\{Person_1\}$) :- people(p-grp$\{Person_1\}$).

sg(sg-grp$\{Person_1,\ Person_2\}$) :- parent($Pname_1$, children$\{Person_1\}$),
 parent($Pname_2$, children$\{Person_2\}$),
 sg(sg-grp$\{Sg\text{-}Person_3\}$),
 in($Pname_1$, sg-grp$\{Sg\text{-}Person_3\}$),
 in($Pname_2$, sg-grp$\{Sg\text{-}Person_3\}$).

sg(sg-grp$\{Person_1,\ Person_2\}$) :- parent($Pname_1$, children$\{Person_1\}$),
 parent($Pname_2$, children$\{Person_2\}$),
 sg(sg-grp$\{Sg\text{-}Person_3\}$),
 sg(sg-grp$\{Sg\text{-}Person_4\}$),
 married($Spouse\text{-}A_5,\ Spouse\text{-}B_5$),
 in($Pname_1$, sg-grp$\{Sg\text{-}Person_3\}$),
 in($Pname_2$, sg-grp$\{Sg\text{-}Person_4\}$),
 in($Spouse\text{-}A_5$, sg-grp$\{Sg\text{-}Person_3\}$),
 in($Spouse\text{-}B_5$, sg-grp$\{Sg\text{-}Person_4\}$). □

3 SQL^*/NR Queries with Aggregates

In this section we discuss the inclusion of aggregate operators, namely avg, min, max, sum, and count, in SQL^*/NR. The inclusion of aggregates in SQL^*/NR offers a wider range of queries in SQL^*/NR that makes the language more appealing to standard SQL users. Furthermore, supporting aggregates in SQL^*/NR also provides a form of data storage reduction by allowing SQL^*/NR users the ability to evaluate aggregates on attributes in a database relation "on the fly" instead of storing the values generated by such operators in the database.

Built-in aggregate operators supported by SQL^*/NR are embedded within a subquery, called *Aggregate-subquery*. Attributes in a nested relation can be chosen as arguments of aggregate operators that apply to different groups of tuples. We define aggregates and a grouping construct[2] in SQL^*/NR queries, and include an approach for transforming an SQL^*/NR query with aggregates into LDL/NR rule expressions.

3.1 Aggregate Subqueries

An *Aggregate-subquery* is an SQL^*/NR subquery as discussed in Section 2.1 with the inclusion of aggregate operators in the **Select** clause and a **group by** construct. An *Aggregate-subquery*, referenced as the Select-From-Where-GroupBy Expression (SFWGB-Expression for short), is of the form:

SELECT attribute-list-with-aggregates
FROM relation-list
[WHERE boolean-expression]
Group By column-names

The **Select** clause in an SFWGB-Expression has the following syntax:

SELECT *atomic-attr$_1$, ..., atomic-attr$_n$, non-atomic-attr$_1$, ...,*
non-atomic-attr$_m$, agg-str$_1$, ..., agg-str$_p$

where *atomic-attr$_i$* $(1 \leq i \leq n)$ and *non-atomic-attr$_j$* $(1 \leq j \leq m)$ are as defined in Section 2.1, and *agg-str$_k$*, $1 \leq k \leq p$, is of the form:

Agg ([DISTINCT] [relation-name.] attribute-name) AS agg-attr

where **Agg** is either **min, max, sum, count** or **avg** which is applied to **attribute-name**. There is an **AS** clause which specifies a reference name (i.e., agg-attr) to the result generated from the aggregation. The **DISTINCT** keyword, which is optional, eliminates duplicates before the aggregate operator is applied.

The difference between an SFW-Expression and an SFWGB-Expression lies in the fact that a **Select** clause in an SFWGB-Expression contains aggregate operators that is applied to one or more attributes on the same level of nesting in a nested relation. For example, to retrieve the average age of the children of each student, where the Student relation in as shown in Figure 1, we enter:

SELECT ... avg(Student.Children.Age) AS Children-Age-Avg,

Furthermore, there exists a *group-by* construct in an SFWGB-Expression but not in any SFW-Expression. In an SFWGB-Expression, the **group-by** construct is of the form:

Group By *atomic-attr$_1$, ..., atomic-attr$_n$*

---

[2] A grouping construct includes a grouping operator, i.e., **group-by**, that collects a number of tuples in a nested relation based on a subset of attributes in these tuples that have the same value.

where all $atomic\text{-}attr_i$, $1 \leq i \leq n$, are atomic attributes at the same level of nesting in a nested relation to which the **Group By** operation is applied. These attributes specified in the **Group By** clause are used to form groups of tuples. Tuples that have the same value on all the attributes $atomic\text{-}attr_1, \ldots,$ $atomic\text{-}attr_n$ specified in the **Group By** clause are placed in one group. These groups are then used by the aggregate operators given in the **Select** clause of an SFWGB-Expression.

3.2 SQL^*/NR Queries with Aggregate Subqueries

An SQL^*/NR query with aggregates is an extended version of the SQL^*/NR query discussed in Section 2.2 with the addition of an *Aggregate-subquery* included in a **THEN INTO** statement. The format of an SQL^*/NR query with aggregates is as follows:

> INSERT INTO R_1
> > *Basis-subquery*
>
> ALSO
> > *Recursive-subquery$_1$*
>
> $\ldots$
>
> ALSO
> > *Recursive-subquery$_n$*
>
> THEN INTO R_2
> > *Aggregate-subquery*

where the **THEN INTO** statement includes the keyword *THEN INTO* and a nested relation R_2 followed by an *Aggregate-subquery* (an SFWGB-Expression). R_1, which is the resultant relation generated by the **INSERT INTO** statement, is treated as a temporary (input) relation to which the *Aggregate-subquery* applies to yield the final desired relation R_2. The following example includes an SQL^*/NR query with aggregates.

Example 10. Let Parent be the nested relation in Figure 5. We specify the following query that generates the relation *Dec-info* in Figure 5. *Dec-info* contains the average age of all the descendants of Daniel.

```
INSERT INTO Temp
   SELECT PName AS A-Name, (SELECT C-Name AS D-Name,
                                   C-Age AS D-Age
                            FROM Children) AS Descendants
   FROM Parent
   WHERE PName = "Daniel"
ALSO
   SELECT T.A-Name, (SELECT C-Name AS D-Name, C-Age AS D-Age
                     FROM Children) AS Descendants
   FROM Parent P, Temp T
```

WHERE P.PName in T.Descendants.D-Name
THEN INTO Dec-Info
    SELECT A-Name, AVG(D-Age) AS D-Avg-Age
    FROM Temp
    Group By A-Name □

Parent		
PName	*Children*	
	C-Name	*C-Age*
Daniel	Tom	60
	Merle	55
Merle	Steve	30
	Ann	25
Steve	Joseph	5

Temp		
A-Name	*Descendants*	
	D-Name	*D-Age*
Daniel	Tom	60
	Merle	55
	Steve	30
	Ann	25
	Joseph	5

Dec-Info	
A-Name	*D-Avg-Age*
Daniel	35

Fig. 5. Relations *Parent*, *Temp*, and *Dec-Info*

As shown in Figure 5, the temporary relation *temp* is generated as a result of evaluating the recursive query which retrieves all the descendants of "Daniel". Applying the *Aggregate-subquery* in the **THEN INTO** statement to *temp* yields the desired resultant relation *Dec-info*.

3.3 Transforming SQL^*/NR Queries with Aggregates into Rule Expressions in LDL/NR

The process of transforming the **From** and **Where** clauses of an SQL^*/NR subquery with aggregates is exactly the same as that for an SQL^*/NR subquery without aggregates. We choose to adopt similar notation used by [SR91, Gel93] to represent the **group by** construct and aggregate operators in an LDL/NR rule expression. The following is the general form of a **group by** subgoal (tuple term) that is to be included in the body of a rule expression RE. This subgoal contains the transformed aggregate constructs specified as arguments in the **Select** clause of the *Aggregate-subquery*:

$$\text{group-by}(r(t), [X_1, \ldots, X_m], [Y_1 = Agg_1 \,([\text{DISTINCT}]\,(Attr_1)), \ldots,$$
$$Y_n = Agg_n \,([\text{DISTINCT}]\,(Attr_n)])$$

where

- r is the relation to which the grouping is applied and is included in the **From** clause of the *Aggregate-subquery*.
- t represents the arguments in r which are either constants, variables, or set terms.
- The grouping list $X_1, \ldots, X_m$ consists of one or more variables that must appear in t and on which the grouping is based. ($X_1, \ldots, X_m$ are the attributes specified in the **group by** clause of the *Aggregate-subquery*.)

- The list of $Y_1, \ldots, Y_n$ is the aggregation list. Each Y_i, $1 \leq i \leq n$, is a new variable (attribute) which is the reference name following an aggregate operator and the AS keyword in the **Select** clause of the *Aggregate-subquery*. $Y_1, \ldots, Y_n$ are included as arguments in the head tuple term of RE.
- $Attr_1, \ldots, Attr_n$ are attribute names in r to which aggregate operators $Agg_1, \ldots, Agg_n$ (e.g. sum and count) are applied, respectively. A built-in predicate for each one of the aggregate operators that is represented by Agg_1, $\ldots, Agg_n$ (e.g., MIN, MAX, SUM, AVG, and COUNT), respectively, in the **group by** subgoal is provided in LDL/NR.
- Within the body of RE, the **group by** subgoal represents a relation over attributes (variables) $X_1, \ldots, X_m$ in the grouping list and attributes (variables) $Y_1, \ldots, Y_n$ in the aggregation list.

Given below is an example to show the rule expressions in LDL/NR transformed from an SQL^*/NR with aggregates.

Example 11. Let Q be the SQL^*/NR query with aggregates given in Example 10. The transformed rule expressions of Q are:

temp($Pname$, descendants{descendants'(C-$Name$, C-Age)}) :-
    parent($Pname$, children{children'(C-$Name$, C-Age)}),
    $Pname$ = "Daniel".

temp(A-$name_2$, descendants{descendants'(C-$Name_1$, C-Age_1)}) :-
    parent($Pname_1$, children{children'(C-$Name_1$, C-Age_1)}),
    temp(A-$Name_2$, descendants{descendants'(D-$Name_2$, D-Age_2)}),
    in($Pname_1$, descendants{descendants'(D-$Name_2$)}).

dec-info(A-$Name$, D-Avg-Age) :-
    group-by(temp(A-$Name$, descendants{descendants'(D-$Name_1$, D-Age_1)}),
        [A-$Name$],
        [D-Avg-Age = AVG(D-Age_1)]). □

4 Summary

We have proposed an extended SQL, called SQL^*/NR, that not only can handle both recursive queries and nested relations, but also allows aggregate operators. An algorithm, which transforms an SQL^*/NR query Q into rules expressions in LDL/NR which can be evaluated to retrieve the desired answers to Q, has also been given in [Qar95]. The proposed SQL^*/NR constructs, which includes query and subquery, have been implemented [Qar95]. The implementation includes accepting an SQL^*/NR query Q (defined according to the syntax described in Appendix A), transforming Q into rule expressions REs in LDL/NR (according to the algorithm described in [Qar95]), and evaluating REs to yield the desired result of Q.

Appendix

A SQL^*/NR BNF

The following is a modified BNF definition of the queries in SQL^*/NR. Non-distinguished symbols are enclosed with "<>". The structure [...] indicates an optional entry, and the structure {...} indicates an additional zero or more repetitions of the entry.

```
<query expression>::-INSERT INTO <nested-relation-name><basis-subquery>
                [{ALSO <recursive-subquery>}]
                [THEN INTO <nested-relation-name><aggregate-subquery>]
<basis-subquery>, <recursive-subquery>::-SELECT <attribute-list>
                                FROM <relation-list>
                                [WHERE <boolean-expression>]
<aggregate-subquery>::-SELECT <attribute-list-with-aggregates>
                FROM <relation-list>
                [WHERE <boolean-expression>]
                GROUP BY <column-name> {<column-name>}
<attribute-list>::-<atomic-attr>{<atomic-attr>} {<non-atomic-attr>} |
                {<atomic-attr>}<non-atomic-attr> {<non-atomic-attr>}
<atomic-attr>::-[<relation-name>.]<attribute-name> [AS <column-name>]
<non-atomic-attr>::-[<rel-name>.]<embedded-relation-name> [AS <column-name>] |
                (<basis-subquery>) AS <column-name>
<relation-name>::-<rel-name>|<rel-name>.<rel-name>
<rel-name>::-<nested-relation-name> |<embedded-relation-name> |<reference-name>
<relation-list>::-<relation> {<relation>}
<relation>::-<nested-relation-name> [<reference-name>] |
        [<nested-relation-name> | <reference-name> .]
        <embedded-relation-name> [<reference-name>] |
        (NEST <nested-relation-name> ON attribute-list AS <column-name>) |
        (UNNEST <nested-relation-name> ON <attribute-list>)
<boolean-expression>::-(<boolean-expression><bool-optr> <boolean-expression>) |
                <LR-OP><Comp-P><RR-OP> |
                <LS-OP><Comp-SetM><RS-OP>
<bool-optr>::-AND | OR
<LR-OP>,<RR-OP>::-<nested-rel-attr>|<constant>
<LS-OP>::-<nested-rel-attr>|<[nested-rel-attr,LS-OP]>
<RS-OP>::-<nested-relation>
<nested-rel-attr>::-<attribute-name> | <rel-name>.<nested-rel-attr>
<nested-relation>::-<rel-name> | <rel-name>.<nested-relation>
<Comp-P>::-< | > | ≤ | ≥ | = | ≠
<Comp-SetM>::-in | notin
<attribute-name>,<column-name>,<embedded-relation-name>,
<nested-relation-name>,<reference-name>::-alphabetic-char {alphanumeric-char}
<attribute-list-with-aggregates>::-<attribute-list>
    <agg>([DISTINCT][<relation-name>.]<attribute-name>) AS <attribute-name>
    {<agg>([DISTINCT][<relation-name>.]<attribute-name>) AS <attribute-name>}
<agg>::-MIN | MAX | SUM | COUNT | AVG
```

References

[CC90]   Q. Chen and W. Chu. *Deductive and Object-Oriented Database*, chapter HILOG: A High-order Logic Programming Language for Non-1NF Deductive Databases, pages 431–452. Elsevier Science Publishers, 1990. W. Kim, et al. (Editors).

[CK91]   Q. Chen and Y. Kambayashi. Nested Relation Based Database Knowledge Representation. In *Proceedings of 1991 ACM SIGMOD International Conference on Management of Data*, pages 328–337. ACM, 1991.

[Gel93]   A. V. Gelder. Foundation of Aggregation in Deductive Databases. In *Proceedings of the 3rd Intl. Conference on Deductive and Object-Oriented Databases*, pages 13–33. Springer-Verlag, 1993. Lecture Notes in Computer Science, 470.

[GM92]   J. Grant and J. Minker. The Impact of Logic Programming on Databases. *Communications of ACM*, 35(3):67–81, March 1992.

[GN90]   H. Gallaire and J.-M. Nicolas. Logic and databases: An assessment. In *the Third International Conference on Database Theory*, pages 177–186. Springer-Verlag, December 1990. Lecture Notes in Computer Science, 470.

[Hul89]   Richard Hull. Four Views of Complex Objects: A Sophisticate's Introduction. In S. Abiteboul, P. Fischer, and H. Schek, editors, *Nested Relations and Complex Objects in Databases*, pages 87–116. Springer-Verlag, New York, 1989. Lecture Notes in Computer Science, 361.

[KC93]   K. Koymen and Q. Cai. SQL^*: A Recursive SQL. *Information Systems*, 18(2):121–128, 1993.

[KS91]   H. Korth and A. Silberschatz. *Database System Concepts, Second Edition*. McGraw-Hill, Auckland, 1991.

[Kup90]   G. M. Kuper. Logic Programming with Sets. *Journal of Computer and System Sciences*, 41:44–64, 1990.

[Llo87]   J. W. Lloyd. *Foundations of Logic Programming, Second, Extended Edition*. Spring-Verlag, New York, 1987.

[LN95]   S. J. Lim and Y-K Ng. Set-Term Matching in a Logic Database Language. In *Proceedings of the 4th International Conference on Database Systems for Advanced Applications*, pages 189–196, Singapore, April 1995.

[Qar95]   N. Qaraeen. SQL*/NR: A Recursive Query Language for Nested Relations. Master's thesis, Brigham Young University, Provo, Utah, July 1995. (Available via http://www.dragon.cs.byu.edu/homepage.html).

[RKB87]   M. Roth, H. Korth, and D. Batony. SQL/NF: A Query Language for ¬1NF Relational Databases. *Information Systems*, 12(1):99–114, 1987.

[RKS88]   M. Roth, H. Korth, and A. Silberschatz. Extended Algebra and Calculus for Nested Relational Databases. *ACM Transactions on Database Systems*, 13(4):389–417, December 1988.

[SR91]   S. Sudarshan and R. Ramakrishman. Aggregation and Relevance in Deductive Databases. In *Proceedings of 17th VLDB.*, pages 501–511, Barcelona, Spain, 1991.

[STZ92]   O. Shmueli, S. Tsur, and C. Zaniolo. Compilation of Set Terms in the Logic Data Language (LDL). *Logic Programming*, 12(1):89–119, 1992.

[Uni91]   UniSQL. *UniSQL/X Database Management System User's Manual*. UniSQL Inc., Austin, Texas, 1991.

Partitioning Pipelines with Communication Costs

Sumit Ganguly, Apostolos Gerasoulis, Weining Wang

Rutgers University, New Brunswick NJ 08904 , USA

Abstract. In this paper, we consider the problem of scheduling a database query execution graph on a parallel machine. Specifically, we consider the problem of data-partitioning pipelined operators with the objective of minimizing response time. This is a basic problem in scheduling database execution trees. Partitioning promises increased parallelism and memory availability at the price of greater communication overhead. Current partitioning methods [BB90, TWPY92, LCRY93, NSHL93] do not consider these trade-offs. We present a mathematical framework within which these alternatives can be quantified for many interesting practical scenarios. We then present an algorithm whose performance is within a factor of 2 of the optimum possible.

1 Introduction

The success of parallel database systems in the commercial and academic world has generated considerable research interest in all aspects of parallel database systems. In particular, query optimization for parallel executions brings a set of new problems to the forefront. These include (a) the design of cost models that can accurately predict the response time of an execution tree, (b) the design of search algorithms that can optimize over a subset of possible execution trees using the cost model, (c) the design of scheduling algorithms for minimizing the response time of a single execution tree and (d) design of dynamic schemes that adapt to skew at run-time.

A consensus among researchers is that the query optimization problem should be approached in two phases: Phase 1 generates a tree and addresses issues (a) and (b) above, and Phase 2 addresses issues (c) and (d) above. The problem that we focus in this paper is issue (c), the design of good scheduling algorithms for a given database execution tree.

Several algorithms have been proposed to schedule general database execution trees [NSHL93, TL94, HCY94]. These algorithms first transform a given tree into a segmented tree, such that each segment is a set of relational operators that can run in a pipelined parallel fashion[Sch90, DeWGra92]. (In the context of Phase 1 of query optimization, the segmentation technique has been used by [Hon91, Hon92, LVZ93, SYT93, ZZBS94]). Each segment, consisting of multiple operators in a pipeline, is then data-partitioned to take advantage of parallelism in the architecture. Partitioning pipelined operators therefore is a basic problem in scheduling database execution trees/graphs.

Current scheduling algorithms for partitioning a pipeline segment typically use the optimal single-shelf greedy algorithm of [TWPY92]. In [LCRY93], the algorithm presented in [TWPY92] has been extended to 2-phase hash-joins. An important cost-factor that the above scheduling algorithms do not consider is the trade-off between partitioning and communication. Increased partitioning promises reduced processing time and increased memory availability. However, increased partitioning also increases the communication overhead. A computation-communication tradeoff for pipelines *without* partitioning was examined in [HM94].

The main contribution of the paper is an algorithm that partitions a set of pipelined operators. This algorithm is proven to be within a factor of 2 of the optimum performance. The conditions for the applicability of the algorithm are as follows.

- The response time of an operator is a non-decreasing function of the number of processors assigned to another operator in the pipelined set.
- The work done by an operator is a non-decreasing function of the number of processors assigned to the operator.

These conditions are sufficiently general to allow a large variety of practical scenarios to be modeled. This includes allocation policies for communication buffer allocation, network congestion and thread startup overhead.

The rest of the paper is organized as follows. Section 2 presents a review of the terminology used in the paper. Section 3 presents the assumptions we make regarding the architecture, data placement and the communication model. In Section 4, we present a cost model for partitioned pipelines. Section 5 presents the partitioning algorithm and Section 6 presents examples comparing our algorithm to existing schemes. Section 7 discusses generalizations to the cost model. Finally, we conclude in Section 8.

2 Review

In this section, we first briefly describe the notions of execution trees and operator trees.

We consider the space of execution trees [SAC+], where each node is annotated to represent a specific method of joining the operands and of accessing the relations (See Figure 1). Each join operation may be a composition of several operations, for example, sorting and merging files, building a hash table and probing it. Given an execution tree, an operator tree [Gan92, GHK92, Hon91, SriEls93] is an expansion of each node of the execution tree into its constituent operations. Data dependency between adjacent operations in the operator tree is placed on the edges of the tree. We consider two kinds of data dependencies between operations: *sequential* and *pipelined*. Sequential dependency arises in executions where the consumer operation cannot begin until the producer operation has completed. Examples of sequential dependencies are (a) the producer or the consumer is a "sort" operation, or (b) the producer is a "build-hash-table" operation and the consumer is a "probe-hash-table" operation. Pipeline dependency arises in situations tuples can be consumed without waiting for the producer to complete. Examples of pipelined dependency are (a) when producer is a "merge" operation and the consumer is a "probe" operation or (b) the producer is a "probe-hash-table" operation and the consumer is the "index-scan" of a "nested-loop" operation. Since properties of relational implementation operators are well known, it is easy to construct a table of data dependencies between all pairs of relational operators [Gan92, GHK92].

We assume that there is a centralized cost model (similar to the System R cost model) that estimates the time of computation of each of the operators in an operator tree on a sequential machine, the sizes of the inputs and the outputs for each operator and the memory requirements of each of the operators.

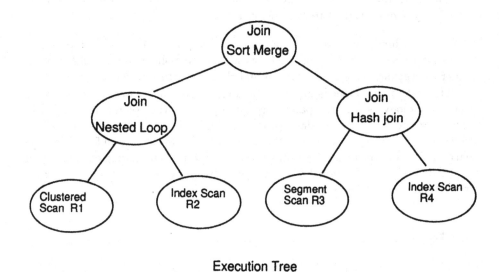

Execution Tree

Fig. 1. Example of Execution tree

3   Architectural Assumptions

In this section, we discuss assumptions about the architecture, data placement and the communication model.

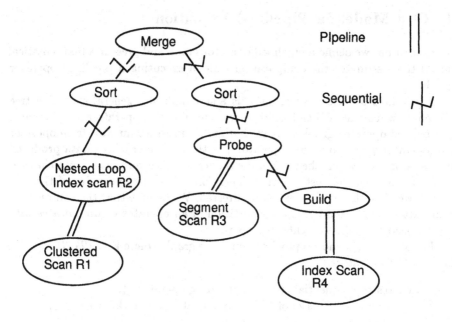

Operator Tree

Fig. 2. The Operator Tree for Figure 1

3.1 Architecture and Data Placement

The results of the paper are applicable to a distributed memory system (also called shared-nothing architecture) in which each database relation is assumed to have been fragmented across all disks. The results in the paper are also applicable to shared-disks architecture.

3.2 Communication Costs

The cost of communicating a single message of M bytes from a sender to a remote receiver is modeled as:

$$t_{send} = t_{receive} = \alpha + M\beta$$

where α is the (constant) operating system call overhead for message startup. and β is the per-byte cost of transmitting the message (message volume component) into network channel. Thus, each send operation involves two components, the *startup* component and the *message-transmit* component.

We assume that the actual transit time for a message is small compared to the processing time required to package the message and hand it to the network. The communication model therefore ignores the number of hops traveled by the message as a cost factor. This assumption is true to a first approximation in fast interconnection networks today.

4 Cost Model for Pipelined Execution

In this section, we define a pipelined operator graph and present a mathematical model that estimates the computation time of executing a pipelined operator graph.

A *pipelined operator tree (graph)* is a subgraph of a general execution tree (graph) in which data dependencies between neighboring operators are pipelined. A pipelined operator graph is a directed graph, in which an edge from operator i to operator j indicates that operator j is the consumer of the data produced by operator i. However, the roles of the sender and receiver may be interchanged without changing the cost function. Therefore, from the viewpoint of the cost model, we view the edges as being undirected. We assume that the communication pattern is all-to-all, i.e., each partition of an operator communicates with all the partitions of its neighboring operator.

Let $G = (V, E)$ denote a pipelined operator graph, where $V = \{T_1, T_2, \ldots, T_n\}$ is the set of operators.

- t_i denotes the sequential cost of computing operator T_i.
- $c_{ij} = c_{ji}$ denotes the size of the communication along the edge $e_{\{i,j\}}$.

Definition 1. Given a pipeline operator graph over n nodes, a partitioning scheme is a vector $\mathbf{x} = (x_1, x_2, \ldots, x_n)$. $\qquad\square$

The variable x_i denotes the number of processors that are assigned to operator T_i and is called the *degree of partitioning* of T_i. Assuming uniform division of load, each partition has a computation time of t_i/x_i units. Each partition of T_i sends/receives a total of c_{ij}/x_i bytes along the undirected edge $e_{\{i,j\}}$. The *message-transmit* cost for all messages sent/received along edge e_{ij} by a partition of operator i is therefore $(c_{ij} \cdot \beta)/x_i$. We now calculate the *startup* component of communication time spent by any partition of operator i while communicating along the edge $e_{\{i,j\}}$. Let λ_{ij} denote the number of messages sent/received from a partition of operator i to an operator in partition j. Since the communication patterns are assumed to be all-to-all, each partition of an operator communicates with all the partitions of its neighboring operator. The *startup* cost incurred by a partition of operator i corresponding to communication along edge $e_{\{i,j\}}$ is $\lambda_{ij} \cdot \alpha \cdot x_j$. For non-communicating pairs of operators i and j, $\lambda_{ij} = \lambda_{ji} = 0$.

The term λ_{ij} is a function of the degrees of partitioning of operators i and j respectively and also on the buffer allocation policy. Section 4.2 details the specific cost formulae for λ_{ij} for various buffer-allocation policies. The total communication overhead incurred by a partition of operator i is

$$t_i^{comm}(\mathbf{x}) = \sum_{j=1}^{n} (\lambda_{ij} \cdot \alpha \cdot x_j) + \sum_{j=1}^{n} (c_{ij} \cdot \beta) / x_i \qquad (1)$$

The computation time incurred by a partition of operator i is

$$t_i^{comp}(\mathbf{x}) = t_i/x_i \qquad (2)$$

Let $\tau_i(\mathbf{x})$ denote the active time spent by a processor in computing a partition of T_i. Then,

$$\tau_i(\mathbf{x}) = t_i^{comp}(\mathbf{x}) + t_i^{comm}(\mathbf{x}) \tag{3}$$

The work done by a partitioned computation is calculated as

$$W(\mathbf{x}) = \sum_{i=1}^{n} x_i \cdot \tau_i(\mathbf{x}) \tag{4}$$

4.1 Steady Pipelines

Intuitively, the time required to complete the execution of a pipeline is determined by the member that takes the longest time. The assumption for this conclusion is that the time taken to fill the pipeline and the time taken to flush the pipeline are negligible when compared to the time taken to complete the pipeline. We refer to such pipelines as *steady* pipelines. Consider what fraction of work is completed by the first node in a pipeline before the last node in the pipeline is ready to start processing? This fraction of work must be at least

$$min\{1/\lambda_{ij} \mid \{i,j\} \text{is an edge}\}$$

For a pipeline to to be steady, the first operator cannot be too far ahead of the final operator. This motivates the following definition.

Definition 2. A pipeline is ϵ-steady if $\lambda_{ij} \geq 1/\epsilon$, for each edge $e_{\{ij\}}$.

For example, suppose that $\epsilon = 0.1$. Then, a pipeline is ϵ-steady, if at least $1/0.1 = 10$ messages are sent between any pair of communicating partitions.

Earlier models for pipelining [HM94] do not consider steady pipelines. The reason for this is that they do not consider partitioning of operators. By partitioning an operator, the available communication buffer size increases, resulting in fewer messages being transmitted between any pair of communicating partitions. This can cause unsteady executions, resulting in a more complicated mathematical model. In order to keep the cost model simple, we make the steadiness assumption on pipeline execution. In practice, the vast majority of database pipelines are steady.

4.2 Buffer Allocation Schemes

In the previous sections, we presented a cost function for computing the response time of partitioned operators. An important parameter in this expression is λ_{ij}, which represents the number of messages that are sent/received between partitions of operator i and j. This parameter is intimately connected with the scheme used to allocate communication buffers to partitioned operators. The goal of this section is to demonstrate that it is very reasonable to assume, under various realistic buffer allocation schemes, that the response time function $\tau_i(x_1, \ldots, x_n)$ satisfies the following two general properties.

1. Constraint 1: $\tau_i(x_1, \ldots, x_n)$ is a non-decreasing function of x_j for all $j = 1 \ldots, n, j \neq i$.

2. Constraint 2: $x_i \cdot \tau_i(x_1, \ldots, x_n)$ is a non-decreasing function of x_i.

Property 1 states that increasing the degree of parallelism of an operator may not decrease the time taken to complete another operator (typically a neighbor). Property 2 states that the total work increases as we increase the degree of parallelism of an operator.

We now discuss various communication buffer allocation schemes. These schemes yield different expressions for λ_{ij}'s and consequently give rise to different optimization problems. We enumerate a few possible buffer allocation schemes and the expressions for λ_{ij} for each of these scenarios. The list of buffer allocation policies considered below are meant only to be representative of the ones possible. In each case, note that the two properties for τ_i above hold. Consider ϵ-steady pipelines.

Scheme A : Allocate a buffer of size b for each communicating pair of partitioned operators. Then

$$\lambda_{ij} = max(c_{ij}/(b \cdot x_i \cdot x_j), 1/\epsilon) \tag{5}$$

Note that by substituting this expression for λ_{ij} into (1), the two properties for τ_i are satisfied.

Scheme B: Suppose an operator T_i has n_i neighbors. Each partition of a operator receives n_i buffers, each of size b. Each buffer is used by the partition for communicating with all partitions of a neighboring operator. The number of messages that must be sent or received by a partition of T_i along the edge $e_{\{i,j\}}$ is $c_{ij}/(b \cdot x_i)$. Hence,

$$\lambda_{ij} = max\left(c_{ij}/(b \cdot x_i), c_{ij}/(b \cdot x_j), 1/\epsilon\right) \tag{6}$$

Note again that τ_i satisfies the two properties enumerated above.

Scheme C: In this scheme each partition is assigned a buffer of size b. The partition divides this buffer in proportion to the size of the data it needs to communicate with its neighbor. Let $i_1, i_2, \ldots, i_{n(i)}$ denote the $n(i)$ neighbors of operator T_i. Let $b_1, b_2, \ldots, b_n$ be the size of the buffers allocated for each of the neighbors. Then

$$\frac{c_{i,i_1}}{b_1} = \frac{c_{i,i_2}}{b_2} = \frac{c_{i,i_3}}{b_3} = \ldots = \frac{c_{i,i_{n(i)}}}{b_{n(i)}} = \mu_i \text{ (say)} \tag{7}$$

subject to the constraint $\sum_{j=1}^{n(i)} b_j = b$. This gives

$$\mu_i = \sum_{j=1}^{n(i)} c_{i,i_j}/b \tag{8}$$

We conclude that

$$\lambda_{ij} = max(\mu_i/x_i, \mu_j/x_j, 1/\epsilon) \tag{9}$$

Note again that τ_i satisfies the two properties.

5 Partitioning Operators

In this section, we present a novel and general scheduling problem and an efficient algorithm that computes a near-optimal schedule for the given problem. The problem formulation is strictly more general than previous models for scheduling partitionable operators, such as [BB90, TWPY92]. It allows the response time of an operator to be dependent on the number of processors assigned to itself and to other operators in the graph, while satisfying certain constraints. The models proposed earlier, such as [BB90, TWPY92], do not allow the response time of a operator to be dependent on the number of processors assigned to other operators. As discussed in Section 4, this feature is essential to modeling response time of partitioned pipelines.

5.1 Problem Formulation

Let $T_1, T_2, \ldots, T_n$ be n given operators. Operators are partitionable, i.e., each operator may be executed using one or more processors. All partitions of all operators must run concurrently. If the total number of partitions exceeds the number of processors p, then multiprogramming is used to switch between partitions of operators assigned to a processor. The overhead of context switch of multiprogramming is assumed to be negligible.

Let x_i denote the number of processors assigned to operator T_i. Associated with each operator T_i is a *response time function* $\tau_i(x_1, x_2, \ldots, x_n)$ that estimates the time required to complete each of the x_i partitions of T_i. We define the notion of a schedule next.

Definition 3. For $i = 1, \ldots, p$, let x_i be the degree of partitioning for operator T_i and let p be the number of processors. For $i = 1, \ldots, n$, a *schedule* S assigns x_i distinct processors to each operator T_i. □

The above definition ensures that partitions of a operator are assigned to distinct processors. This implies that all partitions of operators take equal time to complete. We now define the notion of the load on a processor. The load on a processor is defined as the sum of the times required by each partition that is assigned to it.

Definition 4. Given a schedule of operators $T_1, T_2, \ldots, T_n$ on p processors. The load on a processor j is defined as

$$L_j = \sum \{\tau_i(x_1, x_2, \ldots, x_n) \mid \text{ a partition of } T_i \text{ is assigned to processor } j \quad (10)$$

We are now in a position to define the response time of a schedule.

Definition 5. The response time of a schedule S, denoted by $R(S)$ is the maximum load on any processor.

$$R(S) = max_{1 \leq i \leq p} L_i(S) \quad (11)$$

The work done by a schedule is defined as follows.

Definition 6. The work done by a schedule S, denoted by W^S is defined as

$$W^S = \sum_{i=1}^{n} x_i \cdot \tau_i(x_1, x_2, \ldots, x_n) \qquad (12)$$

We study the problem of finding a schedule with the optimal response time under the following constraints.

1. Constraint 1: $\tau_i(x_1, \ldots, x_n)$ is a non-decreasing function of x_j for all $j = 1 \ldots, n, j \neq i$.
2. Constraint 2: $x_i \cdot \tau_i(x_1, \ldots, x_n)$ is a non-decreasing function of x_i.

The problem of finding an optimal schedule under the above constraints is called the operator partitioning problem (OPP).

Theorem 7. OPP is NP-hard.

Proof. Suppose $\tau_i(x_1, \ldots, x_n) = t_i$. This is the independent operator scheduling problem which is known to be NP-hard. $\qquad\square$

Figure 3 presents the algorithm PPO that produces a schedule whose response time is within a factor of 2 of the optimal response time possible. The algorithm uses the LPT scheduling heuristic [Gra69] used for scheduling independent sets of operators. The LPT heuristic works by having an idle processor execute the largest operator (in terms of time) among the set of operators that are yet to be executed. A widely used property of the LPT heuristic is the following.

Lemma 8 Gra69,BB90. *Let R denote the response time of an LPT schedule of operators with computation times $t_1, t_2, \ldots, t_n$, such that $R > max_{i=1}^{n} t_i$. If the number of processors p is > 1, then $R \leq 2\frac{W}{p}\left(1 - \frac{1}{p+1}\right)$.* $\qquad\square$

We now present the proof of near-optimality of the algorithm. Let x_i^j denote the value of the program variable x[i] at the end of the jth iteration, i.e., when $\sum_{i=1}^{n} x_i - n = j$. Let x^{opt} denote an optimal partitioning vector. The terms $R, R^{new}, oldmax$ etc. refer to the values of the corresponding program variables in the algorithm PPO.

Lemma 9. $R = old\_max$ at the end of all program statements.

Proof. Consider Algorithm PPO presented in Figure 3. If control enters statement 1 of the main program, then $R = old\_max$. Otherwise $R > old\_max$ at the initialization condition and the program would have terminated at $\boxed{T0}$. If control enters statement 1 from statement 9 or from statement 20, the condition follows from the checks in statement 7 and statement 18 respectively. The rest is very straightforward. $\qquad\square$

Algorithm PPO

Input: Operators $T_1, T_2, \ldots, T_n$, response time functions $\tau_i(x_1, x_2, \ldots, x_n)$ for each $i = 1, 2, \ldots, n$ and the number of processors p.

Output: The degree of partitioning x_i for operator T_i, $i = 1, 2, \ldots, n$. The schedule is given by $LPT(x_1, x_2, \ldots, x_n)$.

Initially:

I.1 $x_i = 1$, for $i = 1, 2, \ldots, n$.

I.2 $R = LPT(\mathbf{x})$;   /* R records the height of the current LPT schedule */

I.3 $old\_max = max_{i=1}^n \tau_i(\mathbf{x})$;   /* find highest partition */

I.4   **if** $(R > old\_max)$ **return x;** $\boxed{T0}$

1.   *state 0:* Pick T_i with highest value of $\tau_i(\mathbf{x})$;

2.   **newx** = **x** except **newx** $[i]$ = **x**$[i]$ +1;

3.   **if** (**newx** $[i] > p$) **return x;** $\boxed{T1}$

4. $new\_max = max_{i=1}^n(\mathbf{newx})$;

5.   **if** $(new\_max \geq old\_max)$ **goto** *state 1*;

6. $R^{new} = LPT(\mathbf{newx})$;

7.   **if** $(R^{new} == new\_max)$ {

8.   **x** =**newx**; $new\_max = old\_max$;

9. $R = R^{new}$; **goto** *state 0*;

9'   }

10.   **if** $R^{new} < R$ **return newx;** $\boxed{T2}$

11.   **if** $R^{new} \geq R$ **return x;** $\boxed{T3}$

12.   *state 1:* Pick T_i with highest value of $\tau_i(\mathbf{x})$;

13.   **newx** $[i]$ = **newx**$[i]$ +1;

14.   **if** (**newx** $[i] > p$) **return x;** $\boxed{T4}$

15. $new\_max = max_{i=1}^n(\mathbf{newx})$;

16.   **if** $(new\_max \geq old\_max)$ **goto** *state 1*;

17. $R^{new} = LPT(\mathbf{newx})$;

18.   **if** $(R^{new} == new\_max)$ {

19.   **x** =**newx**; $new\_max = old\_max$;

20. $R + R^{new}$; **goto** *state 0*;

20'   }

21.   **if** $R^{new} < R$ **return newx;** $\boxed{T5}$

22.   **if** $R^{new} \geq R$ **return x;** $\boxed{T6}$

Fig. 3. Algorithm PPO: Partitioning Pipelined Operators

Lemma 10. *For* $1 \leq i \leq n$, $x_i \leq x_i^{opt}$.

Proof. Consider the first iteration $j+1$ at which $x_i^{j+1} = x_i^{opt} + 1$. Let R^j denote the value of R at the end of iteration j. It follows that at iteration j $x_i^j = x_i^{opt}$ and $x_k^j \leq x_k^{opt}$ for $1 \leq k \leq n$ and $R^j = \tau_i(x_1^j, \ldots, x_n^j)$. From constraint 1 on τ_i, it follows that

$$R^j = \tau_i(x_1^j, \ldots, x_n^j) \leq \tau_i(x_1^{opt}, \ldots, x_n^{opt}) \leq R^{opt} \qquad (13)$$

Since the algorithm terminated at an iteration higher than j, the response time of the schedule obtained by the algorithm, say $R^{sch} < R^j$. We conclude that $R^{sch} < R^{opt}$ implying a contradiction. □

Lemma 11. *Let W^{sch} denote the work done by the schedule output by the algorithm PPO. Let W^{opt} be the work done by an optimal schedule. Then $W^{sch} \leq W^{opt}$.*

Proof.

$$W^{sch} = \sum_{i=1}^{n} x_i \tau_i(x_1, \ldots, x_n)$$

From Lemma 3 $x_j \leq x_j^{opt}$. From constraint 1 on τ_i, the expression for W^{sch} satisfies the inequality

$$W^{sch} \leq \sum_{i=1}^{n} x_i \cdot \tau_i(x_1^{opt}, \ldots, x_{i-1}^{opt} x_i, x_{i+1}^{opt}, \ldots, x_n^{opt}) \qquad (14)$$

Since $x_i \leq x_i^{opt}$ from Lemma 3, constraint 2 on τ_i implies that

$$\sum_{i=1}^{n} x_i \cdot \tau_i(x_1^{opt}, \ldots, x_{i-1}^{opt} x_i, x_{i+1}^{opt}, \ldots, x_n^{opt})$$

$$\leq \sum_{i=1}^{n} x_i^{opt} \cdot \tau_i(x_1^{opt}, \ldots, x_{i-1}^{opt}, x_i^{opt}, x_{i+1}^{opt}, \ldots, x_n^{opt}) = W^{opt} \qquad (15)$$

It follows that $W^{sch} \leq W^{opt}$. □

Lemma 12. *If algorithm PPO terminates at $\boxed{T0}$ then $R \leq 2R^{opt}\left(1 - \frac{1}{p+1}\right)$.*

Proof. By Lemma 8, it follows that $R < 2\frac{W^{sch}}{p}\left(1 - \frac{1}{p+1}\right)$. From Lemma 11, it follows that $W^{sch} \leq W^{opt}$ and so, $R < 2\frac{W^{opt}}{p}\left(1 - \frac{1}{p+1}\right) \leq 2\frac{R^{opt}}{p}\left(1 - \frac{1}{p+1}\right)$. □

Lemma 13. *If algorithm PPO terminates at $\boxed{T1}$ then $R = R^{opt}$.*

Proof. Let T_i be the highest operator found at line 1 of Algorithm PPO. Suppose $x_i = x_i^{opt}$. Then

$$R = \tau_i(x_1, \ldots, x_n) \leq \tau_i(x_1^{opt}, \ldots, x_n^{opt}) \leq R^{opt}$$

Thus, if $x_i = x_i^{opt}$, it follows that $R = R^{opt}$. From Lemma 10, it follows that the only other possibility is $x_i < x_i^{opt}$. This however is impossible since $x_i = p$. □

Lemma 14. *If algorithm* PPO *terminates at* $\boxed{T2}$ *then* $R^{new} \leq 2R^{opt}\left(1 - \frac{1}{p+1}\right)$.

Proof. Since, $R^{new} > new\_max$, it follows that $R^{new} \leq 2 \cdot W^{sch}/p$. From Lemma 11, it follows that $W^{sch} \leq W^{opt}$. Hence,

$$R^{new} \leq 2\frac{W^{opt}}{p}\left(1 - \frac{1}{p+1}\right) \leq 2R^{opt}\left(1 - \frac{1}{p+1}\right) \qquad (16)$$

□

Lemma 15. *If algorithm* PPO *terminates at* $\boxed{T3}$, *then* $R \leq 2R^{opt}\left(1 - \frac{1}{p+1}\right)$.

Proof. Suppose that T_i is the highest operator found in the last execution of statement 1. From Lemma 10, there are two cases, namely, either $x_i = x_{opt}$ or $x_i < x_i^{opt}$.

Consider the case when $x_i = x_i^{opt}$. Then, from Constraint 2 and Lemma 10, it follows that

$$R = \tau_i(x_1, x_2, \ldots, x_n) \leq \tau_i(x_1^{opt}, \ldots, x_n^{opt}) \leq R^{opt}$$

This implies that $R = R^{opt}$.

It follows that either $R = R^{opt}$ or $x_i < x_i^{opt}$. Let us now assume that $x_i < x_i^{opt}$ and $R > R^{opt}$. We claim that for all iterations j, $x_j^{new} \leq x_j^{opt}$. Otherwise, there is an iteration $j + 1$ at which $x_k^{new} > x_k^{opt}$. Let R^j denote the response time of the schedule at the end of iteration j. It follows that

$$R \leq R^j = \tau_j(x_1^{new}, \ldots, x_n^{new}) \leq \tau_j(x_1^{opt}, \ldots, x_n^{opt}) \leq R^{opt}$$

This implies that $R = R^{opt}$ contradicting the hypothesis. We therefore conclude that

$$R \leq R^{new} \leq 2\frac{W^{new}}{p}\left(1 - \frac{1}{p+1}\right) \leq 2\frac{W^{opt}}{p}\left(1 - \frac{1}{p+1}\right) \leq 2R^{opt}\left(1 - \frac{1}{p+1}\right) \qquad (17)$$

□

Lemma 16. *If the algorithm* PPO *terminates at* $\boxed{T4}$, $R = R^{opt}$.

Proof. Suppose that T_i is the highest operator found in the last execution of statement 1 . From Lemma 10, there are two cases, namely, either $x_i = x_{opt}$ or $x_i < x_i^{opt}$.

If $x_i = x_i^{opt}$, it follows that $R = R^{opt}$ using an argument similar Lemma 13. Otherwise $x_i < x_i^{opt}$. Following an argument similar to that in Lemma 15, it follows that for all iterations j, $x_k^j \leq x_k^{opt}$ for $k \in \{1, \ldots, n\}$. Since, in the final iteration, $x_k^{new} = p + 1$ for some operator T_k, this would be a contradiction. Hence $R = R^{opt}$. $\qquad\square$

Lemma 17. *If the algorithm* PPO *terminates at* $\boxed{T5}$ *then* $R^{new} \leq 2R^{opt}\left(1 - \frac{1}{p+1}\right)$.

Proof. The proof proceeds along the lines of Lemma 14. $\qquad\square$

Lemma 18. *If the algorithm* PPO *terminates at* $\boxed{T6}$, $R \leq 2R^{opt}\left(1 - \frac{1}{p+1}\right)$.

Proof. The proof proceeds along the lines of Lemma 15. $\qquad\square$

Theorem 19. *Let* R^{sch} *be the response time of the schedule obtained by Algorithm* PPO. *Then* $R^{sch} \leq 2R^{opt}\left(1 - \frac{1}{p+1}\right)$.

Proof. The result follows from Lemmas 12 to 18. $\qquad\square$

The algorithm PPO can be implemented in $O(n^2 \cdot p \cdot log(n \cdot p))$ time. The upper bound presented by the analysis is tight, as can be seen from the following example. Let n be the number of operators and p be the number of processors, such that $p = 2n - 1$. The response time function of the ith operator is defined as

$$\tau_i(x_1, \ldots, x_n) = 1/x_i$$

The optimal schedule partitions each of the n operators onto p processors yielding a response time of $R^{opt} = n/p$. The schedule generated by Algorithm PPO gives 2 processors to $n - 1$ operators and 1 processor to the nth operator. Thus, $R^{sch} = 1$. Thus

$$\frac{R^{sch}}{R^{opt}} = \frac{p}{n} = \frac{2n-1}{n} = 2 - \frac{1}{n} = 2 - \frac{2}{p+1} \qquad (18)$$

6 Examples

In this section, we present two examples that compare the schedule generated by the algorithm presented in this paper against other proposed schemes. The first example compares the performance of the schedule obtained by the algorithm PPO versus an optimal partitioning algorithm that does not consider communication. The second example compares the performance against a more sophisticated partitioning algorithm [BB90].

Example 1. Consider the operator tree shown in Figure 4(a). The figure depicts a 2-way join between three relations R_1, R_2 and R_3. The relations are assumed to be uniformly declustered on the disks; hence the scan operators are not considered for partitioning. The pipeline consists of two operators, with the *merge-join* as the producer node and *nested-loop* join as the consumer node. The *merge-join* operator has an estimated CPU cost of 50 million machine cycles, whereas the *nested-loop* join has an estimated CPU time of 50 billion machine cycles. The output size of the *merge-join* operator is estimated to be 16 MB.

Figure 4(b) shows an optimal processor allocation on 101 processors *without* taking communication into account. This schedule would be produced by [TWPY92] and by [BB90] if communication is not taken into account. Subroutines such as the algorithm proposed by [TWPY92]is used by database scheduling algorithms proposed by [NSHL93, TL94] and with a modification (that does not address communication issues) by [LCRY93]. To simplify the mathematical expressions, we ignore the scan nodes at this moment.

Let us now calculate communication costs for the schedule obtained in Figure 4(b). We assume that the communication parameters are as follows.

- $\alpha = 3000$ machine cycles.
- $\beta = 1$ cycle per byte.

Further, we assume that 100KB of communication buffer is available per partition of each operator. This is identical to *Scheme B* for communication buffer allocation presented in Section 4.2 with $b = 100$KB. Let x_1 (respectively, x_2) denote the number of processors assigned to the merge-join operator (respectively, nested-loop operator) and let τ_1 (respectively, τ_2) denote the number of processors assigned to the merge-join operator (respectively, nested-loop operator). Let $\epsilon = 0.1$ (although the precise value is inconsequential). The expression for τ_1 and τ_2 are as follows.

$$\tau_1(x_1, x_2) = (50 \cdot 10^6/x_1) + (16 \cdot 10^6/x_1) + x_2 \cdot 3000 \cdot max(16 \text{ MB}/(x_1 \cdot 100 \text{ KB}),$$
$$16 \text{ MB}/(x_2 \cdot 100 \text{ KB}), 0.1)$$

$$(19)$$

$$\tau_2(x_1, x_2) = (50 \cdot 10^9/x_2) + (16 \cdot 10^6/x_2) + x_1 \cdot 3000 \cdot max(16 \text{ MB}/(x_1 \cdot 100 \text{ KB}),$$
$$16 \text{ MB}/(x_2 \cdot 100 \text{ KB}), 0.1)$$

$$(20)$$

Figure 4(c) shows the cost of the schedule obtained for the schedule in 4(b), using the above formulae. Figure 4(d) shows the schedule obtained by using the partitioning algorithm PPO proposed in this paper and its cost. Note that the benefit in performance exceeds a factor of 2 over the optimal scheme that does not consider communication. □

Example 2. Figure 5(a) shows a pipeline between two merge-join operators A and B with estimated individual times of 10 million machine cycles each and a data flow of 10MB from A to B. Let the communication buffer size $b = 10$KB, $\alpha = 2000$ cycles and $\beta = 1$ cycle-per-byte. Let x_1 and x_2 represent the degree of

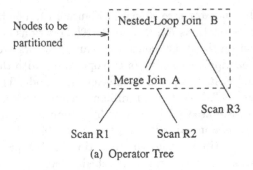

Nodes to be partitioned →

Nested-Loop Join B

Merge Join A

Scan R1   Scan R2   Scan R3

(a) Operator Tree

time (millions of machine cycles)

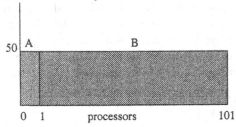

(b) optimal schedule without taking communication
costs into account

time (millions of machine cycles)

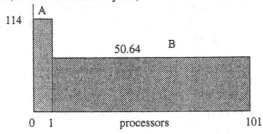

(c) schedule for (b) taking communication costs
into account

time (millions of machine cycles)

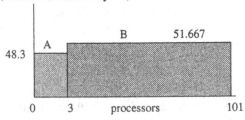

(d) schedule produced by Algorithm PPO

partitioning of operators A and B respectively and let τ_1 and τ_2 represent the response time function of operators A and B respectively. Assume that a buffer of size 10KB is allocated to each partition for communication. Let $\epsilon = 0.1$ (the precise value is inconsequential). The expressions for τ_1 and τ_2 are as follows.

$$\tau_1(x_1, x_2) = (10^7/x_1) + (10 \cdot 10^6 \cdot 1)/x_1 + x_1 \cdot 2000 \cdot max(10 \text{ MB}/(x_1 \cdot 10 \text{ KB}),$$
$$10 \text{ MB}/(x_2 \cdot 10 \text{ KB}), 0.1)$$

$$(21)$$

$$\tau_2(x_1, x_2) = (10^7/x_2) + (10 \cdot 10^6 \cdot 1)/x_2 + x_2 \cdot 2000 \cdot max(10 \text{ MB}/(x_1 \cdot 10 \text{ KB}),$$
$$10 \text{ MB}/(x_2 \cdot 10 \text{ KB}), 0.1)$$

$$(22)$$

Figure 5(b) shows the schedule returned by the algorithm presented in [BB90]. The cause of the termination of this algorithm is depicted in Figure 5(c). The algorithm in [BB90] initially assigns one processor to each operator, to get a schedule shown in Figure 5(b). It then temporarily assigns one more processor to the bottleneck operator (which could be either A or B in this case), say operator A, and recalculates the response time. Since the response time for B increases, the algorithm terminates after rejecting the temporary increase in processor allocation. The schedule produced by the algorithm PPO is shown in Figure 5(d). Note that the response time of the PPO schedule is about a factor of 5 less than the response time of the schedule produced by algorithm in [BB90].

$\square$

7   Generalizations

In this section, we discuss some of the generalities that can be incorporated into the pipeline cost model of Section 4.

Thread Startup and Cleanup Overhead. Optimization and scheduling algorithms were criticized in [DeWGra92] for not taking into account the overhead required to spawn a partitioned operator on a processor and to destroy the thread once it is no longer needed. Let g represent this overhead. Note that by adding g to each of the τ_i's, the new τ_i's continue to satisfy the two constraints of Section 5.1. The algorithm PPO can therefore be used for finding a near-optimal schedule while incorporating thread startup overhead.

Modeling Network Congestion. Network congestion is an important factor that affects the response time of a partitioned parallel computation. As the degree of partitioning of operators is increased, a greater load is placed on the network, which may cause a substantial degradation in performance. Analytical models for predicting network congestion for partitioning database operators have not yet been validated. We believe however, that for a wide variety of practical scenarios, the response time function $\tau_i(x_1, x_2, \ldots, x_n)$ will satisfy the two constraints, namely,

- τ_i is a non-decreasing function of x_j, $j \neq i$.

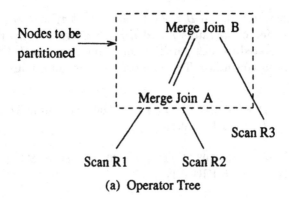

Nodes to be partitioned →

(a) Operator Tree

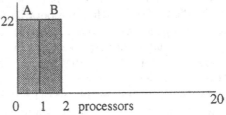

time (millions of machine cycles)

(b) Schedule produced by algorithm
in [BB90]

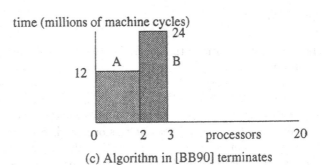

(c) Algorithm in [BB90] terminates

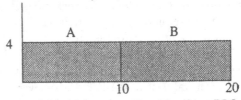

(d) Schedule produced by Algorithm PPO

Fig. 5. Schedule for Example 2

- $x_i \cdot \tau_i$ is a non-decreasing function of x_i.

Since the near-optimality of the proposed algorithm relies only these two properties of τ_i's, we believe that the algorithm PPO can be used in conjunction with refined cost models that take network congestion into account.

8 Conclusions and Future work

In this paper, we have considered the problem of scheduling a set of operators that can run in a pipelined parallel fashion. This problem is a basic component in scheduling database execution trees. Execution trees are scheduled in units of segments after a decomposition phase [NSHL93, HCY94, Hon91, Hon92, SYT93, ZZBS94, LST91], where each segment consists of a pipelinable set of operators. Partitioning promises the benefit of increased parallelism and increased memory availability at the price of increased communication overhead. Current partitioning algorithms [BB90, TWPY92, LCRY93, NSHL93] do not consider these trade-offs in partitioning.

We present an algorithm called PPO for partitioning pipelines with the objective of minimizing the response time. The algorithm PPO yields schedules with response time that are within a factor of 2 of the response time of the optimum schedule. The algorithm PPO is robust, in the sense that its applicability relies on few general constraints which are satisfied in many practical scenarios.

We are currently extending this algorithm to incorporate clustering, i.e., to allow neighboring operators to be scheduled in such a way that there is no communication. We are also extending the ideas in the paper to consider more general forms of data placement.

Acknowledgements: We would like to thank Prof. B.R. Badrinath and Prof. Michiel Noordewier at the Department of Computer Sciences, Rutgers University, for providing many interesting comments on the draft of the paper.

References

[BB90] K.P. Belkhale and P. Banerjee: Approximate Algorithms for the Partitionable Independent Task Scheduling Problem. Proceedings of the 1990 International Conference on Parallel Processing. pp. I-72 - I-75

[Gamma] D.DeWitt, S. Ghandehariziadeh, D. Schneider, A. Bricker, H.Hsiao, R.Rasmussen; The Gamma Database Machine. IEEE Transactions on Knowledge and Data Engineering, March 1990

[DeWGra92] D. DeWitt and J. Gray: The future of high performance database systems. Communications of the ACM, June 1992

[Gan92] S. Ganguly. Parallel Evaluation of Deductive Database Queries. PhD thesis, University of Texas, Austin, 1992

[GHK92] S. Ganguly, W. Hasan and R. Krishnamurthy. Query Optimization for Parallel Executions. Proceedings of the 1992 ACM SIGMOD International Conference on Management of Data

[Gra69] R.L. Graham. Bounds on Multiprocessing Timing Anomalies SIAM J. Appl. Math., 17(1969) 416–429

[HM94] W. Hasan and R. Motwani. Optimization Algorithms for Exploiting the Parallelism-Communication Trade-off in Pipelined Parallelism. Proceedings of the 1994 International Conference on Very Large Databases.

[Hon91] W. Hong and M. Stonebraker. Optimization of Parallel Query Execution Plans in XPRS. Proceedings of the First International Conference on Parallel and Distributed Database Systems. December 1991

[Hon92] W. Hong. Exploiting Inter-Operation Parallelism in XPRS Proceedings of the 1992 ACM SIGMOD International Conference on Management of Data

[LVZ93] R.S.G. Lanzelotte, P. Valduriez and M. Zait. On the Effectiveness of Optimization Search Strategies for Parallel Execution. Proceedings of the 1993 International Conference on Very Large Databases

[LCRY93] M-L. Lo, M-S. Chen, C.V. Ravishankar and P.S. Yu. On Optimal Processor Allocation to Support Pipelined Hash Joins. Proceedings of the 1993 ACM SIGMOD International Conference on Management of Data.

[HCY94] Hui-I Hsiao, M-S. Chen, P.S. Yu. On Parallel Execution of Multiple Pipelined Hash Joins. Proceedings of the 1994 ACM SIGMOD International Conference on Management of Data.

[LST91] H. Lu, M.C. Shan and K.L. Tan. Optimization of Multi-Way Join Queries for Parallel Execution. Proceedings of 1991 International Conference on Very Large Databases

[NSHL93] T.H. Niccum, J. Srivastava, B. Himatsingka, J-Z. Li. A Tree-Decomposition Approach to the Parallel Execution of Relational Query Plans. Technical Report, University of Minnesota at Minneapolis

[PMC$^+$91] H. Pirahesh, C. Mohan, J.Cheung, T.S. Liu and P. Selinger. Parallelism in Relational Database Systems: Architectural Issues and Design Approaches. Proceedings of the 1991 International Conference on Parallel and Distributed Information Systems

[Sch90] D. Schneider. Complex Query Processing in Multiprocessor Database Machines. PhD thesis, University of Wisconsin, Madison, 1990

[SAC$^+$] P. Selinger, M.M. Astrahan, D.D. Chamberlain, R.A. Lorie and T.G. Price. Access Path Selection in a Relational Database Management System. Proceedings of the 1979 ACM SIGMOD International Conference on Management of Data

[SriEls93] Jaideep Srivastava and G. Elsesser. Query Optimization for Parallel Relational Databases. Preliminary version appeared in Proceedings of 1993 International Conference on Parallel and Distributed Information Systems

[SYT93] Eugene J. Shekita, Honesty C. Young and Kian-Lee Tan. Multi-Join Optimization for Symmetric Multiprocessors. Proceedings of the 1993 Conference on Very large Databases

[TL94] K-L. Tan, H. Lu. On resource scheduling of multi-join queries in parallel database systems. Information Processing Letters 48 (1993), 189–195.

[TWPY92] J. Turek, J.L. Wolf, K.R. Pattipati and P.S. Yu. Scheduling Parallelizable Tasks: Putting it All on the Shelf. Proceedings of the 1992 ACM Sigmetrics Conference

[ZZBS94] M. Ziane, M. Zait, and P. Borla-Salamet. Parallel Query Processing in DBS3. In Proceedings of the 1993 International Conference on Parallel and Distributed Information Systems

Index of Authors

Springer-Verlag
and the Environment

We at Springer-Verlag firmly believe that an international science publisher has a special obligation to the environment, and our corporate policies consistently reflect this conviction.

We also expect our business partners – paper mills, printers, packaging manufacturers, etc. – to commit themselves to using environmentally friendly materials and production processes.

The paper in this book is made from low- or no-chlorine pulp and is acid free, in conformance with international standards for paper permanency.

Lecture Notes in Computer Science

For information about Vols. 1–928

please contact your bookseller or Springer-Verlag

Vol. 964: V. Malyshkin (Ed.), Parallel Computing Technologies. Proceedings, 1995. XII, 497 pages. 1995.

Vol. 965: H. Reichel (Ed.), Fundamentals of Computation Theory. Proceedings, 1995. IX, 433 pages. 1995.

Vol. 966: S. Haridi, K. Ali, P. Magnusson (Eds.), EURO-PAR '95 Parallel Processing. Proceedings, 1995. XV, 734 pages. 1995.

Vol. 967: J.P. Bowen, M.G. Hinchey (Eds.), ZUM '95: The Z Formal Specification Notation. Proceedings, 1995. XI, 571 pages. 1995.

Vol. 968: N. Dershowitz, N. Lindenstrauss (Eds.), Conditional and Typed Rewriting Systems. Proceedings, 1994. VIII, 375 pages. 1995.

Vol. 969: J. Wiedermann, P. Hájek (Eds.), Mathematical Foundations of Computer Science 1995. Proceedings, 1995. XIII, 588 pages. 1995.

Vol. 970: V. Hlaváč, R. Šára (Eds.), Computer Analysis of Images and Patterns. Proceedings, 1995. XVIII, 960 pages. 1995.

Vol. 971: E.T. Schubert, P.J. Windley, J. Alves-Foss (Eds.), Higher Order Logic Theorem Proving and Its Applications. Proceedings, 1995. VIII, 400 pages. 1995.

Vol. 972: J.-M. Hélary, M. Raynal (Eds.), Distributed Algorithms. Proceedings, 1995. XI, 333 pages. 1995.

Vol. 973: H.H. Adelsberger, J. Lažanský, V. Mařík (Eds.), Information Management in Computer Integrated Manufacturing. IX, 665 pages. 1995.

Vol. 974: C. Braccini, L. DeFloriani, G. Vernazza (Eds.), Image Analysis and Processing. Proceedings, 1995. XIX, 757 pages. 1995.

Vol. 975: W. Moore, W. Luk (Eds.), Field-Programmable Logic and Applications. Proceedings, 1995. XI, 448 pages. 1995.

Vol. 976: U. Montanari, F. Rossi (Eds.), Principles and Practice of Constraint Programming — CP '95. Proceedings, 1995. XIII, 651 pages. 1995.

Vol. 977: H. Beilner, F. Bause (Eds.), Quantitative Evaluation of Computing and Communication Systems. Proceedings, 1995. X, 415 pages. 1995.

Vol. 978: N. Revell, A M. Tjoa (Eds.), Database and Expert Systems Applications. Proceedings, 1995. XV, 654 pages. 1995.

Vol. 979: P. Spirakis (Ed.), Algorithms — ESA '95. Proceedings, 1995. XII, 598 pages. 1995.

Vol. 980: A. Ferreira, J. Rolim (Eds.), Parallel Algorithms for Irregularly Structured Problems. Proceedings, 1995. IX, 409 pages. 1995.

Vol. 981: I. Wachsmuth, C.-R. Rollinger, W. Brauer (Eds.), KI-95: Advances in Artificial Intelligence. Proceedings, 1995. XII, 269 pages. (Subseries LNAI).

Vol. 982: S. Doaitse Swierstra, M. Hermenegildo (Eds.), Programming Languages: Implementations, Logics and Programs. Proceedings, 1995. XI, 467 pages. 1995.

Vol. 983: A. Mycroft (Ed.), Static Analysis. Proceedings, 1995. VIII, 423 pages. 1995.

Vol. 984: J.-M. Haton, M. Keane, M. Manago (Eds.), Advances in Case-Based Reasoning. Proceedings, 1994. VIII, 307 pages. 1995.

Vol. 985: T. Sellis (Ed.), Rules in Database Systems. Proceedings, 1995. VIII, 373 pages. 1995.

Vol. 986: Henry G. Baker (Ed.), Memory Management. Proceedings, 1995. XII, 417 pages. 1995.

Vol. 987: P.E. Camurati, H. Eveking (Eds.), Correct Hardware Design and Verification Methods. Proceedings, 1995. VIII, 342 pages. 1995.

Vol. 988: A.Ù. Frank, W. Kuhn (Eds.), Spatial Information Theory. Proceedings, 1995. XIII, 571 pages. 1995.

Vol. 989: W. Schäfer, P. Botella (Eds.), Software Engineering — ESEC '95. Proceedings, 1995. XII, 519 pages. 1995.

Vol. 990: C. Pinto-Ferreira, N.J. Mamede (Eds.), Progress in Artificial Intelligence. Proceedings, 1995. XIV, 487 pages. 1995. (Subseries LNAI).

Vol. 991: J. Wainer, A. Carvalho (Eds.), Advances in Artificial Intelligence. Proceedings, 1995. XII, 342 pages. 1995. (Subseries LNAI).

Vol. 992: M. Gori, G. Soda (Eds.), Topics in Artificial Intelligence. Proceedings, 1995. XII, 451 pages. 1995. (Subseries LNAI).

Vol. 993: T.C. Fogarty (Ed.), Evolutionary Computing. Proceedings, 1995. VIII, 264 pages. 1995.

Vol. 994: M. Hebert, J. Ponce, T. Boult, A. Gross (Eds.), Object Representation in Computer Vision. Proceedings, 1994. VIII, 359 pages. 1995.

Vol. 995: S.M. Müller, W.J. Paul, The Complexity of Simple Computer Architectures. XII, 270 pages. 1995.

Vol. 996: P. Dybjer, B. Nordström, J. Smith (Eds.), Types for Proofs and Programs. Proceedings, 1994. X, 202 pages. 1995.

Vol. 997: K.P. Jantke, T. Shinohara, T. Zeugmann (Eds.), Algorithmic Learning Theory. Proceedings, 1995. XV, 319 pages. 1995.

Vol. 998: A. Clarke, M. Campolargo, N. Karatzas (Eds.), Bringing Telecommunication Services to the People – IS&N '95. Proceedings, 1995. XII, 510 pages. 1995.

Vol. 999: P. Antsaklis, W. Kohn, A. Nerode, S. Sastry (Eds.), Hybrid Systems II. VIII, 569 pages. 1995.

Vol. 1000: J. van Leeuwen (Ed.), Computer Science Today. XIV, 643 pages. 1995.

Vol. 1004: J. Staples, P. Eades, N. Katoh, A. Moffat (Eds.), Algorithms and Computation. Proceedings, 1995. XV, 440 pages. 1995.

Vol. 1005: J. Estublier (Ed.), Software Configuration Management. Proceedings, 1995. IX, 311 pages. 1995.

Vol. 1006: S. Bhalla (Ed.), Information Systems and Data Management. Proceedings, 1995. IX, 321 pages. 1995.

Vol. 1007: A. Bosselaers, B. Preneel (Eds.), Integrity Primitives for Secure Information Systems. VII, 239 pages. 1995.

Vol. 1008: B. Preneel (Ed.), Fast Software Encryption. Proceedings, 1994. VIII, 367 pages. 1995.

Vol. 1009: M. Broy, S. Jähnichen (Eds.), KORSO: Methods, Languages, and Tools for the Construction of Correct Software. X, 449 pages. 1995.